IEE RADAR, SONAR, NAVIGATION AND AVIONICS SERIES 7

Series Editors: Professor E. D. R. Shearman
P. Bradsell

ULTRAWIDEBAND RADAR MEASUREMENTS

analysis and processing

Other volumes in this series:

Volume 1 **Optimised radar processors** A. Farina (Editor)
Volume 2 **Radar development to 1945** R. W. Burns (Editor)
Volume 3 **Weibull radar clutter** M. Sekine and Y. H. Mao
Volume 4 **Advanced radar techniques and systems** G. Galati (Editor)
Volume 5 **Strapdown inertial navigation technology** D. H. Titterton and J. L. Weston
Volume 6 **Surface-penetrating radar** D. J. Daniels

ULTRAWIDEBAND RADAR MEASUREMENTS analysis and processing

L. Y. Astanin and A. A. Kostylev

Translated by L. N. Smirnova

The Institution of Electrical Engineers

Published by: The Institution of Electrical Engineers, London, United Kingdom

The Institution of Electrical Engineers,
Michael Faraday House,
Six Hills Way, Stevenage,
Herts. SG1 2AY, United Kingdom

British Library Cataloguing in Publication Data

A CIP catalogue record for this book
is available from the British Library

ISBN 0 85296 894 9

Typeset by Alden Bookset, Didcot
Printed by Bookcraft, Bath

Contents

Introduction

Ultrawideband (UWB) signals came into use as a result of researchers' efforts to learn as much as possible about the radar target. The initial desire to discriminate between two closely flying planes changed, with the development of electronic technologies, to the quest for constructing a three-dimensional image of a target. The potential for direct reduction of the incident pulse duration was soon exhausted and followed by a detailed analysis of target-reflected signals. It became clear that the most important changes in a target response occurred during a transient process of 1–2 oscillations' duration. This fact in itself led to the idea of using UWB signals of this duration without energy expenditure for steady oscillation transmission.

The current state of UWB radar science appears to be following a third wave in researchers' interest in UWB signal application. This manifests itself in the relative proportions of theoretical models and technological capabilities. The first steps towards wideband pulse technology, originally known as milli-microsecond range technology, were made in the 1950s and resulted from the progress in electrovacuum and transistor physics. At that time, radio pulses of nanosecond duration attracted the attention of few workers, who used exotic methods for the shaping and registration of such signals.

The appearance in the 1970s of sampling oscilloscopes with frequency bandwidths of over 10 GHz greatly stimulated the development of fast response (picosecond range) transistors. The simultaneous transition to a new generation of computers provided signal processing capabilities which were earlier unthinkable. High resolution radars were designed for the recognition of airborne and space targets, and UWB signals were adapted to conventional radars and theoretical models. It became clear, eventually, that the information potential of these signals could only be exploited by new approaches to their theoretical treatment, signal processing and representation of measurements. These lines of research developed rapidly in the early 1990s. Today, interest in UWB radar measurements is growing. It is accompanied by the appearance of new applications, such as ground penetrating radars, construction control, etc. The number of publications in these fields is rapidly increasing, and the overview edited by Taylor [100] deserves special attention.

The authors of the present book have made an attempt to express their point of view concerning some principal issues in both the theoretical description of UWB radar signals and their computer processing. The focus is on the possibility of modelling and measuring UWB signal waveforms as a principal source of information about a target. The scope of the issues is much greater than in previous publications [1, 85]. The authors have become aware that narrowband theory based on signal representation in the complex plane with the Hilbert transform can be applied to UWB signals with reservations. The principal task of theoretical analysis, to the authors' minds, should be the transient process and what is discarded in the analytical signal mapping: discontinuity points in the function and its derivatives. For this reason, a UWB signal is to be analysed in the phase plane treated as a generalised complex plane. This explains the frequent use of the word *generalised* in this book. This term primarily refers to the concept of frequency, which becomes devoid of its averaged parameter period to be treated as the angular velocity of the vector end that can be defined and measured at any moment of time. This is the basic idea of UWB waveform analysis in the time domain. The concepts of generalised signal envelope, phase and instantaneous frequency as a measure of the waveform informative value are introduced, and the latter concept is extended to a polarised vector signal. In line with this approach is the widely used wavelet transform analysis in the time domain. A time domain version of the geometrical diffraction theory is considered as being applicable to the solution of UWB electrodynamic problems.

This book presents a detailed discussion of mathematical models for radar targets and the statistical processing of scattered signals. The efficiency of the regularising algorithms suggested is illustrated with a description of experimental data processing.

Further progress in UWB radar systems is closely associated with the design of fast response (picosecond-range) devices, efficient antennas and analogue-to-digital converters for a frequency of 10 GHz. These devices can considerably improve UWB radar performance.

The authors are grateful to V.I. Norkin (Section 1.4), D.V. Sajapin (Sections 1.5, 3.4), S.E. Shaldayev (Section 1.6), Yu.N. Kalinin (Section 5.7) and A.F. Kononov (Section 5.8) who provided them with their results.

Chapter 1

Ultrawideband signals: definition, modelling and measurement

1.1 Definition and modelling problems

The basic function of radar is to obtain information about the object being tracked (radar target) by measuring the characteristics of the electromagnetic field scattered by the target. Originally, radar was used to measure the coordinate parameters defining the target location and movement. With improvements in hardware and software, and the introduction of new design principles, radar precision and resolving power became high enough to obtain the noncoordinate data necessary for target classification and for the measurement of the target size, shape, orientation and state, allowing target recognition and imaging.

It is well known that the precision and resolving power of radar measurements are directly related to the parameters of the signals used. The desire to broaden the scope of information to be extracted made radar engineers turn to signals with a wide frequency bandwidth. These provide a space resolution of the order of a fraction of the target size and permit measurement of responses from local scattering centres.

Wideband radar measurements became possible due to the application of fast response equipment, which could generate radar pulses of nano- and picosecond duration. The new radar tracking standards necessitated the use of radar characteristics which could adequately describe the properties of wideband signals and responses, permitting their classification in terms of informative features.

To measure target coordinates, conventional radar used narrowband signals with a small ratio of the signal bandwidth, Δf, to the carrier frequency, f_0, i.e. with a fractional bandwidth $\Delta f / f_0 \ll 1$. Signals with a bandwidth of a few hundreds of megahertz provide a resolution and range estimate accuracy within a few tens of centimetres, which is sufficient for separate measurement of the local responses that are to be modelled by local radar characteristics. It is expedient to call such signals wideband or high resolution signals, although their fractional bandwidth for, say, the X-band frequencies is still low (although the inequality $\Delta f/f_0 \ll 1$ is less strong). Both narrowband and wideband radar signals can be formed as simple pulses or as signals with fill-in-pulse modulation (with a spread spectrum). Since the inequality $\Delta f/f_0 \ll 1$ holds true, complex envelopes [35, 114] are widely used for an analytical description of these signals.

The maximum body of information is obtained from measurements of signals whose bandwidth, Δf, and centre (carrier) frequency, f_0, are comparable, i.e.

$$\Delta f/f_0 \approx 1 \tag{1.1}$$

These are referred to as ultrawideband (UWB) signals [1].

Noncoordinate data can be obtained by measuring target responses in a frequency range wide enough for radar characteristics to vary considerably with frequency. This dependence is known to be very essential in the resonance frequency range. A wide bandwidth provides the high range resolution necessary for separate measurement of responses from local scattering centres. At the same time, a comparatively low centre frequency under the condition $\Delta f/f_0 \approx 1$ gives information in the frequency range resonant with respect to the size of the scattering centres proper. This information is extracted from the waveform variation of the scattered radio signal; therefore, the characteristics of UWB signals must describe this variation.

According to expr. 1.1 given in the frequency domain, UWB signals are essentially nonmonochromatic, since they consist of components with very different frequencies. Fig. 1.1 illustrates the spectral density function for a UWB signal.

The signal bandwidth is defined as $\Delta f = f_H - f_L$ and the centre frequency as $f_0 = (f_H + f_L)/2$; the signal bandwidth factor, as a measure of fractional

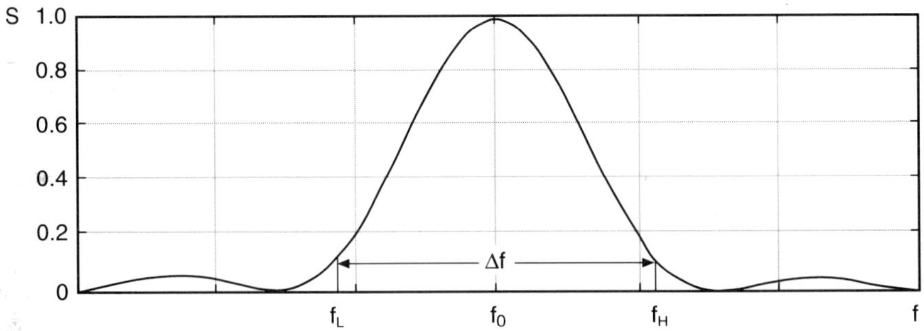

Fig. 1.1 UWB signal spectral density function

bandwidth, is

$$\mu = \frac{\Delta f}{f_0} = 2\frac{f_H - f_L}{f_H + f_L} \tag{1.2}$$

where f_H and f_L are the highest and the lowest frequencies in the signal bandwidth, respectively. In the limiting case, the signal bandwidth, Δf, is comparable with its centre frequency, so that $\mu \approx 1$. With such nonmonochromatic radar signals, a target response can be investigated over a wide frequency range.

Since the response of an object is dependent on its geometric size, a, and the signal wavelength, λ, the radar signal parameters are chosen in accordance with the class of the target being tracked. For wavelengths $\lambda \ll a$, responses are said to be of a high frequency or optical nature; for $\lambda \approx a$ they are called resonance responses and for $\lambda \gg a$ they are low frequency or Rayleigh responses. The investigation of radar target responses over a wide frequency range implies the extraction of information about the target in all three ranges simultaneously, using a nonmonochromatic radar signal with a bandwidth containing the necessary frequencies such that

$$f_H \gg \frac{c}{a} \gg f_L \tag{1.3}$$

where c is the speed of light.

In Fig. 1.2, the modulus of the frequency response, $H(ka) = H(2\pi a/\lambda)$, from a typical target (sphere) is shown against the spectral density functions of narrowband, high resolution and UWB signals. Here $k = 2\pi/\lambda$ is a wavenumber and ka is a target electrical size. The condition $\Delta ka/k_0 a \ll 1$ for a signal spectral density function $S^{(A)}$ to be narrowband is met, since the bandwidth $\Delta ka^{(A)}$ is narrow. Then the frequency response of a target is simulated by a complex number, $H(k_0 a)$, in the resonance and Rayleigh frequency regions. The spectral density function $S^{(B)}$ is also narrowband, despite its large bandwidth; i.e. $\Delta ka/k_0 a \ll 1$ due to the high magnitude of the centre frequency, ω_0, which corresponds to $(k_0 a)^{(B)}$.

In radar performance modelling, the narrowband nature of $S^{(B)}$ is evident in that the frequency response, $H(ka)$, is uniform within the spectrum bandwidth and can be simulated by a constant quantity, $H(k_0 a) = H_0$.

For UWB signals, the spectral density function, $S^{(C)}(ka)$, includes components of both the resonance and low frequency regions. In the resonance region $ka = 2\pi a/\lambda = 0.5, \ldots, 5$; the object's size is comparable with the wavelength, so the field diffraction in the shadow region and re-reflections between local scattering centres are significant. The interference of waves reflected by the object and those bending around it causes oscillations of the frequency response even for smooth bodies like a sphere (see Fig. 1.2). The oscillation period provides information on the target size.

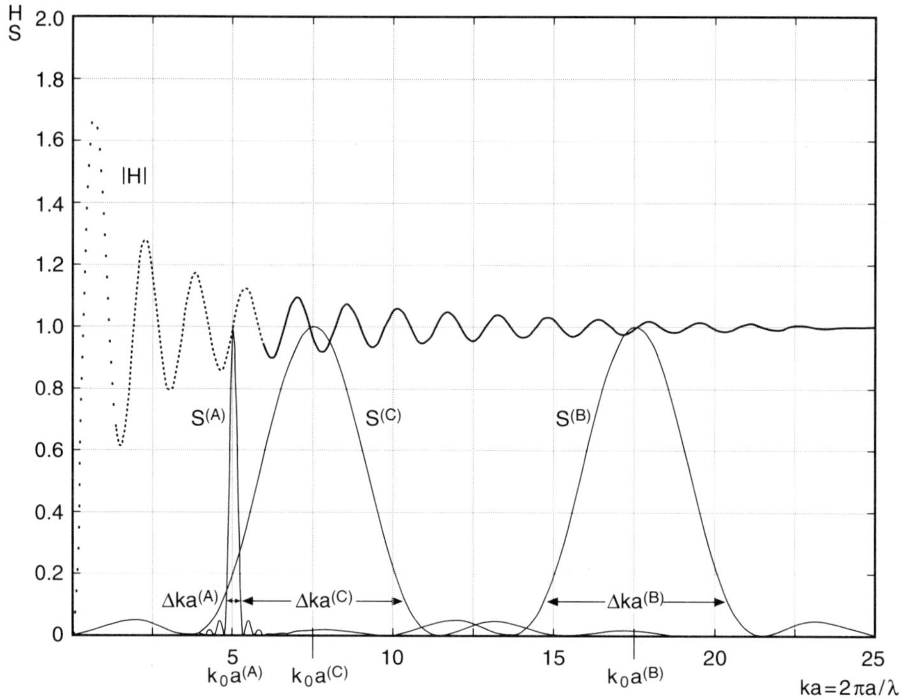

Fig. 1.2 Three classes of radio signal spectra

Finally, the bandwidth of the UWB signal spectrum contains components of the low frequency, Rayleigh region where $ka \ll 1$. These components are usually insignificant when compared with the frequency response of the target as a whole (Fig. 1.2). Nevertheless, for targets of complex geometry, the resolving power of UWB signals is sufficient to single out return signals from local scattering centres, whose size is a small fraction of the total target size. With respect to these scatterers, the spectral density function of UWB signals lies in the Rayleigh region and it therefore contains information about their properties such as size, orientation of the principal axes, etc. [12, 13]. The high information value of the low frequency region, in combination with its high resolution, is a specific feature of UWB signals which may be termed as a frequency dualism.

In any case, when reflected by a target, a UWB signal changes its spectral density function, depending on the centre frequency, f_0. This changes the signal waveform in the time domain, $s(t)$, providing information about the target's characteristics at various frequencies. For this reason, when measuring UWB radar signals, one has to analyse the time domain structure of the return signal and to find its correlation with the target parameters.

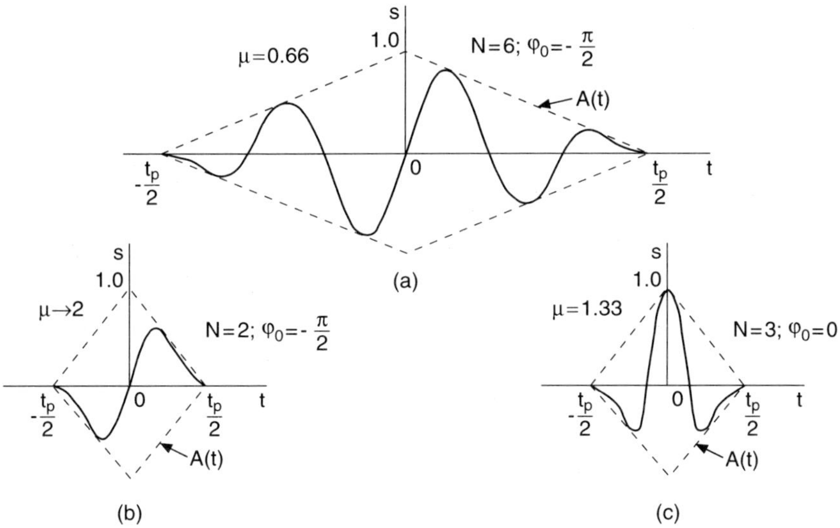

Fig. 1.3 Real models of UWB radio pulses

a The envelope, A(t), mainly reflects the signal shape
b, c The phase, φ_0, mainly determines the signal shape

Spectral representation may be valid for UWB signals with various time domain structures. At present, UWB radar systems use radio pulses of nano and picosecond duration, signals with linear frequency modulation and discrete, multifrequency signals, etc.

Consider models of UWB finite radio pulses with duration t_p, N quasihalf periods (time sidelobes) of the high frequency fill-in and centre frequency of signal spectral density function, $f_0 = 1/T_0$ (Fig. 1.3). We may estimate the pulse bandwidth to be $\Delta f = f_H - f_L = 2/t_p \approx 4/NT_0$. Then the bandwidth factor may be taken as

$$\mu = \Delta f / f_0 \approx 4/N \quad (1.4)$$

For the radio pulse shown in Fig. 1.3a, the representation with an envelope $A(t)$, say, of triangular shape (dashed lines) and initial phase φ_0, gives a rather clear picture of the signal. But when the number of time sidelobes, N, decreases, the envelope does not clearly reflect the signal waveform (Figs. 1.3b,c). Therefore, the conventional model of UWB signals, $s(t) = A(t)\cos(\omega_0 t + \varphi_0)$, with a bandwidth factor greater than $\mu \approx 0.66, \ldots, 1.0$, is not fully appropriate, since the functions $A(t)$ and $\varphi(t)$ are then incompatible with the voltage measured at the amplitude and phase detector outputs. For this reason a specific feature of UWB radar design is that radio signals are not detected when received. The task is to work out generalised characteristics capable of describing UWB waveforms and to offer processing procedures for their measurement. We shall discuss these problems in some detail below.

It is important that the spectral density function, $S(\omega)$, of any UWB radio signal, $s(t)$, at the null frequency ($\omega = 0$) should have a zero value, i.e.

$$S(\omega) = \int_{-t_p/2}^{t_p/2} s(t) \exp(-i\omega t)\, dt \,|_{\omega=0}$$

$$= \int_{-t_p/2}^{t_p/2} s(t)\, dt = 0 \tag{1.5}$$

This condition (Rayleigh's postulate) corresponds to the requirement that the lines of force of any electromagnetic field in free space should be closed, or that a finite-dimensional radiator should not emit a steady current field. All models of incident and reflected radar signals must meet the requirement of eqn. 1.5, and this defines a special class of models which can be correctly called radio signal models. Strictly, narrowband signals with $\Delta f / f_0 \ll 1$ also belong to this class, but since their structure contains a large number of time sidelobes, $N \gg 1$, the condition of eqn. 1.5 is usually satisfied rather accurately and does not need to be considered. For a UWB signal with $N = 6$ (Fig. 1.3a) eqn. 1.5 is satisfied exactly for $\varphi_0 = -\pi/2$ and approximately for other values of φ_0, while for a signal with $N = 2, \ldots, 3$ (Figs. 1.3b,c) it may be violated depending on the model phase, φ_0. When the condition of eqn. 1.5 is met, the lower frequency bound, f_L, is positive and the signal bandwidth factor is restricted to $\mu < 2$. This distinguishes UWB radio signals from video pulses, for which $f_L = 0$ (DC current) and $\mu = 2$, while radio signals permit radiation and waveguide transmission in accordance with eqn. 1.5.

The specificity of UWB signals associated with the condition of eqn. 1.5 will be discussed in Section 1.2.

An important feature of radar signals is their coherence, commonly attributed to their strict phase stability because of the narrow bandwidth. For UWB signals, the concept of coherence must be generalised to mean the exact repeatability of the signal waveform, as is illustrated by the $s(t)$ pulse waveform in Fig. 1.4: $s_T(t) = s(t \pm nT)$, where $n = 0, 1, 2 \ldots$, and $T = 1/F$ is the repetition rate. A strict definition of coherence implies an absolutely stable repetition rate of radio pulses, T, and on this condition is the concept of coherence equivalent to the requirement for all components of a periodic UWB signal to be coherent. A weaker requirement for coherence simply reduces to the same pulse waveform, $s(t)$ (Fig. 1.4), in each period T relative to the reference time $t \pm nT = 0$, where T is given by the main UWB radar oscillator.

This property of UWB pulses can be used successfully for coherent pulse storage and for target detection and estimation of its speed and angle [46, 47]. Modern fast response electronic devices used in UWB radars provide the necessary precision of signal shaping and processing. Conventional descriptions of narrowband radar systems widely apply the ambiguity

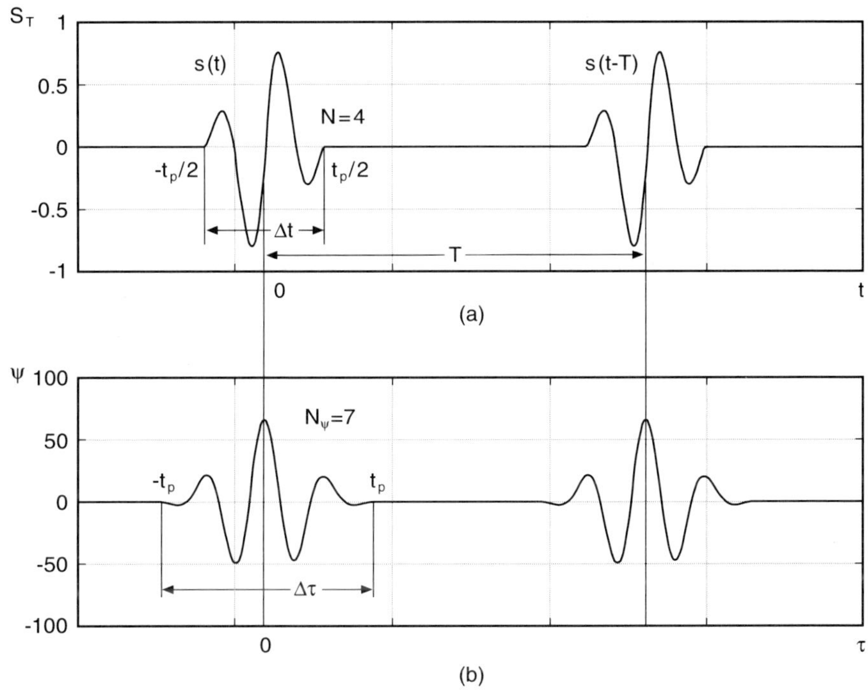

Fig. 1.4 *a* Periodic UWB radio pulses
b Their correlation function

function [71]

$$X(\tau, \Omega) = \int_{-\infty}^{+\infty} x(t) y^*(t, \tau, \Omega)\, dt \tag{1.6}$$

This equation represents the cross correlation function of the complex-valued envelope of the incident signal, $x(t) = A(t)\exp(\mathrm{i}\varphi(t))$, and of the signal reflected by a moving target:

$$y(t - \tau, \Omega) = A(t - \tau)\exp[\mathrm{i}\varphi(t - \tau)]\exp[\mathrm{i}\Omega(t - \tau)]$$

The function $X(\tau, \Omega)$ is usually represented as a three-dimensional diagram which permits estimation of the distance (time delay, τ) and speed (Doppler frequency, Ω) of the target.

To consider the applicability of the ambiguity function to the description of UWB signals, we should discuss the following questions. The first is whether we can design a complex-valued envelope that would adequately evaluate the properties of a UWB signal. This issue is related to the problems of signal amplitude and phase detection already mentioned.

The second question to be examined is that of a valid model to describe the waveform change when the signal is reflected by a moving target, since the Doppler frequencies are quite different for different signal frequency components. Both problems can be resolved by introducing appropriate changes in signal modelling. These issues will be discussed later in this chapter.

The third question concerns eqn. 1.6 and is of greater importance. The principal cross section of the ambiguity function (at $\Omega = 0$) is the correlation function of the signal, with a main lobe width which defines the potential accuracy of the parameter, τ, being estimated. The correlation function is known to be the output response of a filter matched with the incident radar pulse waveform. Therefore, the use of this function *a priori* implies matched signal processing. Such processing is optimal for narrowband signals in the problems of target detection and position measurement, providing optimisation of the signal-to-noise ratio at the given signal energy. Matched filtering is necessary in narrowband systems for signals with fill-in-pulse modulation (linear frequency modulation, phase manipulation, etc.), because it provides a gain in the signal amplitude and compression of its duration to 60 dB and more. The gain from matched filtering of simple pulses has been found to be only about 10%; however, the signal waveform changes appreciably.

Matched filtration is necessary for UWB signals with fill-in-modulation for the same reasons as for narrowband signals. The correlation function of such a UWB signal represents a short, simple radio pulse which is to be treated as an incident UWB signal. When such UWB signals are used to investigate radar target performance in a wide frequency range, the reflected waveform may provide information as to the target characteristics. Therefore, matched filtering of UWB radio pulses, which were noted above to change their shape, is of no use, and the measured parameters should be estimated directly from the received signal waveform, $s(t)$.

Fig. 1.4 shows periodic UWB radio pulses, $s(t)$, and their correlation function, $\Psi(\tau)$. The ability to detect separate signals with parameters differing by no less than the minimum resolvable interval (specifically, the interval $\Delta\tau$ of the time delay τ) is usually called the resolving power of radar measurement.

One can see from Fig. 1.4 that the correlation function $\Psi(\tau)$ has a larger number of time sidelobes, N_ψ, than the original radio pulse, $s(t)$, namely $N_\psi = 2N - 1$. According to the function in eqn. 1.4, it possesses a smaller bandwidth factor, μ, and a larger duration, $\Delta\tau$, i.e. worse resolution. This, too, confirms the inapplicability of matched filtration to highly informative UWB signals, which should be measured directly from the received target response. Direct measurement of target responses would be possible in terms of a time domain model of the UWB radio pulse, $s(t)$, involving a parameter to describe the target speed. In contrast to the function in eqn. 1.6, the processing algorithms of this model do not have to use integral transformations in order to provide further detailed analysis of the radio pulse waveform, $s(t)$. This concerns not only the correlation integral of the function in eqn. 1.6 but any other commonly used integral transformation. We shall discuss this problem below.

1.2 UWB signal representation with real-time functions

Let us examine some properties of the UWB signal model in terms of the real finite function, $s(t)$.

One can estimate the total signal duration using the number of half-periods, N, of centre frequency, $f_0 = 1/T_0$, as $\Delta t = NT_0/2$. Then N may be treated as a discrete measure of signal duration. Similarly, the approximate formula in eqn. 1.4 estimates the bandwidth factor of a signal from the number of its time sidelobes, N. The bandwidth factor of the correlation function in Fig. 1.4 is reduced due to its large length. In the general case, the correlation function of a signal comprising N sidelobes consists of $2N - 1$ sidelobes. With known signal length, it is possible to find its bandwidth

$$\Delta f = \mu f_0 = 4f_0/N = 2/\Delta t \tag{1.7}$$

and to use this value for designing the equipment.

It is known that the minimal variance in the parameter σ_τ^2 in a conventional radar measurement is defined by the properties of the ambiguity function, $X(\tau)$, and that the ratio of the total signal energy, E, to the noise spectral density, N_0, is

$$\sigma_\tau^2 = \frac{N_0}{2E\left(-\dfrac{d^2X}{d\tau^2}\right)\Bigg|_{\tau=0}} \tag{1.8}$$

The magnitude $d^2X/d\tau^2$ is equivalent to the notion of effective frequency determined from the spectral density function, $S(f)$, of the signal $s(t)$

$$f_e^2 = \int_{-\infty}^{+\infty} f^2 |S(f)|^2 \, df \Bigg/ \int_{-\infty}^{+\infty} |S(f)|^2 \, df$$

The potential accuracy here is

$$\sigma_\tau^2 = \frac{N_0}{2E(2\pi f_e)^2} \tag{1.9}$$

We shall use Fig. 1.5, depicting a radio pulse, $s(t)$, with centre frequency $f_0 = 1/T_0$, to find the accuracy of delay, τ_0, estimation for UWB signals without detection and matched filtration. The presence of noise, with a standard value of σ_N, produces an estimation error

$$\sigma_\tau = \sigma_N/(A_m 2\pi f_0)$$

or

$$\sigma_\tau^2 = \frac{P_N}{P_s(2\pi f_0)^2}$$

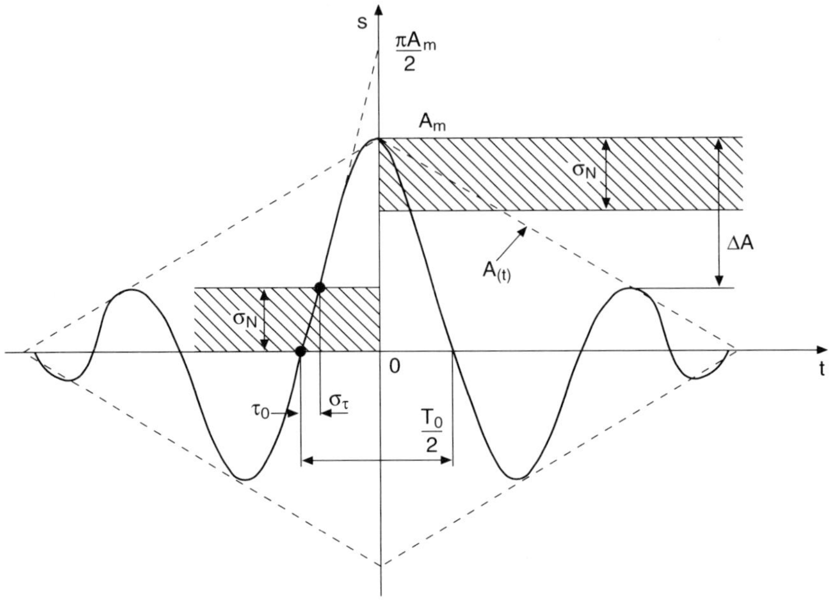

Fig. 1.5 Time domain measurement accuracy using UWB radio signal

where $P_s = A_m^2$ is the maximum instantaneous power of the signal, and $P_N = \sigma_N^2$ is the noise power.

Let us define in more detail the expression for f_e of a UWB signal with arbitrary μ. For this purpose, we represent the signal as $s(t) = A(t)\cos\omega_0 t$. By double differentiation we get

$$\begin{aligned}(2\pi f_e)^2 &= -\frac{1}{A_m}\frac{d^2 s}{dt^2}\bigg|_{t=0} \\ &= -\frac{1}{A_m}\frac{d^2 A}{dt^2} + (2\pi f_0)^2 \\ &= (2\pi F_e)^2 + (2\pi f_0)^2\end{aligned}$$

where F_e is the effective frequency of the signal envelope, $A(t)$ and $A_m = A(0)$. In the flat spectral density approximation the radio signal bandwidth may be estimated [1] as $\Delta f = 2\sqrt{3}F_e$, i.e.

$$\begin{aligned}(2\pi f_e)^2 &= (2\pi f_0)^2\left[1 + \left(\frac{\Delta f}{2\sqrt{3}f_0}\right)^2\right] \\ &= (2\pi f_0)^2\left[1 + \left(\frac{\mu}{2\sqrt{3}}\right)^2\right]\end{aligned}$$

Thus, the potential measurement accuracy of time-dependent parameters, when a UWB signal is used without detection, is practically determined by its centre frequency, f_0, and somewhat exceeds it.

The use of UWB signals not only implies unambiguous parameter measurement within the period T_0, but also provides unambiguous signal peak counting to achieve the measurement accuracy σ_T. It is clear from Fig. 1.5 that unambiguous counting of the parameter t within the signal, $s(t)$, is only possible when two neighbouring peaks of the signal (or its correlation function) are clearly discernible, which implies a sufficient fall of the signal envelope, $A(t)$, in the period of high frequency fill-in, i.e. $\Delta A > \sigma_N$. Using the approximation $A(t) = A_M(1 - kt/t_p)$, we get

$$\sqrt{\frac{P_s}{P_N}} \geqslant \frac{2}{k\mu}$$

where k is the envelope form factor. Therefore, in order to provide unambiguous counting within a UWB signal at $\mu \approx 1$, we do not impose strong constraints on the signal-to-noise ratio, while for narrowband signals and $\mu \to 0$ the constraints become essential and cannot be practically attained.

Real models of narrowband signals are inconvenient owing to their oscillating nature and are meaningless since their high frequency fill-in contains little information. On the contrary, if the UWB signal consists of a few time sidelobes, the real models may become convenient and adequate.

For the total signal, $s(t)$, the condition of eqn. 1.5 for its mean value to be zero is met, i.e.

$$\int s(t)\,dt = 0 \tag{1.10}$$

where the integration limits must correspond with the duration of a finite radio signal.

Let us assume that the condition imposed by eqn. 1.10 defines the class $\{R\}$ of finite radio signals. Models within this class will be distinguished by the nature of their spectral density functions, $S(\omega)$, in the neighbourhood of point $\omega = 0$. In the class $\{R\}$, we single out the subclasses $\{R^{(\nu)}\}$ with the zero derivative

$$d^{(k)}S(\omega)/d\omega^k = 0 \tag{1.11}$$

for $\omega = 0$ and $0 \leqslant k \leqslant \nu$.

By the properties of the Fourier transform, eqn. 1.11 implies that the time moment of the signal, $s(t)$, is equal to zero up to the νth order

$$\int t^k s(t)\,dt = 0$$

$$\text{at} \quad 0 < k \leqslant \nu. \tag{1.12}$$

We represent the spectral density function of a signal, $s(t)$, at the point $\omega = 0$ by a Taylor series, i.e., $S(\omega) = a_0 + a_1\omega + a_2\omega^2 + \ldots$. From eqn. 1.11, for a signal from the subclass $\{R^{(\nu)}\}$ we have $a_i = 0$ for $0 \leqslant i \leqslant \nu$ or

$$S(\omega) = a_{\nu+1}\omega^{\nu+1} + a_{\nu+2}\omega^{\nu+2} + \ldots \tag{1.13}$$

For the spectral density function of the signal $s_I(t) = \int_{-\infty}^{t} s(t)\,dt$ obtained by integrating $s(t)$, we have $S_I(\omega) = S(\omega)/i\omega$ or, according to eqn. 1.13

$$S_I(\omega) = -ia_{\nu+1}\omega^{\nu} - ia_{\nu+2}\omega^{\nu+1} + \ldots \qquad (1.14)$$

We can formulate this result as a theorem: every integration transfers a finite radio signal from the subclass $\{R^{(\nu)}\}$ into the subclass $\{R^{(\nu-1)}\}$. Therefore, any signal of the subclass $\{R^{(\nu)}\}$ allows ν consecutive integrations, still remaining in the class of finite radio signals $\{R\}$, when eqn. 1.10 is satisfied.

Narrowband radio signals can be approximately attributed to the subclasses $\{R^{(\nu)}\}$ with large values of ν. This is so because if the finite function $s(t)$ in eqn. 1.12 is highly oscillatory, the integral is approximately equal to zero whatever the value of ν.

For a model of a radio signal with a rectangular envelope

$$s_r(t) = A_0 \cos(\omega_0 t + \varphi_0) \cdot [1(t - t_1) - 1(t - t_2)]$$

where $1(t)$ is a step function, the condition of eqn. 1.10 is met for the expression

$$\omega_0(t_2 - t_1) = 2\pi k \qquad (1.15)$$

where $k = 1, 2, \ldots$ and $t_2 - t_1 = t_p$ is a pulse duration. Such signal models belong to the subclass $\{R^{(0)}\}$ of radio signals.

In order to extract the subclass $\{R^{(1)}\}$ from the set of these signals, in addition to eqn. 1.10, we equate to zero the first moment of signal model, $s(t)$

$$\int_{t_1}^{t_2} t \cdot s(t)\, d(t) = 0$$

This is equivalent to the zero derivative of the spectral density function, $dS(\omega)/d\omega = 0$, at the null frequency and leads to the equation $t_1 \sin(\omega_0 t_1 + \varphi_0) - t_2 \sin(\omega_0 t_2 + \varphi_0) = 0$, with the roots

$$\omega_0 t_{1,2} + \varphi_0 = k\pi \qquad (1.16)$$

where $k = 0, 1, 2, \ldots$

Eqn. 1.15 defines the relationship between the pulse length, t_p, and frequency, ω_0, invariantly with respect to the initial phase, φ_0, of the high frequency fill-in, while eqn. 1.16 defines this phase. Since there are no other parameters in the model of $s_r(t)$, we cannot single out the subclass $\{R^{(2)}\}$. This is clearly demonstrated by the absence of roots in the second moment of the signal model.

The signal $s_r^{(0)}(t)$ of subclass $\{R^{(0)}\}$ stops being a radio signal after the first integration, because the function $s_1^{(0)}$ in Fig. 1.6a is no longer valid for eqn. 1.10. Double integration violates the finite nature of the original signal (function $s_2^{(0)}$ in Fig. 1.6a). On the other hand, the signal $s_r^{(1)}(t)$ of subclass $\{R^{(1)}\}$ after the first integration still remains in the class $\{R\}$, because the function $s_1^{(1)}$ (Fig. 1.6b) meets the requirement that the mean value in eqn. 1.10 is zero. Double integration violates this condition and, if integrated three times, the signal stops being finite.

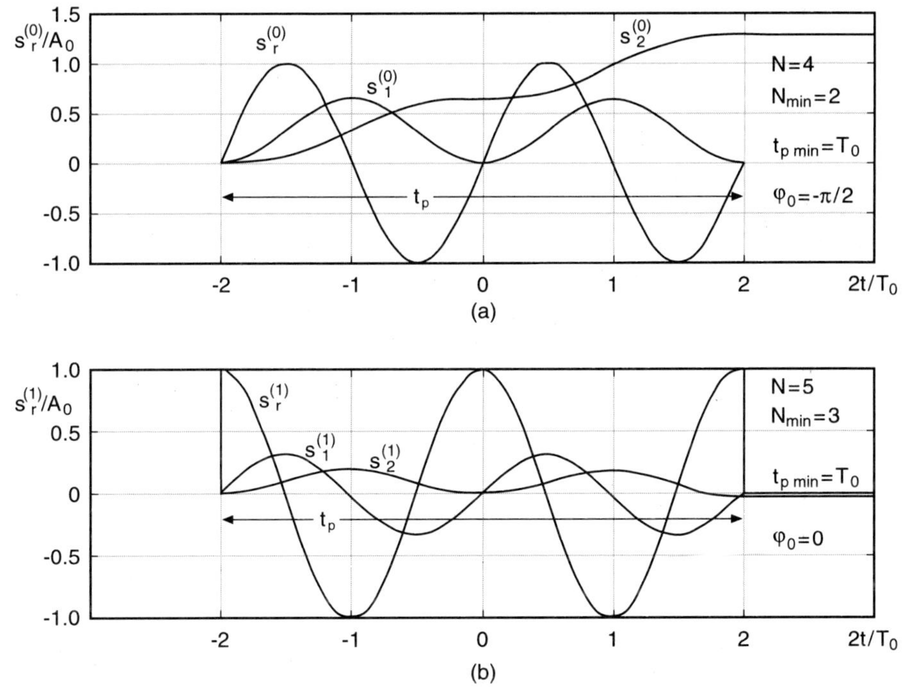

Fig. 1.6 UWB radio signal waveform transformations by consecutive integrations

a Rectangular pulse of the subclass $\{R^{(0)}\}$
b Rectangular pulse of the subclass $\{R^{(1)}\}$

For a radio pulse model $s_T(t) = A_T(t)\cos(\omega_0 t + \varphi_0)$ with a triangular envelope $A_T(t)$, carrier frequency, ω_0, and the initial phase of high frequency fill in, φ_0, eqn. 1.10 yields the condition $\varphi_0 = \pm\pi/2$ (odd function $s_T(t)$) sufficient for the signal to belong to the subclass $\{R^{(0)}\}$. Such a signal is depicted in Fig. 1.6c for a minimum number of time sidelobes, $N_{min} = 2$. For $\omega_0 t_p/2 = 2k\pi$ $(k = 1, 2, 3 \ldots)$, i.e. for an even number of periods of the high frequency fill-in, T_0, the radio signal should be attributed to the subclass $\{R^{(1)}\}$ (Fig. 1.6d). When the conditions $\omega_0 t_p/2 = 2k\pi$ and $\varphi_0 = \pi/2$ are met simultaneously, the signal belongs to the class $\{R^{(2)}\}$ (Fig. 1.6e). In the triangular pulse model we allow for all degrees of freedom; therefore, the third derivative, $d^{(3)}S(\omega)/d\omega^3$, cannot vanish, and there is no triangular radio pulse in the subclass $\{R^{(3)}\}$.

In addition, one can see that UWB signals in Figs. 1.3a,b belong to the subclass $\{R^{(0)}\}$, while the signal in Fig. 1.3c does not belong to the class of radio signals.

Finite radio pulses within the class $\{R\}$ have a generic property that the number of time sidelobes of the pulse decreases by one with every integration (compare the signals $s_T^{(2)}$, $s_1^{(2)}$ and $s_2^{(2)}$ in Fig. 1.6e). We shall show that this may serve as a discriminating feature for the type of scattering

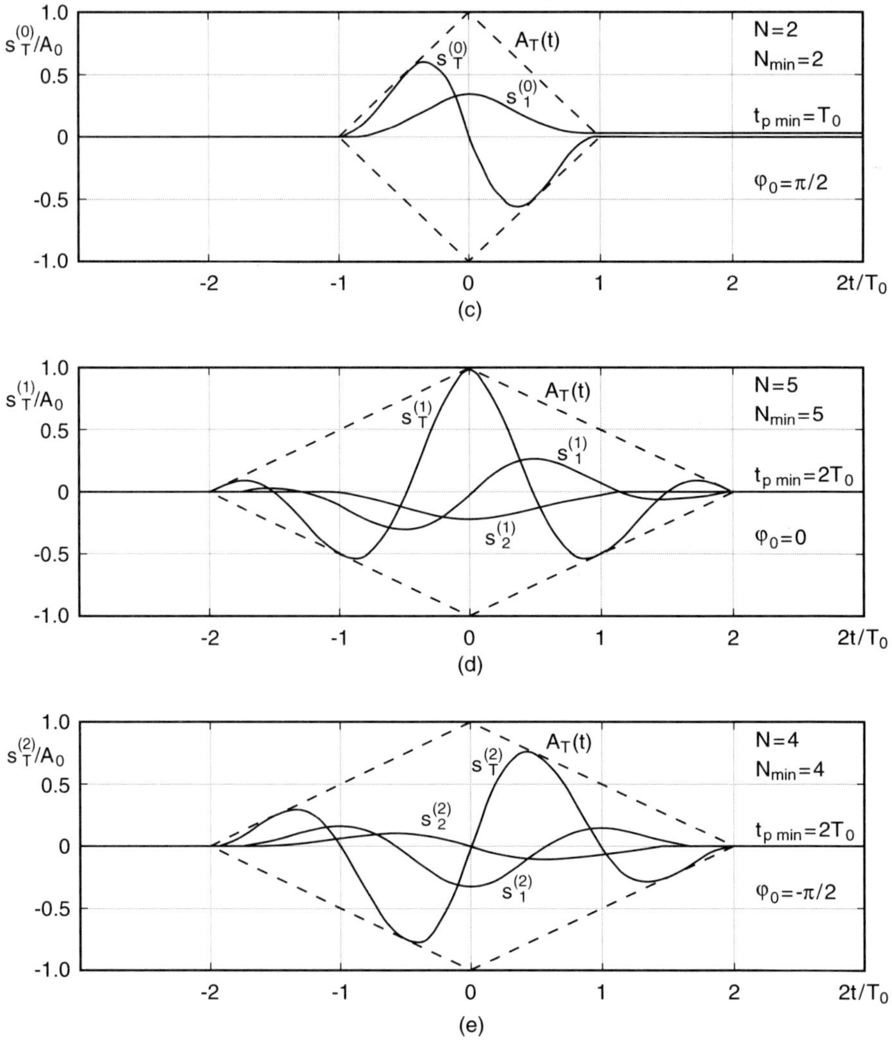

Fig. 1.6 Continued

c Triangular pulse of the subclass $\{R^{(0)}\}$
d Triangular pulse of the subclass $\{R^{(1)}\}$
e Triangular pulse of the subclass $\{R^{(2)}\}$

which shows, for example, how many times a radar signal is re-reflected by a target.

We can use eqn. 1.13 to show that the first differentiation of any finite signal transfers it into the class of radio signals, and that the number of time sidelobes increases by one with every differentiation. For narrowband signals the relative energy of the initial and mean timelobes is insignificant, and these features need not be allowed for; however, their role is essential for UWB signals.

Recall that we have restricted ourselves to a one-parameter model of the UWB signal. But even this simplest case illustrates its main properties, which influence the choice of methods for the simulation and estimation of signal parameters. This justifies the use of a special term – ultrawideband radio signals.

1.3 The applicability of conventional modelling

1.3.1 Frequency-domain representation of UWB signals

Frequency-domain representation of radar signals and responses is an effective technique that has found wide application. In this method a real signal model, $s(t)$, is expressed by its Fourier transform

$$S(\omega) = \int_{-\infty}^{+\infty} s(t) \exp(-\mathrm{i}\omega t)\, dt$$

This integral is the total signal representation, in which the value of the spectral density, $S(\omega)$, at any frequency, ω, is defined by all instantaneous values of the $s(t)$ function. The integral model has stimulated the use of integral parameters such as a video pulse form factor, k, conventionally employed for the estimation of pulse risetime, t_r

$$t_r = \frac{k}{\Delta f} \tag{1.17}$$

where Δf is the pulse bandwidth.

Similarly, integral (moment) evaluations are made of the spectral density function, $S(f)$, for estimation of effective frequency, f_e, represented as

$$f_e = \sqrt{\int f^2 |S(f)|^2\, df} \Big/ \sqrt{\int |S(f)|^2\, df} \tag{1.18}$$

In accordance with eqn. 1.9, this defines the potential measurement accuracy in the time domain. By analogy, the effective signal duration is estimated as

$$t_e = \sqrt{\int t^2 s^2(t)\, dt} \Big/ \sqrt{\int s^2(t)\, dt} \tag{1.19}$$

These characteristics are considered to be resolving power estimates in the time and frequency domains, respectively.

Theoretically, this spectral model allows restoration of the time domain signal structure. However, if a signal has discontinuities in the $s(t)$ function or its derivatives, the $S(\omega)$ spectral density extends to infinity with respect to frequency ω. So it becomes necessary to truncate the signal's spectral density at certain frequencies, $\pm\omega_H$, because of the restrictions imposed by the computational and instrumental feasibility of measurement. As a result, the inverse transformation to the time domain can be made only approximately. The restored signal bandwidth shows transient processes in the vicinity of discontinuity points having a finite rise time and oscillation distortions of the signal structure (Gibbs phenomenon). For finite signals this occurs at their origins and end points, leading to the violation of the causality principle and impeding the attainment of information on the event onset, i.e. on the time moment of response arrival. The use of smooth truncation of the signal spectral density (for example, of the Fejer window) may remove oscillations in a transient characteristic, but the finite rise time and the causality violation still remain.

The duration of a transient process, t_r, is defined by the value ω_H of the upper bandwidth frequency. By measuring video pulses with Δt duration this problem can be solved for a much wider frequency bandwidth than is normally used in the treatment of short duration transient processes, i.e. $t_r \ll \Delta t$. However, in the UWB model the duration of a transient process is comparable with the centre frequency period, $T_0 = 2\pi/\omega_0$, namely with the duration, t_p, of the UWB pulse proper (or its correlation function). In this case, measurement errors may be appreciable. Excessive frequency, $\omega_H \gg \omega_0$, may be very difficult to create owing to the increased noise and because the spectral density of UWB signals occupies the whole of the technologically possible frequency bandwidth.

Clearly, a frequency domain representation of UWB responses involves serious challenges associated with the integral character of transformations, which obscure the signal bandwidth details. One can resolve these difficulties by representing individual segments of a signal, using the concepts of instantaneous and current spectra [54, 71] by which the signal is represented in the time-frequency plane. These issues will be discussed below.

1.3.2 A signal complex model with the Hilbert transform

In order to extract information from received radar signals, the signals are processed to give them a form suitable for measurement. Any measurement involves comparison with a certain standard, say, by calculating the correlation integral. In this way the conditions for signal detection and for the measurement of the target position and speed are optimised. To obtain noncoordinate information, one should analyse the time domain structure of the response, which can generally be written as

$$\begin{aligned} s(t) &= A(t)\cos\Psi(t) \\ &= A(t)\cos[\omega_0 t + \varphi(t)] \end{aligned} \tag{1.20}$$

where $\Psi(t)$ is the full phase of the oscillating factor.

For narrowband signals, the linear component of the full phase, $\omega_0 t$, adds no information, while the functions $A(t)$ and $\varphi(t)$ mainly give information of different physical nature – dissipation and dispersion characteristics, respectively. To find these functions, one must construct a complex signal

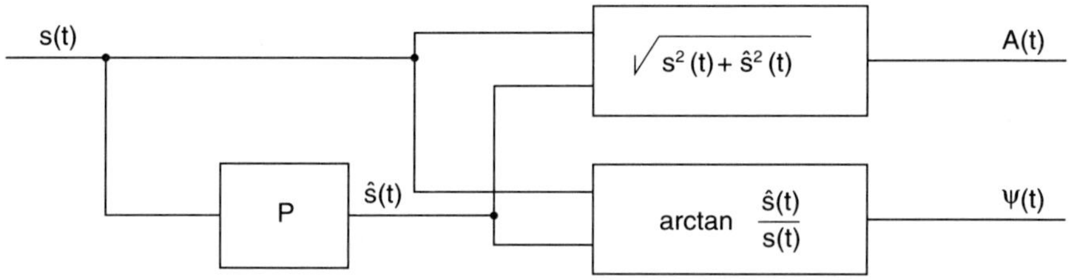

Fig. 1.7 Structure of complex-valued signal shaper

as a pair of functions in two quadrature channels and perform detection, considerably reducing the requirements for the processing time. Then the functions $A(t)$ and $\varphi(t)$ are calculated by processing the complex signal in both quadrature channels.

When dealing with UWB signals the centre frequency, ω_0, should not be ignored. Consider the applicability of the conventional complex pulse model for the description of UWB responses.

A complex UWB signal is commonly represented as a pair of functions, say, $s(t)$ and $\hat{s}(t)$ in Fig. 1.7

$$u(t) = s(t) + i\hat{s}(t) \tag{1.21}$$

This is convenient because of the use of conventional mathematical apparatus for complex-valued functions. If the operator P of the $s(t) \rightarrow \hat{s}(t)$ transformation has been chosen, then any $s(t)$ uniquely corresponds to the complex signal described by eqn. 1.21. In the usual quadrature processing, a narrowband signal, $u(t)$, can be written representatively as

$$u(t) = A(t) \exp i\Psi(t) \tag{1.22}$$

where envelope $A(t)$ and phase $\Psi(t)$ can be derived from Fig. 1.7.

Radio engineering theory commonly employs the formation of complex signals, $u(t)$, with the Hilbert transform of $s(t)$ as $\hat{s}(t)$

$$\begin{aligned} \hat{s}(t) = s_H(t) &= \frac{1}{\pi}\int_{-\infty}^{+\infty} \frac{s(\tau)}{t-\tau}\, d\tau \\ &= s(t) * \frac{1}{t} \end{aligned} \tag{1.23}$$

representing the convolution of the real function, $s(t)$, with the discontinuity function, $1/t$. It has been shown [114] that the Hilbert operator is the only one in the class of linear homogeneous continuously differentiable operators that can provide the envelope $A(t)$ and phase $\Psi(t)$ of the complex model (eqn. 1.22), satisfying some physically reasonable requirements. In particular, they coincide, in the limiting case, with the parameters of a monochromatic signal, $s(t) = A_0 \cos \omega_0 t$.

It is important for our treatment of UWB signals that the Hilbert imaginary component, $s_H(t)$, of the complex signal, $u(t)$, is an infinitely differentiable function; therefore, it does not satisfy the physical principle of causality even if the original signal, $s(t)$, is finite. For this reason, the complex signal

$$u(t) = s(t) + \mathrm{i}s_H(t) \tag{1.24}$$

formed with the operator of eqns. 1.21 and 1.23 is known as the analytical signal, and does not possess causality in time.

Examine in this connection the analytical modelling of UWB radio pulses. We shall make the Hilbert transformation for an arbitrary (also UWB) radar signal

$$\begin{aligned} s(t) &= A(t)\cos\Psi(t) \\ &= A(t)\cos[\omega_0 t + \varphi(t) + \varphi_0] \end{aligned} \tag{1.25}$$

With $\cos x = 1/2[\exp(\mathrm{i}x) + \exp(-\mathrm{i}x)]$, we write for $s(t)$ spectral density

$$S(\omega) = 1/2\exp(\mathrm{i}\varphi_0)F(\omega - \omega_0) + 1/2\exp(-\mathrm{i}\varphi_0)F^*[-(\omega + \omega_0)] \tag{1.26}$$

where

$$F(\omega) = \int_{-\infty}^{+\infty} A(t)\exp[\mathrm{i}(\varphi(t) - \omega t)]\,dt$$

The frequency response of the shaper making the Hilbert transformation is known to correspond to an ideal phaser

$$H(\omega) = -\exp(\mathrm{i}\pi/2)\mathrm{sgn}\,\omega \tag{1.27}$$

Then the Hilbert signal spectral density can be written as

$$S_H(\omega) = S(\omega)H(\omega)$$

The spectral density function, $S(\omega)$, and the frequency response, $H(\omega)$, are shown in Fig. 1.8a,b. For a narrowband signal (dashed lines), the calculation of $S_H(\omega)$ reduces to finding the product $S(\omega)\cdot H(\omega)$ for the frequency values $\omega > 0$ and $\omega < 0$ independently. For a UWB signal with a large value of $\mu = \Delta\omega/\omega_0$, one should approximately assume the absence of streaks in the spectral density functions, $F(\omega \pm \omega_0)$, for the frequency $\omega = 0$ (shaded regions in Fig. 1.8a). In this case, the spectrum, $F(\omega)$, must be treated as finite

$$|F(\omega)| = 0 \tag{1.28}$$

with $|\omega| > \omega_0$.

Then the spectral density of the Hilbert-conjugated signal is defined as

$$\begin{aligned} S_H(\omega) = {} & 1/2\exp[\mathrm{i}(\varphi_0 - \pi/2)]F(\omega - \omega_0) \\ & + 1/2\exp[-\mathrm{i}(\varphi_0 - \pi/2)]F^*[-(\omega + \omega_0)] \end{aligned} \tag{1.29}$$

and as a result

$$s_H(t) = A(t)\sin[\omega_0 t + \varphi(t) + \varphi_0] \tag{1.30}$$

This proves that application of the Hilbert transformation to a radar response (eqn. 1.25) with arbitrary modulation, as well as to a UWB signal,

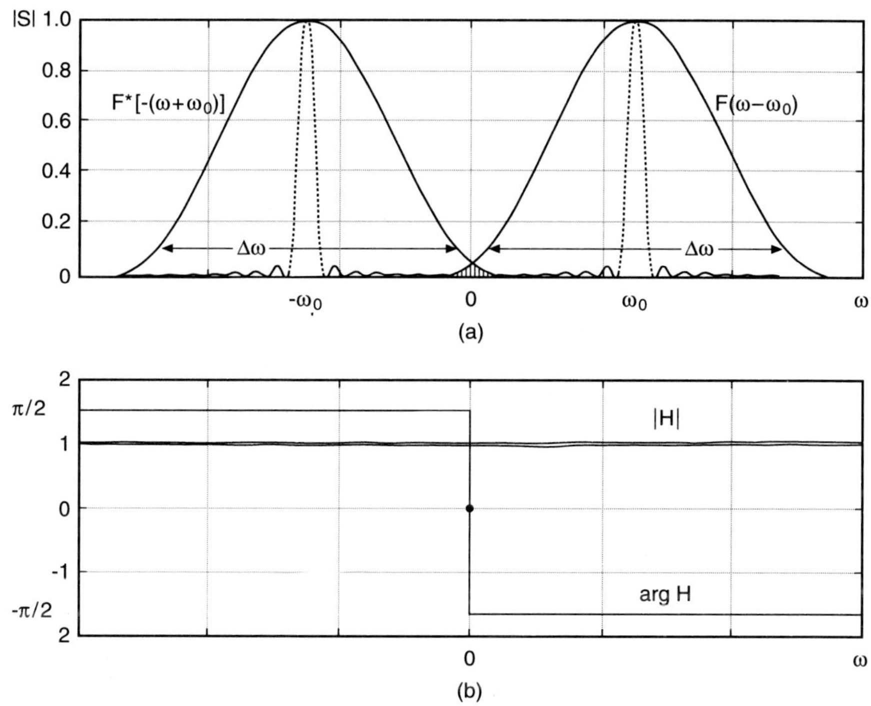

Fig. 1.8 *a* Radio signal spectral density functions
b Frequency response of Hilbert transform shaper

yields exactly the expression (eqn. 1.30) representing the original signal quadrature on the single condition (eqn. 1.28). In this case the complex signal described by eqn. 1.24 can be exactly represented in the exponential form (eqn. 1.22) and is an analytical signal.

Let us use the quadrature component, $s_H(t)$, to construct a complex signal and then define the complex envelope $f(t)$ (with the spectral density function $F(\omega)$)

$$\begin{aligned} u_H(t) &= s(t) + \mathrm{i}s_H(t) \\ &= A(t)\exp[\mathrm{i}(\omega_0 t + \varphi(t))] \\ &= f(t)\exp(\mathrm{i}\omega_0 t) \end{aligned} \quad (1.31)$$

where $A(t) = \sqrt{s^2(t) + s_H^2(t)}$

$$\omega_0 t + \varphi(t) = \arctan\frac{s_H(t)}{s(t)}$$

$$f(t) = A(t)\exp(\mathrm{i}\varphi(t))$$

In keeping with our considerations (Section 1.3.1), the original spectrum bandwidth truncation, as in the expression of eqn. 1.28, leads to an error

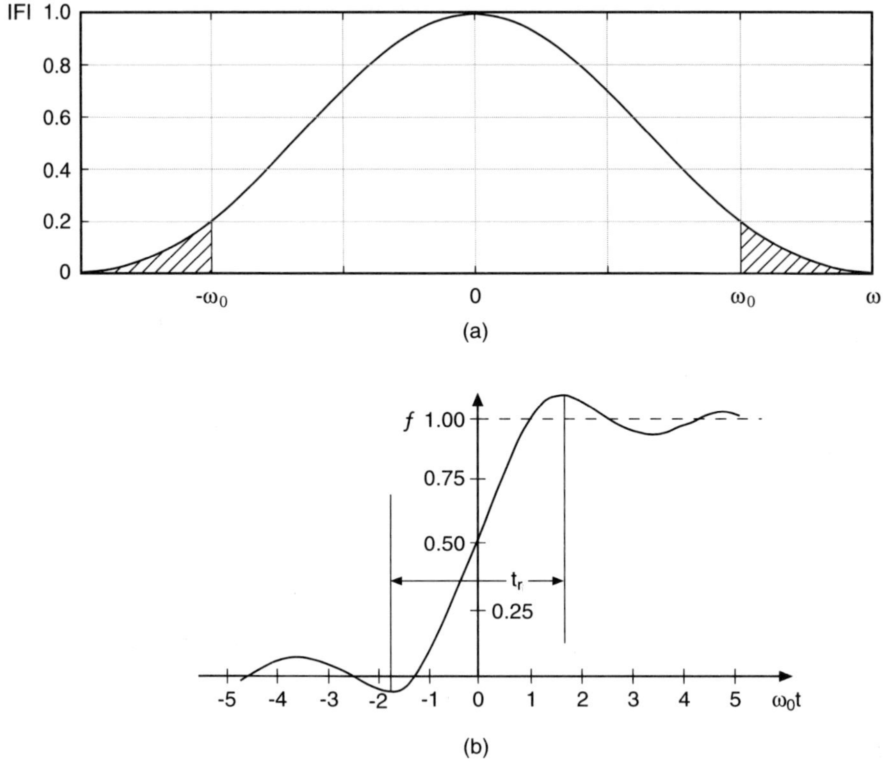

Fig. 1.9 *a* Complex envelope spectrum truncation
b Its effect on transient processes

in the analytical signal envelope, as compared to the original envelope of $s(t)$, when we calculate the Hilbert transform of $s(t)$. For example, for a signal representing the switching of harmonic oscillation

$$s(t) = \cos(\omega_0 t + \varphi_0)1(t) \tag{1.32}$$

with an envelope in the form of a step function, $1(t)$, the truncation of $F(\omega)$ at frequencies $\pm\omega_0$ (Fig. 1.9a) results in the transient process, $f(\omega_0 t) = \mathrm{Si}(\omega_0 t)$, shown in Fig. 1.9b. Since the truncated sections of the $F(\omega)$ bandwidth have the highest frequencies, the influence of truncation reveals itself in the vicinity of the envelope discontinuity point, $\omega_0 t = 0$. The characteristic errors are manifested as a violation of the signal causality ($f(\omega_0 t) \neq 0$ at $t < 0$) and as the finite rise time of the envelope

$$t_r \cong \pi/\omega_0 \tag{1.33}$$

This parameter is comparable with the total duration of a UWB radio pulse.

These factors should be considered as being inadequate to UWB signal and radar response modelling problems, because one cannot obtain a high measurement accuracy in time intervals comparable with the period of the centre frequency, ω_0, or within the transient process time, t_r (eqn. 1.33). Note also that this process has a Gibbsian nature, since the only product $\omega_0 t$ is the argument of $f(\omega_0 t)$.

Let us estimate the error in the analytical signal calculation associated with the spectrum bandwidth truncation, $S(\omega)$, of the original signal, $s(t)$, having discontinuity points. For this, we define the Hilbert transform by a one-sided inverse Fourier transform of the exact expression for $S(\omega)$ [35]

$$u_H(t) = \frac{1}{\pi}\int_0^\infty S(\omega)\exp(\mathrm{i}\omega t)\,d\omega \tag{1.34}$$

This expression yields an analytical signal with a complex envelope, $f_H(t)$, which differs from that of the original signal, $s(t)$, by the presence, in the integral of eqn. 1.34, of the portion $F^*[-(\omega+\omega_0)]$ crossing the point $\omega = 0$ and by the absence of a similar portion, $F(\omega-\omega_0)$ (see Fig. 1.8a). Owing to the Fourier transform linearity, the difference between the envelope $f_H(t)$ of the complex signal $u(t)$ and that of the original signal, $s(t)$, can be represented by the Fourier transform of these spectral portions

$$\begin{aligned}\Delta(t) = &-\frac{1}{2\pi}\exp(\mathrm{i}\omega_0 t)\int_{-\infty}^{-\omega_0} F(\omega)\exp(\mathrm{i}\omega t)\,d\omega\\ &+\frac{1}{2\pi}\exp(-\mathrm{i}\omega_0 t)\int_{-\infty}^{-\omega_0} F^*(\omega)\exp(-\mathrm{i}\omega t)\,d\omega\end{aligned}$$

As was mentioned above, the error is due to the high frequency components of the spectral density function, $F(\omega)$ (see Fig. 1.9a); therefore, its influence is essential in the vicinity of discontinuity points of the signal $s(t)$ or of its derivatives.

For the signal described by eqn. 1.32 with a stepped envelope, $1(t)$, Fig. 1.10a,b shows the components $\Gamma(\omega_0 t)$ and $\Phi(\omega_0 t)$ of the analytical signal complex envelope

$$f_H(\omega_0 t) = \Gamma(\omega_0 t)\exp[\mathrm{i}\Phi(\omega_0 t)] \tag{1.35}$$

obtained with $\Delta(t)$ for various φ_0 values of the model of eqn. 1.32. It follows from Fig. 1.10 that, in spite of the exact calculation of the Hilbert transform for $s(t)$, the envelope causality is still violated and there is a transient oscillation in the time interval comparable with the period $T = 2\pi/\omega_0$. The analogy with the Gibbs phenomenon is preserved.

Similar calculations for the $s(t)$ signal containing discontinuities in the first envelope derivative demonstrate a smaller error in the analytical envelope, as is seen from Fig. 1.10c, where N is the unit of gradient for the original $s(t)$ envelope. In this case, too, the signal causality has been lost, but the degree of manifestation of the error depends on N, which means that the analogy with the Gibbs phenomenon is absent. There is a region, though very narrow, with a complex envelope discontinuity at $t = 0$.

Thus, refined calculations of the Hilbert transform for discontinuous signals do not eliminate errors in the analytical signal in the vicinity of the discontinuity points of signal $s(t)$ and its derivatives.

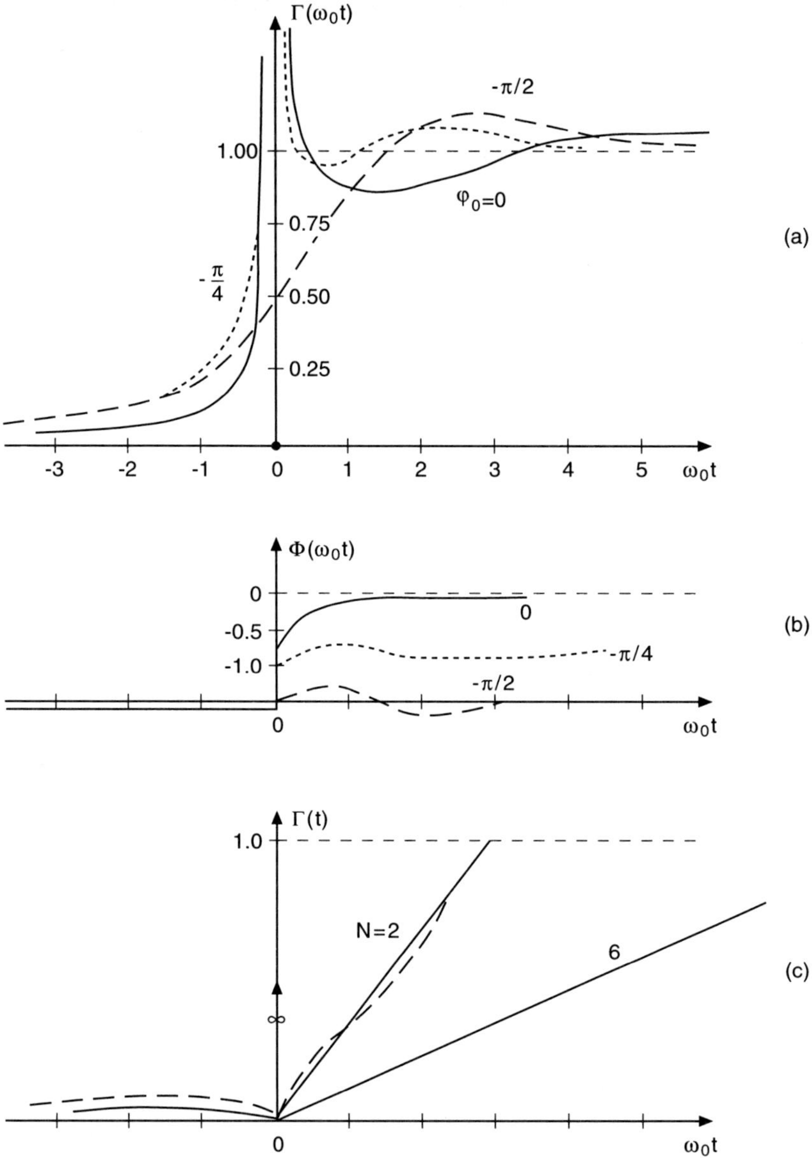

Fig. 1.10 Components of the analytical signal

a, b With stepped envelope
c With ramp envelope

Turn now to the properties of the $A(t)$ envelope in regions with no $s(t)$ discontinuities. We shall restrict ourselves to the case of amplitude modulation, $s(t) = A(t)\cos\omega_0 t$. The usual requirement on the $A(t)$ envelope is that it must coincide with the oscillating $s(t)$ signal at points near its maximum values, s_{max}, and that the tangents at these points must also coincide

$$\frac{dA}{dt} = \frac{ds}{dt} \tag{1.36}$$

with $s \approx s_{max}$.

By differentiating the expression for the complex signal envelope

$$A(t) = \sqrt{s^2(t) + \hat{s}^2(t)} \tag{1.37}$$

we obtain

$$A(t)\frac{dA}{dt} = s(t)\frac{ds}{dt} + \hat{s}(t)\frac{d\hat{s}}{dt} \tag{1.38}$$

According to eqn. 1.37, we must have $\hat{s}(t) = 0$ at the coincidence points, $A(t) = s(t)$. Then

$$\frac{dA(t)}{dt} = \frac{ds(t)}{dt}$$

Therefore, the derivatives of $A(t)$ and $s(t)$ at the tangency points are equal at any $\hat{s}(t)$.

At the same time, for the coincidence point $A(t) = s(t)$ to be the signal peak, s_{max}, the function $\hat{s}(t)$, according to eqn. (1.38), must be proportional to the signal derivative and, hence, turn to zero at $s(t) = s_{max}$

$$\hat{s}(t) = k\frac{ds}{dt} \tag{1.39}$$

Let us show that the function $s_H(t)$, representing the Hilbert transform of $s(t)$, meets the condition of eqn. 1.39 only approximately, so that the coincidence of the envelope $A(t)$ and the real signal maximum, s_{max}, is also approximate.

Consider an amplitude-modulated radio signal

$$s(t) = A(t)\cos\omega_0 t = (1 + m\cos\Omega_m t)\cos\omega_0 t \tag{1.40}$$

Such a signal is continuously differentiable and has a finite bandwidth at $\omega_0 \pm \Omega_m$; therefore, at $\Omega_m \leqslant \omega_0$ the exact expression for the Hilbert transform of $s(t)$, according to eqn. 1.30 has the form

$$s_H(t) = (1 + m\cos\Omega_m t)\sin\omega_0 t \tag{1.41}$$

Compare this expression with the exact magnitude of the derivative ds/dt of the signal, $s(t)$ (eqn. 1.40)

$$\frac{ds}{dt} = -\omega_0\left[(1 + m\cos\Omega_m)\sin\omega_0 t + m\frac{\Omega_m}{\omega_0}\sin\Omega_m t\cos\omega_0 t\right] \tag{1.42}$$

Assuming the modulation to have a narrowband nature

$$\Omega_m/\omega_0 \ll 1$$

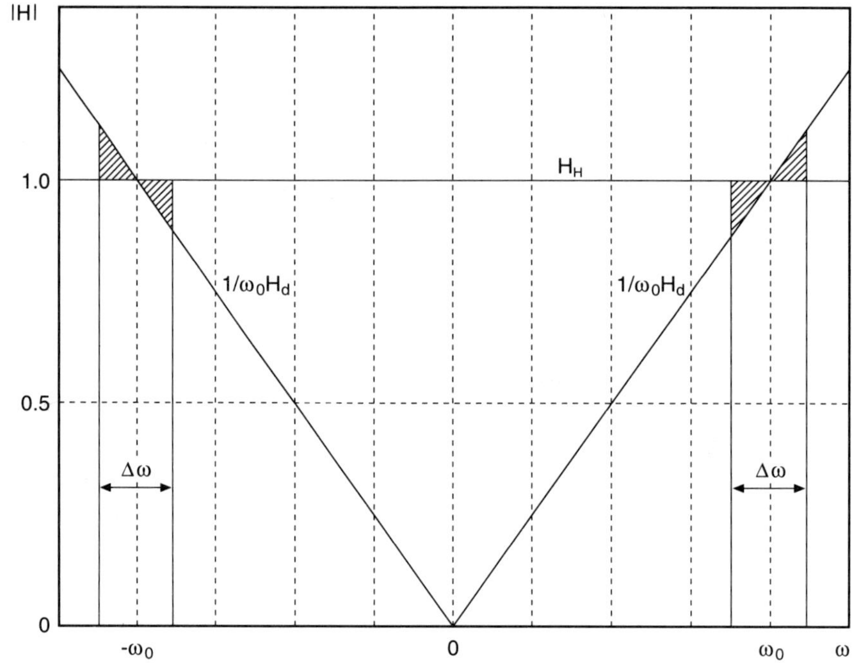

Fig. 1.11 Comparison of Hilbert transformer and differentiator frequency responses

we arrive at an expression coinciding with eqn. 1.41, with the exception of the factor $-\omega_0$. Therefore, the Hilbert transform of the $s(t)$ signal can be regarded as a narrowband approximation of the differentiation procedure for this signal.

A comparison of the Hilbert transform with the differentiation of a signal in the frequency domain leads one to a similar conclusion. Fig. 1.11 gives the amplitude-frequency responses of the Hilbert transformer, $H_H(\omega) \equiv 1$, and of the differentiator, $1/\omega_0 H_d(\omega)$. The regions of difference are shaded, and for a narrowband signal with $\Delta\omega/\omega_0 \ll 1$ they may be neglected. The phase responses of the Hilbert transformer and the differentiator are known to be identical for the whole frequency domain.

Fig. 1.12 shows that the instantaneous maximum value, s_m, of an amplitude-modulated signal, $s(t)$ (eqn. 1.40), coincides neither with the value of eqn. 1.37 for the envelope $A(t_m)$ nor with that of the carrier oscillation, $\cos\omega_0 t_m$, and is achieved at a moment t_m different from $t = 0$. It is easy to obtain the expressions

$$\omega_0 t_m = m\frac{\Omega_m}{\omega_0}\sin\Omega_m t$$

$$s_m = 1 + \frac{1}{2}\left(\frac{m\Omega_m}{\omega_0}\sin\Omega_m t\right)^2$$

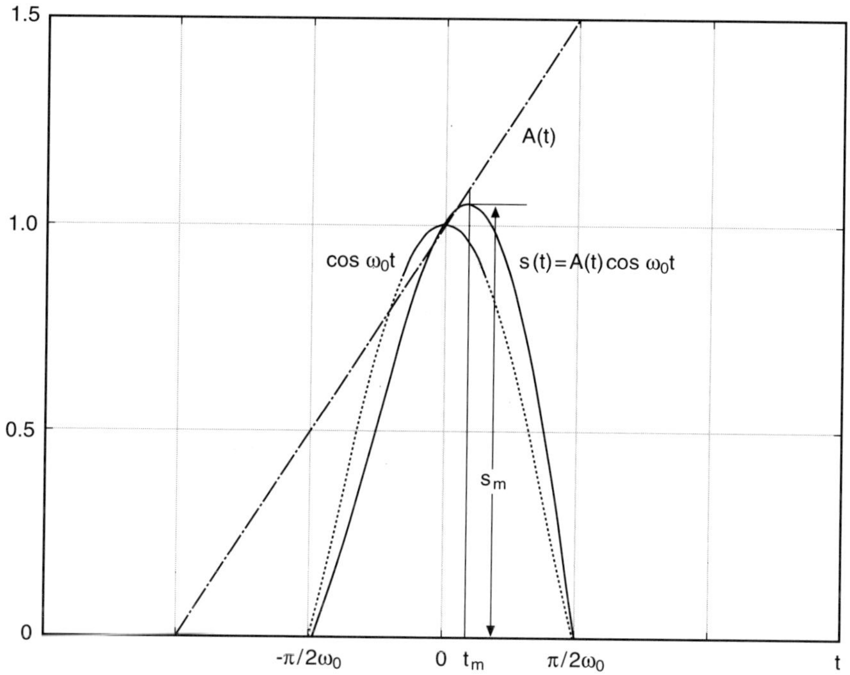

Fig. 1.12 Time sidelobe of UWB amplitude-modulated signal

showing that the error magnitudes depend on how wide the fractional bandwidth, Ω_m/ω_0, is. In UWB modulation ($\Omega_m/\omega_0 \approx 1$) and in high modulation ($m \approx 1$) at the points of maximum envelope steepness ($\sin \Omega_m t \approx 1$), the errors may be quite appreciable, $\omega_0 t_m \approx 1$ and $s_m \approx 1.5$. Then the time sidelobe of the amplitude-modulated oscillation in Fig. 1.12 differs remarkably from the cosine curve halfperiod.

Therefore, in addition to errors in the vicinity of the discontinuity points of an $s(t)$ radio signal, the Hilbert complex model introduces errors in the signal representation at the sites where a smooth envelope shows abrupt changes. The asymptotic similarity of the Hilbert transformation and differentiation of $s(t)$ functions for narrowband signals may become a basis for the extension of the concept of the complex envelope to UWB signals.

1.3.3 Errors in UWB signal description

The specific features of the complex model using the Hilbert transform that we have discussed so far are responsible for errors in the description of UWB signals. It was pointed out in Section 1.3.2 that errors must appear at the sites of abrupt signal change and around the points of discontinuity of the $s(t)$ function. Let us examine the nature of such errors, using the usual complex envelope method.

The properties of radio channel circuits can be defined in terms of their frequency response, $H_0(\omega)$, which, after a shift by ω_0, is expressed in a Taylor series

$$\begin{aligned} H(\omega) &= H_0[(\omega + \omega_0)] \\ &= \sum_{n=0}^{\infty} a_n (\mathrm{i}\omega)^n \end{aligned} \tag{1.43}$$

The filter input signal, $s_i(t)$, is given by the complex model $u(t) = f(t)\exp(\mathrm{i}\omega_0 t)$ with a complex envelope, $f(t)$, and a corresponding spectral density function, $F(\omega)$. We can find the complex envelope $g(t)$ of the output signal $s_0(t)$ as

$$g(t) = \frac{1}{2\pi}\int_{-\infty}^{+\infty} F(\omega)H(\omega)\exp(\mathrm{i}\omega t)\,d\omega$$

Substituting eqn. 1.43 into the above equation, we find that the output signal envelope is in the form of a series in the moments of its spectrum

$$g(t) = \frac{1}{2\pi}\left[a_0\int_{-\infty}^{+\infty} F(\omega)\exp(\mathrm{i}\omega t)\,d\omega + a_1\int_{-\infty}^{+\infty} \omega F(\omega)\exp(\mathrm{i}\omega t)\,d\omega + \dots\right]$$

From the theorem for the spectral density of a derivative, we have

$$g(t) = \sum_{n=0}^{\infty} a_n(-\mathrm{i})^n \frac{d^{(n)}f(t)}{dt^n} \tag{1.44}$$

This relationship describes the response of a linear circuit to an arbitrary input excitation without integral Fourier transform or convolution.

Note that eqn. 1.44 is of a local, i.e. high frequency, nature. The complex envelope of the transformed signal, $g(t)$, is defined by the complex envelope of the input signal, $f(t)$, and its derivatives at the same moment of time.

Let us consider some transformations of the complex envelope in typical radio engineering circuits. A nondistorting circuit has a steady amplitude-frequency response, $|H(\omega)| = A$, and a linear phase-frequency response, namely

$$\begin{aligned} H(\omega) &= A\exp[-\mathrm{i}(\omega + \omega_0)t_d] \\ &= A\exp(-\mathrm{i}\omega_0 t_d)\left(1 - \mathrm{i}t_d\omega - \frac{1}{2!}t_d^2\omega^2 + \dots\right) \end{aligned}$$

According to eqn. 1.44, the transformation of a continuously differentiable complex envelope, $f(t)$, consists in its delay by the time t_d

$$\begin{aligned} g(t) &= A\exp(-\mathrm{i}\omega t_d)\left[f(t) - t_d\frac{df}{dt} + \frac{1}{2!}t_d^2\frac{d^2f}{dt^2} + \dots\right] \\ &= A\exp(-\mathrm{i}\omega t_d)f(t - t_d) \end{aligned}$$

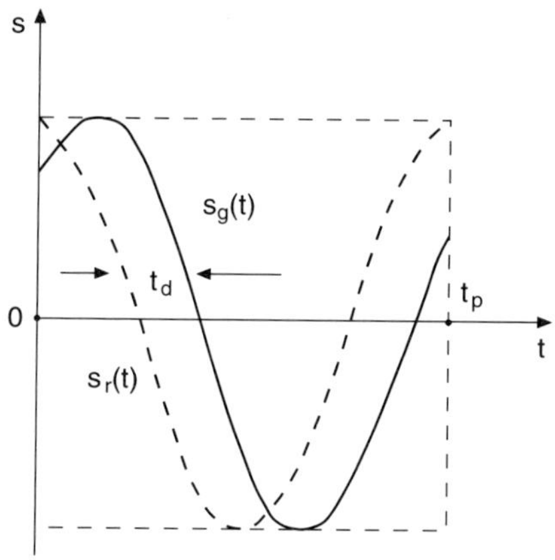

Fig. 1.13 Representation of delayed UWB radio pulse using the complex envelope method

Therefore, the application of the complex envelope approach to a continuously differentiable signal has yielded an exact description.

If the function $f(t)$ or its derivatives have discontinuities, the spectral function of the signal for $\omega = 0$ is obviously nonzero, and the use of eqn. 1.44 introduces an error in the description of signal transformation. Consider a real radio pulse with a rectangular envelope of duration t_p

$$s_r(t) = [1(t) - 1(t - t_p)] \cos \omega_0 t$$

where $1(t)$ is the Heaviside function (step function). We take the first two terms of the series in eqn. 1.44 and find the complex envelope of the signal at the output of a nondistorting circuit

$$g(t) = [1(t) - 1(t - t_p)] \exp(-i\omega t_d) + i\omega_0 t_d^2 [\delta(t) - \delta(t - t_p)]$$

The real output signal, $s_g(t) = [1(t) - 1(t - t_p)] \cos \omega_0 (t - t_d)$, corresponding to $g(t)$, has a high frequency fill-in, delayed by the time, t_d, and an envelope that has no delay (Fig. 1.13). This means that methods using complex envelopes cannot be applied to problems dealing with transformations of discontinuous signals; in other words, these methods are poor when applied to descriptions of transient processes.

Better results can be obtained in a model of signals with a weak discontinuity. If a real signal has a trapezoidal envelope, s_T, the smoothed envelope gives a satisfactory representation of the signal transformation (dotted line in Fig. 1.14).

Note that the local character of eqn. 1.44 permits the description of specific transformations of continuously differentiable UWB signals. For example, the differentiator has a frequency response $H(\omega) = H_0[(\omega + \omega_0)] = i\omega_0(1 + \omega/\omega_0)$ exactly defined by the two terms in eqn. 1.43,

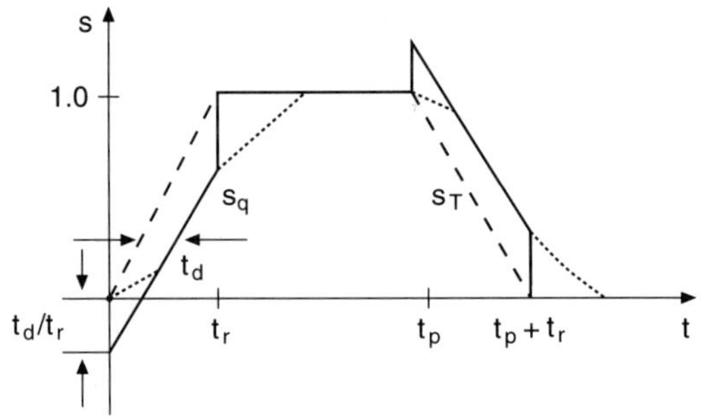

Fig. 1.14 Complex envelope of delayed trapezoidal UWB radio pulse

$a_0 = \mathrm{i}\omega_0$ and $a_1 = \mathrm{i}$. Hence, we get from eqn. 1.44

$$g(t) = \mathrm{i}\omega_0 f(t) + df/dt$$

Here, the first term describes the differentiation of a narrowband radio signal, which changes the high frequency fill-in phase by $\pi/2$ without changing the complex envelope magnitude. The second term accounts for the wideband properties of the signal, and its value is defined by the complex envelope variation rate and is proportional to the bandwidth factor $\mu = \Delta\omega/\omega_0$. On differentiation of the trapezoidal radio signal (Fig. 1.15), the output signal is narrowband in the flat part of the pulse, while the regions with well-defined UWB properties, i.e. with abrupt signal changes, are local and adjoin the points of discontinuity of the signal model.

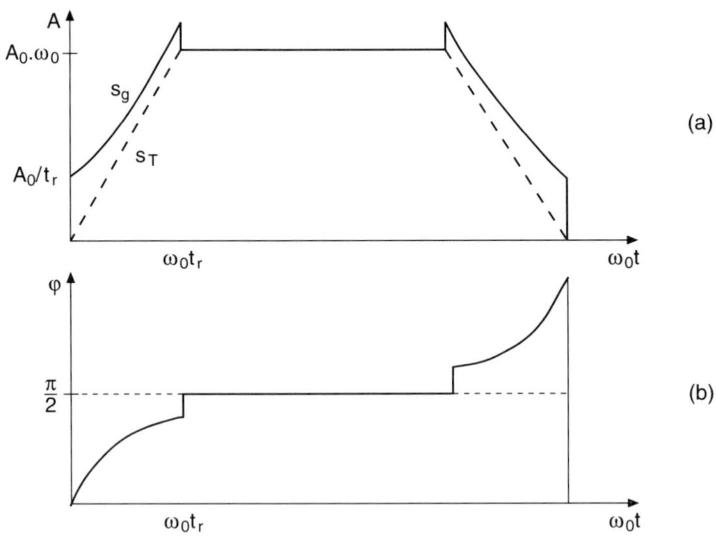

Fig. 1.15 UWB radio pulse complex envelope after differentiation

1.4 Generalised modelling of UWB waveforms

1.4.1 The generalised concept of frequency

We have pointed out that the conventional modelling and processing methods involving Fourier, Hilbert and other integral transformations can satisfactorily describe the general properties of radio signals via, say, integral estimations of errors in the frequency and time domains. Such methods, however, average and smooth out estimates in the vicinity of signal discontinuities, which actually provide information on the origin and end point of the target response, permit separate observations of responses from different scatterers, and so on. The time interval around a discontinuity point for UWB signals is comparable with the total duration of a signal (or of its correlation function), thus producing appreciable estimation errors. Here we consider possible approaches to the description of the UWB signal waveform without integral transformation.

Let us start with a concept fundamental to radio engineering – the concept of frequency, ω_0, of a monochromatic oscillation, $s(t) = A_0 \sin(\omega_0 t + \varphi_0)$. A usual way of finding ω_0 is by measuring the signal period, T, and calculating its frequency, $\omega_0 = 2\pi/T$. For this reason, frequency is believed to be an integral signal characteristic similar to the Fourier transform calculation [71]. On the other hand, the monochromatic oscillation function, $s(t) = A_0 \sin(\omega t + \varphi_0)$, is known to represent the solution to the differential equation of an oscillator

$$\frac{d^2 s(t)}{dt^2} + \omega_0^2 s(t) = 0$$

and at any moment, t_0, the frequency, ω, can be determined as an invariant quantity with respect to the initial oscillation phase

$$\omega = \sqrt{-\frac{d^2 s(t)/dt^2}{s(t)}} = \omega_0$$

This allows one to treat frequency as an instantaneous, or structural, characteristic of a signal. The notion of instantaneous frequency is widely used in analyses of narrowband signals; this quantity is then defined as a derivative of the full phase of the model (see eqn. 1.20)

$$\omega = \frac{d\Psi}{dt} = \omega_0 + \frac{d\varphi}{dt}$$

We extend this concept to an arbitrary signal with an arbitrary amplitude and phase modulation.

Thus, by representing a monochromatic oscillation $s(t) = A_0 \sin(\omega_0 t + \varphi_0)$ in the phase plane $\{s; s_1\}$, as in Fig. 1.16, where

$$s_1(t) = \frac{1}{\omega_0} \frac{ds}{dt}$$

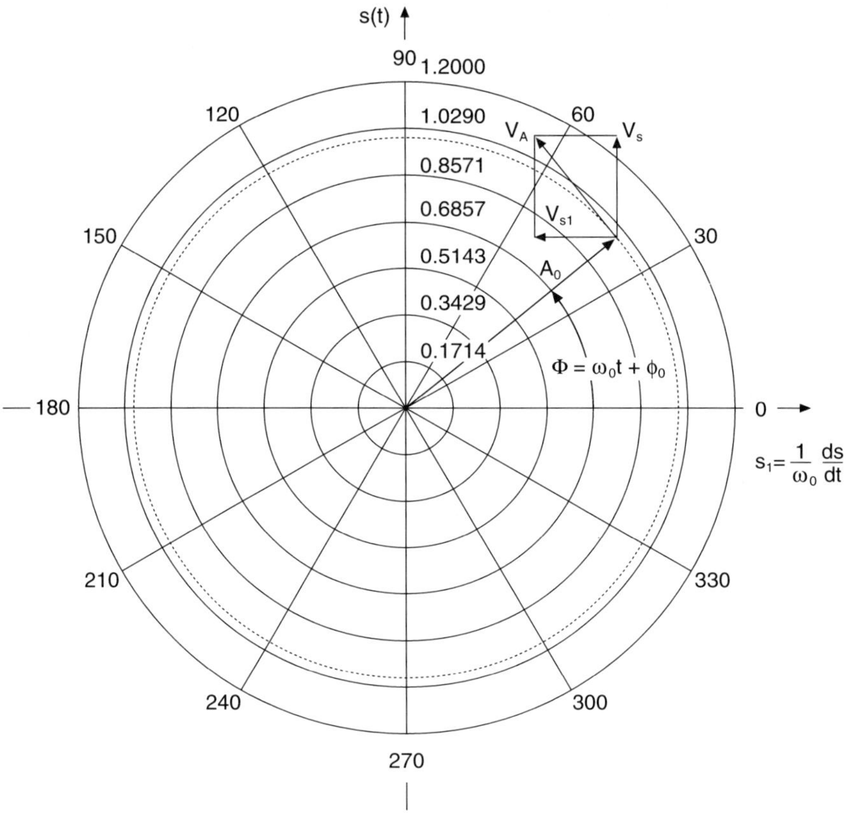

Fig. 1.16 Monochromatic oscillation in the phase plane

one can define the frequency as the velocity of the A_0 vector end normalised to the vector length

$$\omega = \frac{V_A}{A_0} = \frac{\sqrt{V_s^2 + V_{s1}^2}}{A_0}$$

where

$$A_0 = \sqrt{s^2 + s_1^2}$$

$$V_s = \frac{ds}{dt}$$

$$V_{s1} = \frac{ds_1}{dt} = \frac{1}{\omega_0}\frac{d^2 s}{dt^2}$$

For monochromatic oscillation $s(t) = A_0 \sin(\omega_0 t + \varphi_0)$ we get

$$\omega = \frac{1}{A_0}\sqrt{\left(\frac{ds}{dt}\right)^2 + \frac{1}{\omega_0^2}\left(\frac{d^2 s}{dt^2}\right)^2} = \omega_0$$

Then, the frequency can be treated as a measure of the information content in the signal or as a scalar measure of the velocity variation of the vector A representing the signal in the phase plane. It is clear that at $\omega_0 = \text{const}$, a single measurement at an arbitrary moment of time can provide all possible information about a monochromatic signal.

Note that the model in the phase plane uses a two-dimensional signal representation, very much like a conventional complex signal model with the imaginary component as the Hilbert transform of the signal.

With the notion of information content one should, naturally, make allowance for the noise background of measurements. This can be done by using the vast theoretical arsenal of statistical radio engineering. The present analysis, however, focuses on UWB waveform modelling, so that the problems of noise are addressed only when it is absolutely necessary.

As the information content of radar measurements increases, knowledge of the waveform characteristics of signals finds an ever growing application. Earlier, A. Kharkevich [54] analysed some variants of signal representation by using integration in the time-dependent limits for the description of nonstationary processes. He employed the concept of current spectrum

$$S_t(\omega, t) = \int_0^t s(\tau) \exp(-\mathrm{i}\omega\tau)\, d\tau \tag{1.45}$$

and that of instantaneous spectrum

$$S_i(\omega, t) = \frac{dS_t(\omega, t)}{dt} \tag{1.46}$$

Lately, there has been much interest in various methods for representation of nonstationary signals, in particular in the time-frequency plane. This problem will be discussed after the analysis of representation based on the concept of generalised frequency.

1.4.2 Generalised waveform representation in the phase plane

The construction of a two-dimensional signal model in the complex plane implies the creation of a second component, $\hat{s}(t)$, related to the real model, $s(t)$, by a linear, continuous homogeneous operator

$$\hat{s}(t) = \mathrm{L}\{s(t)\}$$

These conditions are met by the Hilbert transform used in the complex signal model, which was shown above to be unsuitable for the description of discontinuities in UWB signals.

Let us represent the signal model $s(t)$ in the phase plane as in Fig. 1.16. For this we attribute the role of the second component in the two-dimensional signal model to its time derivative normalised to the centre frequency, ω_0

$$\hat{s}(t) = s_1(t) \\ = \frac{1}{\omega_0}\frac{ds}{dt} \tag{1.47}$$

This procedure has been widely used in oscillation theory to represent solutions to differential equations and for a qualitative analysis of motion. The components $s(t)$ and $s_1(t)$ for conservative systems, unaffected by external forces, are continuously differentiable state variables of a system. When a system experiences external effects, there appear discontinuity points in the phase plane, requiring a generalised solution to the differential equations. Identification of such points is part of the problem of radar response resolution.

Let us show that the two-dimensional model of a UWB signal in the phase plane, as well as the conventional complex model, permits the unique representation of a real signal, $s(t)$, through its envelope and phase. We shall try to formulate for them the concept of generalised instantaneous frequency.

Examine a finite UWB radio pulse model (Fig. 1.17), consisting of, say, four halfperiods of frequency ω_0 (time sidelobes), given in the interval $(0, 4\pi)$ by the expression

$$s(t) = A(t)\sin[\omega_0 t + \varphi(t)] \\ = A(t)\sin\Phi(t) \tag{1.48}$$

The components $A(t)$ and $\Phi(t) = \omega_0 t + \varphi(t)$ of the signal model can be derived by the Hilbert transform or just chosen for the sake of an illustrative presentation of the signal $s(t)$. With use of a narrowband approximation, the component $\hat{s}(t)$ of the two-dimensional signal model in accordance with eqn. 1.30, is a quadrature of the real signal $s(t)$ (eqn. 1.48): $\hat{s}(t) = s_H(t) = A(t)\cos\Phi(t)$. For any $\hat{s}(t)$, a signal can be represented by the path of the vector end

$$A(t) = \sqrt{s(t)^2 + \hat{s}(t)^2}$$

in the plane $\{s(t), \hat{s}(t)\}$. Following Hamilton, this path may be termed the hodograph of a signal $s(t)$. The velocity module variation of the vector end $A(t)$ along the hodograph is determined by any change in the waveform of a real signal, $s(t)$, which distinguishes this signal from a monochromatic oscillation. The hodograph of an arbitrarily modulated signal is not a circle, in contrast to Fig. 1.16. For example, the hodograph of a UWB pulse (Fig. 1.17) in the phase plane $\{s(t), \hat{s}(t)\}$ is shown in Fig. 1.18.

To represent exactly the finite UWB signal (Fig. 1.17) in the phase plane we take, according to eqn. 1.47, the component $\hat{s}(t) = s_1(t)$ in the form of the $s(t)$ signal derivative normalised to the centre frequency, ω_0.

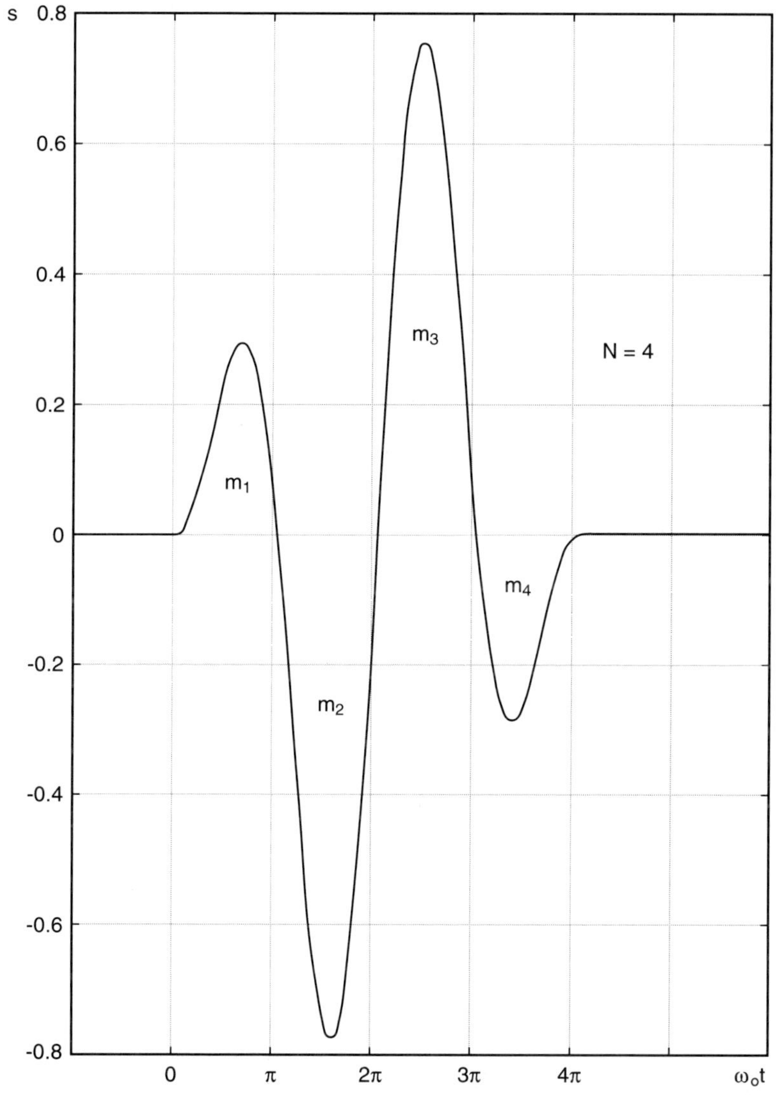

Fig. 1.17 Finite UWB radio pulse model

A precise description of $s_1(t)$ can be made by differentiating eqn. 1.48 with allowance for the variation in both the amplitude and the phase of the signal model, $s(t)$

$$\begin{aligned} s_1(t) &= \frac{1}{\omega_0}\frac{ds}{dt} \\ &= \frac{1}{\omega_0}\left\{\frac{dA}{dt}\sin[\omega_0 t + \varphi(t)] + A(t)\left(\omega_0 + \frac{d\varphi}{dt}\right)\cos[\omega_0 + \varphi(t)]\right\} \end{aligned} \tag{1.49}$$

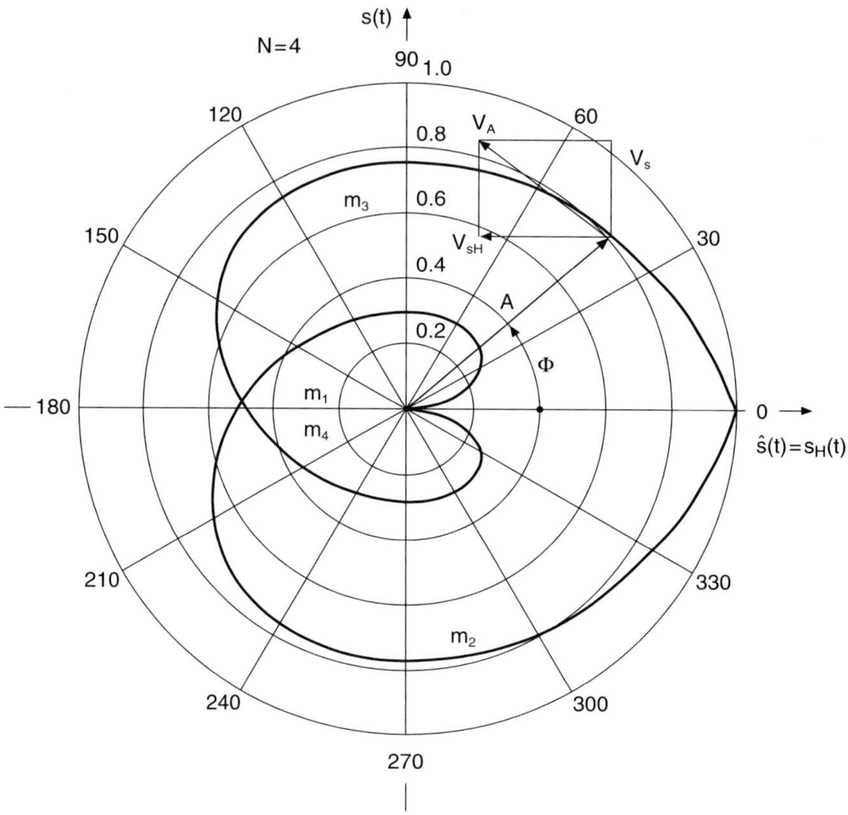

Fig. 1.18 Narrowband approximation of UWB radio pulse hodograph in the phase plane

With the polar coordinates $\{M, \Theta\}$ in the $\{s; s_1\}$ plane, the UWB signal can be represented as

$$s(t) = M(t)\sin\Theta(t) \tag{1.50}$$

This model provides a unique description of the waveform characteristics of an arbitrary signal given by the real function $s(t)$ at any point t. The signal itself can be given in arbitrary ways: by a formula like eqn. 1.48, a table, experimental readings and so on. The signal representation in the phase plane permits definition of the model components (eqn. 1.50) which will be taken to be the generalised envelope, $M(t)$, and phase, $\Theta(t)$

$$M(t) = \sqrt{s^2(t) + \frac{1}{\omega_0^2}\left(\frac{ds}{dt}\right)^2} \tag{1.51a}$$

$$\Theta(t) = \arctan\left[s(t) \Big/ \frac{1}{\omega_0}\frac{ds}{dt}\right] \tag{1.51b}$$

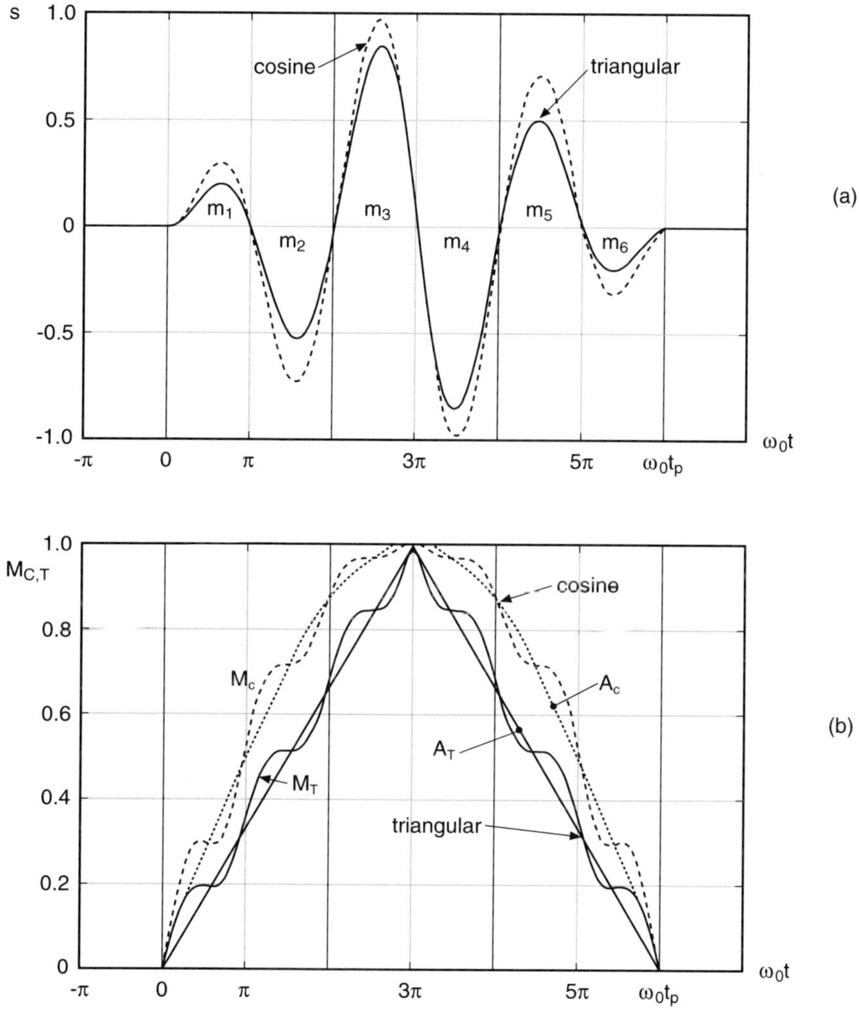

Fig. 1.19 *a* UWB radio signal models
b Their generalised envelopes $M_{C,T}$ and narrowband approximations $A_{C,T}$ (N = 6)

Fig. 1.19 presents the generalised envelopes, $M_T(t)$, and their narrowband approximations, $A_T(t)$, for UWB radio pulses consisting of six quasihalf-periods of the centre frequency, ω_0. The full lines show the signal, $s(t)$, and its generalised envelope, $M_T(t)$, for the model (eqn. 1.48) with a triangular envelope, $A_T(t)$. The broken line is a signal model with a cosine (relative to the maximum) envelope $A_C(t)$ and generalised envelope $M_C(t)$

$$s(t) = \sin\left(\frac{1}{N}\omega_0 t\right)\sin\omega_0 t\left[1(t) - 1\left(t - \frac{N\pi}{\omega_0}\right)\right] \tag{1.52}$$

We mentioned in Section 1.3.2 (Fig. 1.12) that the function $A_T(t)$ represents the signal waveform in the vicinity of local $s(t)$ extrema with an uncontrollable error. The oscillating term of the generalised envelopes, $M_C(t)$, provides information on the time domain positions and magnitudes of the extrema of the oscillating functions, $s(t)$, for both its polarities. This is manifested as the $M_C(t)$ envelopes' oscillations with approximately $2\omega_0$ frequency.

Using the concept of frequency as a characteristic of the signal waveform (Section 1.4.1), we can now define the generalised instantaneous frequency of a UWB signal as the velocity modulus of the end of vector M in the $\{s; s_1\}$ plane, normalised to its length $M(t)$

$$\Omega(t) = \frac{V_M(t)}{M(t)}$$

The hodograph of the UWB radiopulse with its cosine envelope (Fig. 1.19) in the phase plane $\{s(t), \hat{s}_1(t)\}$ is shown in Fig. 1.20 (the full line), where $\hat{s}(t) = s_1(t)$ is defined by eqn. 1.49 and for narrowband approximation $\hat{s}(t) = s_H(t)$ (dotted line).

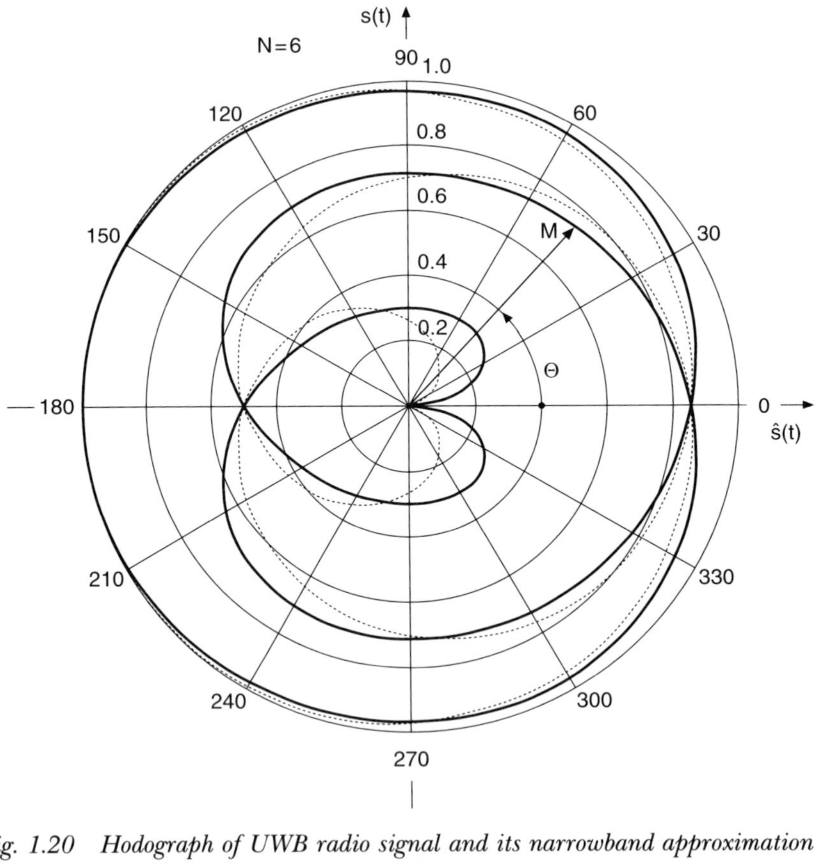

Fig. 1.20 Hodograph of UWB radio signal and its narrowband approximation

In accordance with Fig. 1.20, we can write

$$\Omega(t) = \frac{1}{M(t)}\sqrt{\left(\frac{ds}{dt}\right)^2 + \left(\frac{ds_1}{dt}\right)^2} = \frac{\sqrt{\left(\frac{ds}{dt}\right)^2 + \frac{1}{\omega_0^2}\left(\frac{d^2s}{dt^2}\right)^2}}{\sqrt{s^2(t) + \frac{1}{\omega_0^2}\left(\frac{ds}{dt}\right)^2}} \tag{1.53}$$

The equation for $\Omega(t)$ obtained with the chosen centre frequency, ω_0, is only defined by the waveform of the real function, $s(t)$, and uniquely represents the UWB signal waveform.

Note that the concept of generalised frequency defined as the signal variation rate in the phase plane, $\{s, ds/dt\}$, is closely related to the majority of estimates, made by S. Bernstein (see *Selected Works*, vol. 1, Nauka, Moscow, 1952), of the derivative maxima for the functions with a finite bandwidth bounded by the frequency ω_H, namely

$$\left|\frac{ds}{dt}\right|_{max} \leqslant \omega_H \frac{s_{max}}{\sqrt{3}}$$

and

$$\omega_H \geqslant \sqrt{3}\frac{\left(\frac{ds}{dt}\right)_{max}}{s_{max}}$$

These expressions show that if the upper frequency, ω_H, is bounded, say, by the discretisation conditions, there will be no possibility of restoring waveform features with steepness greater than $\{ds/dt\}_{max}$, when it is normalised to the signal amplitude, s_{max}.

Eqn. 1.53, in contrast to the mentioned estimates, allows us to define the instantaneous frequency $\Omega(t)$, depending on the signal waveform, $s(t)$, and on the time interval in the vicinity of a discontinuity point.

We shall now examine how the signal bandwidth behaves in the calculation of $M(t)$, $\Theta(t)$ and $\Omega(t)$. For this, we shall find the dependences of the components $M(t)$ and $\Theta(t)$ in the generalised model (eqn. 1.50) on the components $A(t)$ and $\Phi(t)$ in the model with the Hilbert transform in narrowband approximation (eqn. 1.48). By substituting eqn. 1.48 into eqns. 1.51, we obtain

$$M(t) = \sqrt{A^2(t)\sin^2\Phi(t) + \frac{1}{\omega_0^2}\left[\frac{dA}{dt}\sin\Phi(t) + A\frac{d\Phi}{dt}\cos\Phi(t)\right]^2}$$

$$\Theta(t) = \arctan\frac{A(t)\sin\Phi(t)}{\frac{1}{\omega_0}\left[\frac{dA}{dt}\sin\Phi(t) + A\frac{d\Phi}{dt}\cos\Phi(t)\right]} \tag{1.54}$$

where $\Phi(t) = \omega_0 t + \varphi(t)$.

Eqn. 1.49 for the $s_1(t)$ coordinate in the phase plane can be written as

$$s_1(t) = A(t)\left\{\frac{dA/dt}{A(t)\omega_0}\sin[\omega_0 + \varphi(t)] + \left(1 + \frac{d\varphi/dt}{\omega_0}\right)\cos[\omega_0 + \varphi(t)]\right\} \tag{1.55}$$

Here, the signal bandwidth factor, $\mu = \Delta\omega/\omega_0$, is expressed as the ratio of the variation rates of the normalised amplitude $(dA/dt)/A(t)$ and phase $d\varphi/dt$ to the centre frequency, ω_0. For a narrowband signal, one should put

$$\frac{dA/dt}{A(t)\omega_0} \ll 1$$

and

$$\frac{d\varphi/dt}{\omega_0} \ll 1 \tag{1.56}$$

then eqn. 1.55 will represent the quadrature component of the $s(t)$ signal

$$\begin{aligned} s_1(t) = s_H(t) &= A(t)\cos[\omega_0 t + \varphi(t)] \\ &= A(t)\cos\Phi(t) \end{aligned} \tag{1.57}$$

which is employed in the conventional complex model when the signal bandwidth is finite and bounded by the frequency $\pm\omega_0$. As was demonstrated in Section 1.3.2, this model is unsuitable for the description of UWB signals with waveform discontinuities.

In the narrowband approximation (eqn. 1.56), the relationships in eqn. 1.54 demonstrate a coincidence of the signal models in the complex and phase planes

$$M(t) = A(t) \quad \text{and} \quad \Theta(t) = \Phi(t)$$

or

$$\begin{aligned} s(t) &= M(t)\sin\Theta(t) \\ &= A(t)\sin\Phi(t) \end{aligned} \tag{1.58}$$

This result indicates that the UWB signal representation in the phase plane has a generalised nature, because it allows, in the limiting case, a reduction in the conventional narrowband relationships.

For comparison, in Fig. 1.19 are shown the narrowband approximations of generalised envelopes $A_T(t)$ and $A_C(t)$, and in Fig. 1.20 is depicted the narrowband approximation of a signal model hodograph (the broken line).

To demonstrate the difference between the generalised characteristics $M(t)$, $\Theta(t)$ and the conventional complex model components, we shall consider an amplitude-modulated signal

$$s(t) = (1 + m\cos\Omega_m t)\sin\omega_0 t \tag{1.59}$$

At $m < 1$ and $\Omega_m < \omega_0$, the signal bandwidth is finite and the quadrature component

$$s_H(t) = (1 + m\cos\Omega_m t)\cos\omega_0 t$$

has been shown to represent exactly the Hilbert transform. In the complex model, we have

$$\begin{aligned} A(t) &= \sqrt{s^2(t) + s_H^2(t)} \\ &= 1 + m\cos\Omega_m t \end{aligned}$$

In order to represent the $s(t)$ signal in the phase plane, one should use the component

$$\begin{aligned} s_1(t) &= \frac{1}{\omega_0}\frac{ds}{dt} \\ &= \left[(1 + m\cos\Omega_m t)\cos\omega_0 t - \frac{m\Omega_m}{\omega_0}\sin\Omega_m t\sin\omega_0 t\right] \end{aligned} \tag{1.60}$$

In contrast to the quadrature component $s_H(t)$ of the $s(t)$ signal, this expression contains a term proportional to the variation rate and is expressed as the ratio of the modulation frequency, Ω_m, and the percentage modulation, m, to the centre frequency, ω_0. The factor $m\Omega_m/\omega_0$ can be regarded as the bandwidth factor of the signal, $s(t)$.

Thus, the difference between the signal representation in the complex and phase planes also reveals itself in the case of UWB modulation ($\Omega_m \approx \omega_0$) for continuously differentiable signals, such as amplitude-modulated oscillations. This difference is represented as a continuously differentiable function of the bandwidth factor, $m\Omega_m/\omega_0$.

Using eqn. 1.60 we can write a formula for the generalised envelope of an amplitude-modulated signal

$$M(t) = \sqrt{\begin{aligned} &(1 + m\cos\Omega_m t)^2\sin^2\omega_0 t \\ &\quad + \left[(1 + m\cos\Omega_m t)\cos\omega_0 t - \frac{m\Omega_m}{\omega_0}\sin\Omega_m t\sin\omega_0 t\right]^2 \end{aligned}}$$

Neglecting the term of the order $(m\Omega_m/\omega_0)^2$, we obtain the formula

$$M(t) = (1 + m\cos\Omega_m t)\sqrt{1 + \frac{m\Omega_m}{\omega_0}\frac{\sin\Omega_m t}{1 + m\cos\Omega_m t}\sin 2\omega_0 t} \tag{1.61}$$

This result shows the generalised envelope, $M(t)$, to be different from the complex envelope, $A(t) = 1 + m\cos\Omega_m t$, by the value of the term proportional to the bandwidth factor, $m\Omega_m/\omega_0$. This additional term oscillates with frequency $2\omega_0$.

Let us now estimate some asymptotic properties of the instantaneous frequency function, $\Omega(t)$. First, the generalised envelope, $M(t)$, for a radio pulse approaches zero at the signal origin and endpoint. The value of instantaneous frequency tends to infinity, indicating a high information content in the waveform for these events. This can be used for solving the problems of radar response resolution. According to eqn. 1.53, a similar

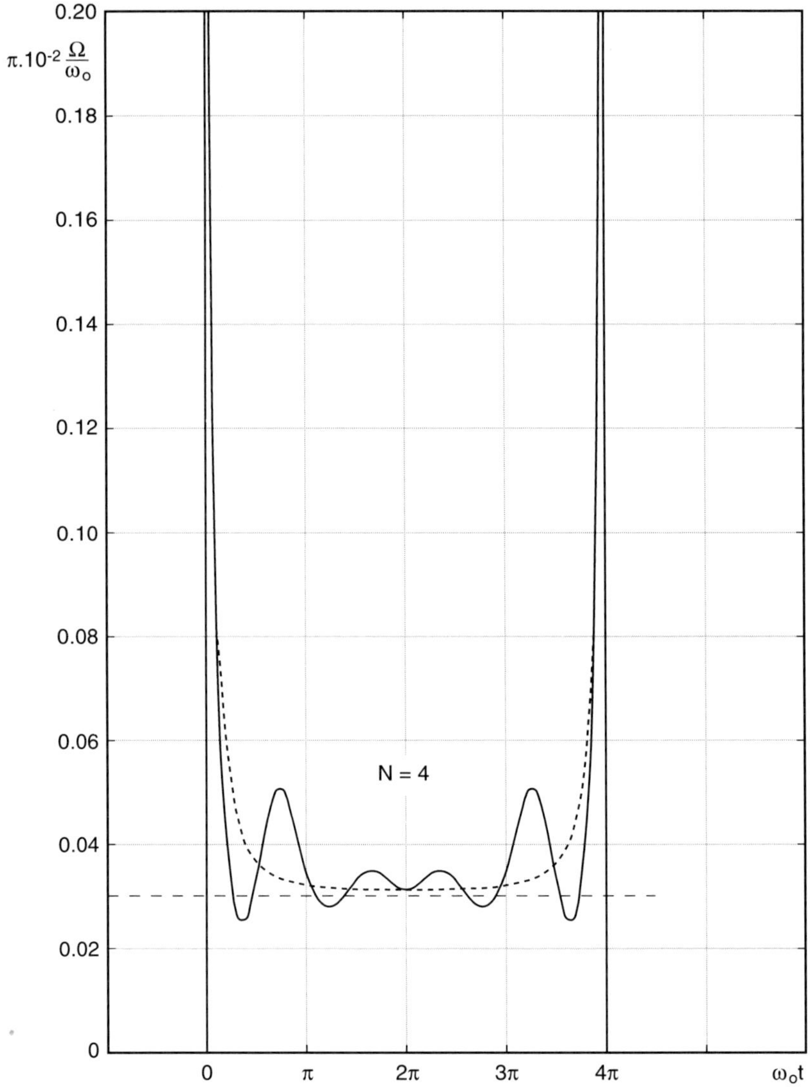

Fig. 1.21 Instantaneous frequency for a UWB radio pulse and its narrowband approximation

situation arises in the vicinity of the discontinuity points of $s(t)$, where the first and the second derivatives tend to show a step function behaviour. On the other hand, at the sites of small deviations of the amplitude and phase of the $s(t)$ signal model (eqn. 1.48) from the monochromatic oscillation, the hodograph in Fig. 1.18 is represented as a circumference, and the instantaneous frequency, $\Omega(t)$, approaches its minimum, Ω_{min}, close to the centre frequency of the signal model

$$\Omega_{min} \approx \omega_0$$

Fig. 1.21 gives an instantaneous frequency plot for the UWB radio pulse shown in Fig. 1.17. Here, the instantaneous frequency, $\Omega(t)$, acquires an infinitely large value at the signal origin and end point and approaches the frequency ω_0 in the middle of the pulse, since the waveform in this region is close to the monochromatic oscillation. Any deviation of $\Omega(t)$ from ω_0 reflects nonstationary properties of the UWB pulse waveform. Therefore, the oscillations of the function $\Omega(t)$ (Fig. 1.21) with a frequency of about $2\omega_0$, approximately reproduce, as for the generalised envelope $M(t)$ (Fig. 1.19b), the positions of extrema of the real function $s(t)$.

From eqn. 1.53 we have

$$\Omega(t) = \omega_0 \sqrt{\left(\frac{dA/dt}{\omega_0 A(t)}\right)^2 + \left(1 + \frac{d\varphi/dt}{\omega_0}\right)^2} \tag{1.62}$$

If the condition of expr. 1.56 is met, we can eliminate from eqn. 1.55 the small terms of the second order to get the equation for a narrowband signal

$$\Omega(t) = \omega_0 + d\varphi/dt$$

coinciding with the conventional definition of instantaneous frequency as a derivative of the full signal phase, $\psi(t) = \omega_0 t + \varphi(t)$, which does not take account of the variation rate of the envelope, $A(t)$.

The generalised frequency diagram built up with the Hilbert transform on the assumption of a finite signal waveform is shown by a dashed line in Fig. 1.21. It has no $2\omega_0$ oscillations and, therefore, cannot reveal the waveform variation. A comparison of the signal hodographs built on the phase planes exactly and by the narrowband approximation (Fig. 1.20) demonstrates similar differences.

1.4.3 Potentialities and some results for generalised UWB waveform modelling

In Section 1.3.2 we described a complex signal model, pointing out that its components, $A(t)$ and $\Psi(t)$, were derived with the Hilbert transform to get a two-dimensional signal to be further processed in terms of the scheme in Fig. 1.7.

Similar processing procedures for obtaining the components $M(t)$, $\Theta(t)$ and $\Omega(t)$ of the generalised model are shown in Fig. 1.22; here the component $s_1(t)$ is formed with the differentiation operator, D. This procedure may be described as generalised UWB signal detection.

Consider the possibility of defining the parameter ω_0 used in the operator D

$$D\{s(t)\} = s_1(t) = \frac{1}{\omega_0}\frac{ds}{dt}$$

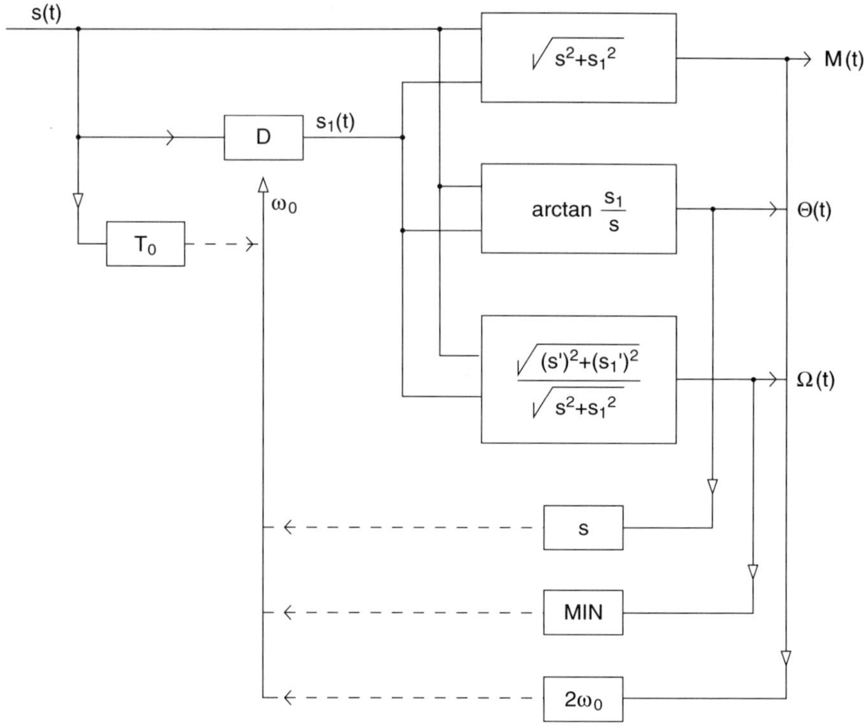

Fig. 1.22 UWB radio signal generalised detection

By definition, ω_0 represents the centre frequency of the signal bandwidth, $S(\omega)$. The quantity ω_0 can be found by various integral estimations, for example by that described in Section 1.1

$$\omega_0 = \frac{\omega_H + \omega_L}{2}$$

or by the moment estimation [46]

$$\omega_0 = \frac{\int \omega |S(\omega)|^2 \, d\omega}{\int |S(\omega)|^2 \, d\omega} \tag{1.63}$$

In addition to the difficulty of determining the frequency bounds, $\omega_{H,L}$, and the integration limits in eqn. 1.63, these estimations require the calculation of the spectral density, $S(\omega)$, involving considerable difficulties (see Section 1.3.1). Besides, the above estimations of ω_0 are integral procedures conceptually inadequate for the problems of finding UWB waveform characteristics.

On the other hand, the use of the UWB signal model in the form of the real function $s(t)$, can provide the estimation of ω_0 as $\omega_0 = 2\pi/T_0$, where T_0 is an average quantity or the value of the principal period of the oscillating function $s(t)$. Since a UWB radio impulse contains only a few oscillation quasiperiods, T_0 can be calculated fairly easily. This is shown by the operator T_0 in Fig. 1.22.

Another way of evaluating ω_0 is to identify the linear term in the phase characteristic

$$\Theta(t) = \omega_0 t + \varphi(t)$$

for example, by expanding the function $\Theta(t)$ into power series. This procedure is presented by the operator S in Fig. 1.22.

If the envelope $M(t)$ changes smoothly in the vicinity of the principal maximum, then the quantity $\Omega(t)$ approaches its minimum, $\Omega(t) \approx \omega_0$, in agreement with eqn. 1.53. Therefore, ω_0 may be estimated from the minimum frequency, $\Omega(t)$, as shown by the operator *MIN* in Fig. 1.22.

Finally, an exact account of the UWB properties in the signal model, $s(t)$, gives rise to oscillations in the function $M(t)$ with a frequency of about $2\omega_0$ (Section 1.4.2). The measurement of the oscillation period (the operator $2\omega_0$ in Fig. 1.22) is also a way of finding the parameter ω_0 of the generalised UWB signal.

The procedure of finding the characteristics M, Θ and Ω in terms of Fig. 1.22 is performed as an iterative process starting with a preliminary estimation of ω_0, say, by the operator T_0. One may believe that the convergence will be satisfactory in the general case of UWB signals.

We shall illustrate this, using V.I. Tikhonov's example [114] of finding the generalised envelope, $M(t)$, of the UWB signal, $s(t) = A(t)\cos\omega_0 t$, with a preset component parameter $s_1(t) = (1/\omega_1)(ds/dt)$ and an error $\omega_1 \neq \omega_0$. Using eqn. 1.51 and neglecting the summands of the order $((dA/dt)/A\omega_0)^2$, we have

$$M(t) = A(t)\sqrt{1 - \frac{1-\alpha^2}{2}\sin^2\omega_0 t - 2\frac{dA/dt}{A(t)\omega_0}\sin 2\omega_0 t}$$

where $\alpha = \omega_0/\omega_1$. For a monochromatic oscillation, $A(t) = A_0$ and inaccurate ω_0 leads to the following expression for $A(t)$ in the A_0 estimate

$$A(t) = A_0\sqrt{1 - \frac{1-\alpha^2}{4}(1 - \cos 2\omega_0 t)}$$

One can see that the error is not large at $\alpha \approx 1$. This gives us hope for a rapid convergence in defining ω_0 for the algorithms in Fig. 1.22. Below, we shall discuss the effect of the quantity ω_0 on generalised modelling of UWB waveforms.

Now let us examine, in detail, some results of the calculation of generalised UWB waveform characteristics. Fig. 1.23a shows a UWB radio pulse with maximum amplitude $s_{max} = 1$ and a rectangular envelope with four time sidelobes, $N = 4$. According to eqn. 1.4, the bandwidth factor of the radio pulse is $\mu = \omega/\omega_0 = 4/N = 1$. The horizontal axis scale in Fig. 1.23

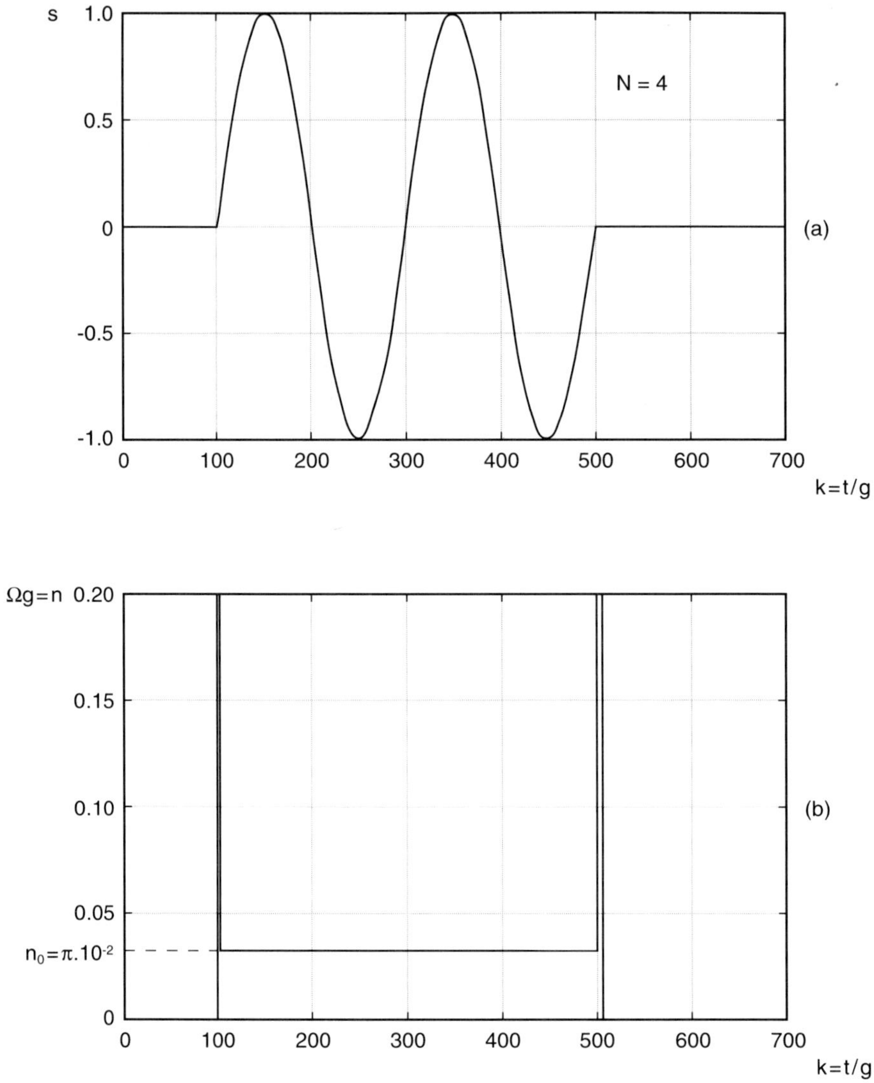

Fig. 1.23 *a* UWB radio signal with rectangular envelope
b Its generalised instantaneous frequency

and in the subsequent diagrams is given by the number of signal counts used in the calculations. For example, the radio pulse in Fig. 1.23a is represented by 400 counts, sufficient for an adequate representation of the signal waveform. The generalised frequency, $\Omega(t)$, variation of this radio pulse, calculated from eqn. 1.53, is given in Fig. 1.23b.

As was indicated in Section 1.4.2, the origin and the end point of the signal are presented by infinite values of the generalised instantaneous frequency, $\Omega(t)$. Within the pulse duration, the radio signal with a rectangular envelope exactly coincides with monochromatic oscillation, so that the instantaneous frequency is exactly equal to ω_0 at any moment of time. In order to define the value of instantaneous frequency, one should set the time scale of the horizontal axis by the quantity g measured as seconds/count. Then the scale of instantaneous frequency, Ω, along the vertical axis in Fig. 1.23b will be $n \cdot 1/g$. For instance, if the duration of the pulse in Fig. 1.23a is 4 ns, the coefficient $g = 4 \times 10^{-9}/400 = 10^{-11}$. Within the pulse duration $n_0 = \pi \cdot 10^{-2}$ or $\Omega(t) = n_0/g = 3.14 \times 10^9 = \omega_0 = 2\pi f_0$; hence, $f_0 = 500\,\text{MHz}$, corresponding to the carrier frequency of the radio pulse in Fig. 1.23a with the period $T_0 = 2\,\text{ns}$.

Fig. 1.24a shows a UWB radio signal as the sum of two radio pulses with a rectangular envelope and maximum amplitude, $s_{max} = 1$. The second pulse is delayed by 20 counts, or by 1/20 of the pulse duration. But the time profile of the total signal in Fig. 1.24a does not give a clear idea of the two signal components, the more so as we are discussing the model problem with zero noise. In contrast, the plot of generalised frequency, $\Omega(t)$, in Fig. 1.24b clearly shows the origin and the end point of each of the two pulses as instantaneous frequency overshoots (peaks). The first overshoot amplitude is infinite, because it has been found from eqn. 1.53 at the zero value of the generalised envelope, $M(t)$. The amplitudes of the second and third overshoots have been calculated at a finite value of the total signal. They have been found to be $n = 0.4$ along the generalised frequency axis, approximately ten times the value of ω_0. The fourth overshoot is calculated by the same algorithm in the presence of the second radio pulse only. But since the calculations are made without the last nonzero value of the signal, the pulse amplitude is large but not infinite, $n = 10^{13}$.

This analysis demonstrates a possible time resolution of overlapping UWB radio pulses delayed by 1/20 of their duration. It should be recalled that, in the conventional approach, the quantity resolvable in radar measurements is the total pulse duration.

The calculation of generalised frequency, $\Omega(t)$, from eqn. 1.53 includes division by the generalised envelope, $M(t)$, resulting in infinite Ω at all points beyond the pulse extent. To regularise the result of the profile calculation in Figs. 1.23b and 1.24b, a simultaneous fulfilment of the condition $M(t) = 0$ was checked at several neighbouring readout points. If this condition was met, it was assumed that $M(t) = 0$ and $\Omega(t) = 0$. In some cases the problem may become ill-posed. For example, Fig. 1.25 presents the sum of two UWB pulses with a rectangular envelope, each having four time sidelobes and the second delayed by 1/2 the centre frequency period, i.e. by one time sidelobe. In this case, the signal in the count range $k = 200–500$ is represented as the difference between the counts and is equal to zero only approximately if the counts were taken independently. This makes the value of $\Omega(t)$ incorrect in this range (Fig. 1.25b).

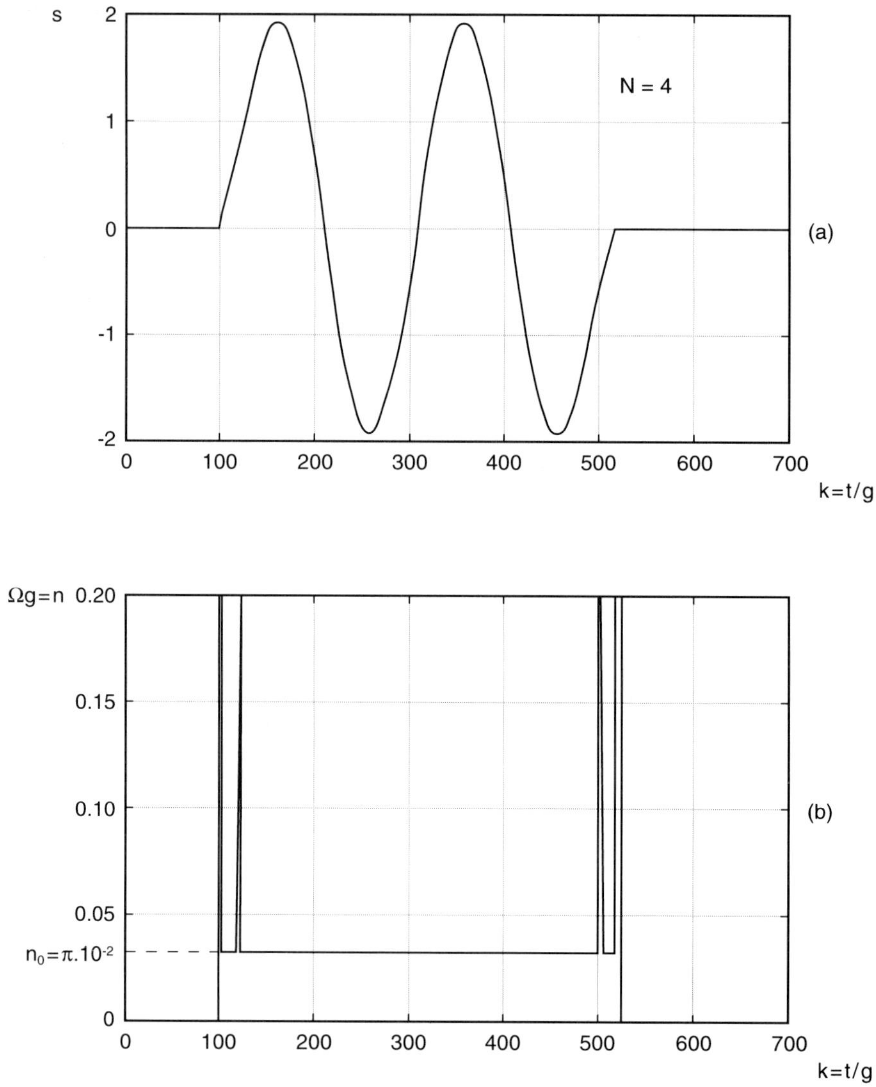

Fig. 1.24 *a* Sum of two delayed UWB radio pulses
b Its generalised instantaneous frequency

Analyse now the possibilities of representing the UWB signal waveform in terms of generalised instantaneous frequency, $\Omega(t)$. Fig. 1.26a shows a radio pulse model with a cosine envelope, which has four sidelobes and belongs to the $\{R^{(2)}\}$ subclass of radio signals (Section 1.2). Fig. 1.26b showing this generalised instantaneous frequency $\Omega(t)$, similar to Fig. 1.21, has characteristic oscillations, with maxima in the time domain which have the same positions as the extrema of the time function $s(t)$ in Fig. 1.26a, revealing the details of the signal waveform.

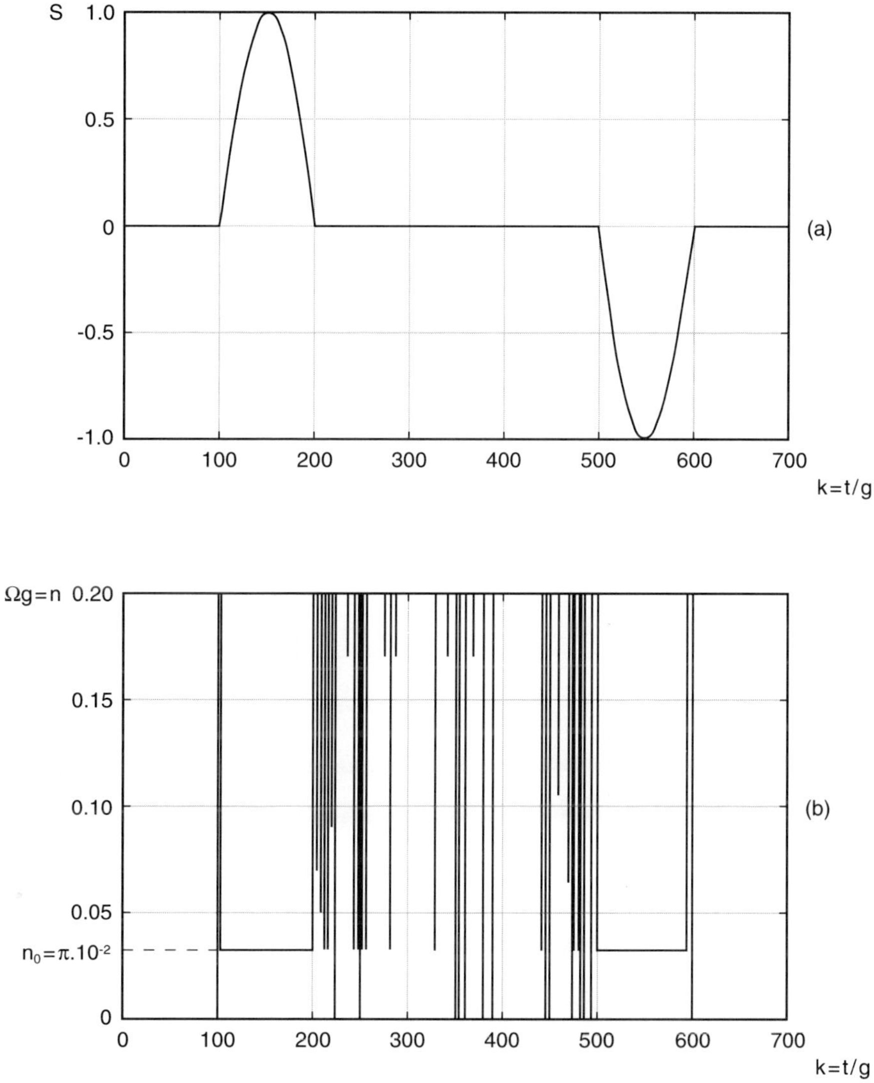

Fig. 1.25 Ill-posed problem in the generalised instantaneous frequency calculation

a Sum of two delayed radio pulses
b Its generalised instantaneous frequency

Fig. 1.27b gives the generalised frequency profile for a signal which is the sum of two pulses in Fig. 1.26a with identical amplitudes and a 20 count delay. It demonstrates the possible time resolution of signals, namely, the ability to identify the time positions of the origin of the second and the end point of the first pulses, indicated by the steps in the $\Omega(t)$ profile. The peak amplitude of a step is the measure of admissible noise level in the problem

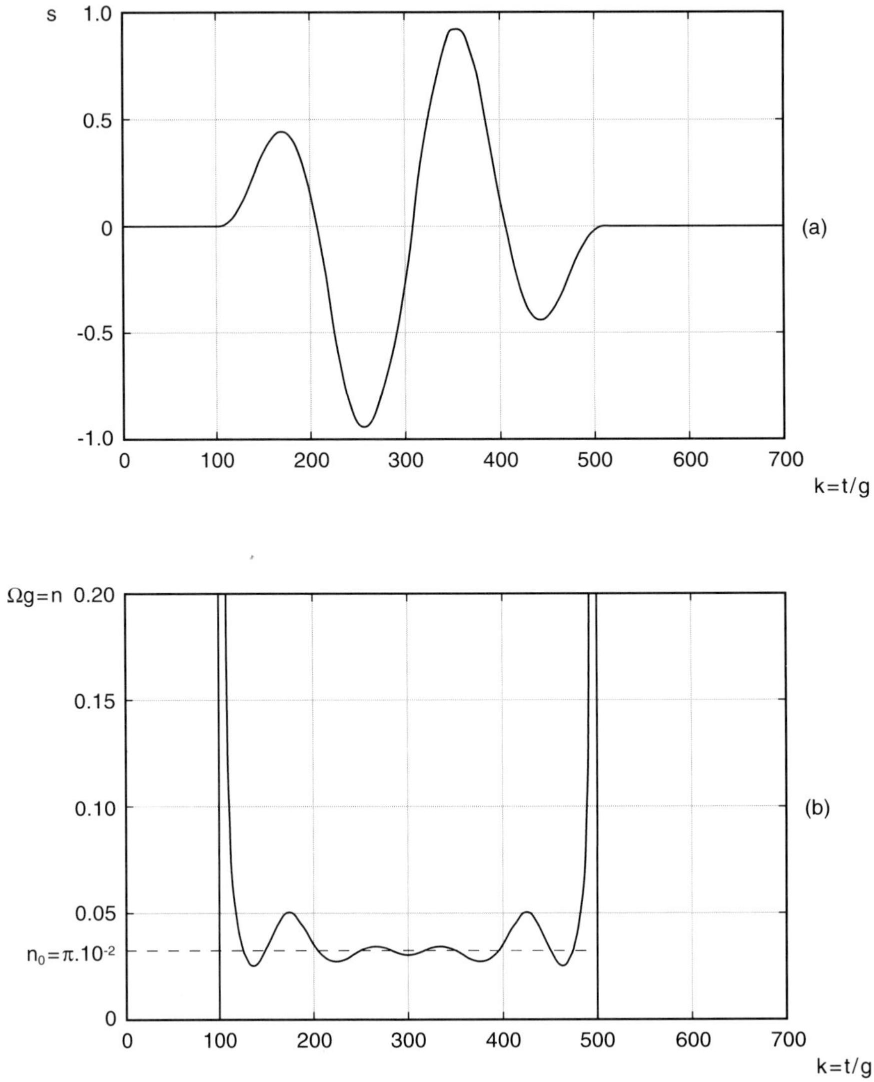

Fig. 1.26 *a* UWB radio pulse with cosine envelope
b Its generalised instantaneous frequency

solution. Note that the signals are practically unresolvable in the total $s(t)$ profile in Fig. 1.27a.

Fig. 1.28 presents similar dependences for the sum of two pulses when the second pulse amplitude is one half of the first. These pulses are also resolvable, but the peak amplitude of the step at the second pulse origin is lower in the presence of the first one than at the end point of the first pulse in the presence of the second one, having a lower amplitude.

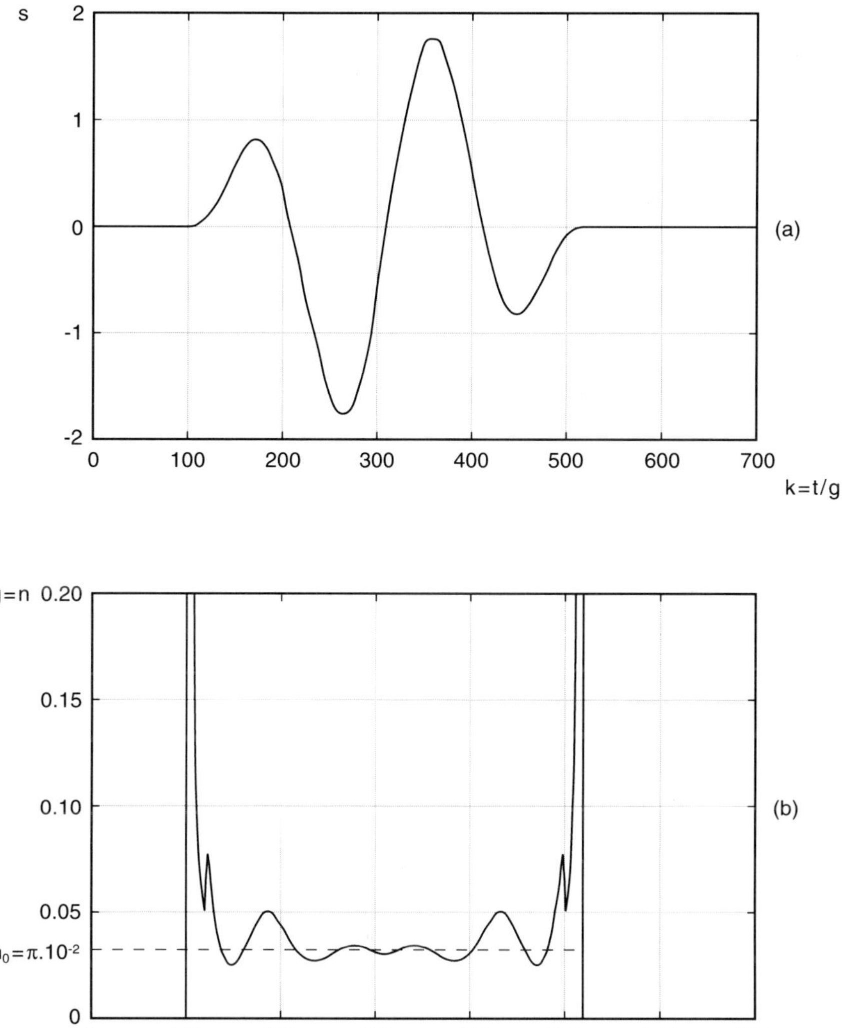

Fig. 1.27 *a* Sum of two delayed UWB radio pulses
b Their time resolution with use of generalised instantaneous frequency plot

Evidently, the limits of admissible noise levels must be defined under the worst conditions.

This type of signal model affects the peak value of the step in the function $\Omega(t)$. Fig. 1.29b gives the generalised instantaneous frequency, $\Omega(t)$, for the sum of two UWB signals at $N = 4$ with a triangular complex envelope and a 20 count time delay. The $\Omega(t)$ step is somewhat larger than that in Fig. 1.27,

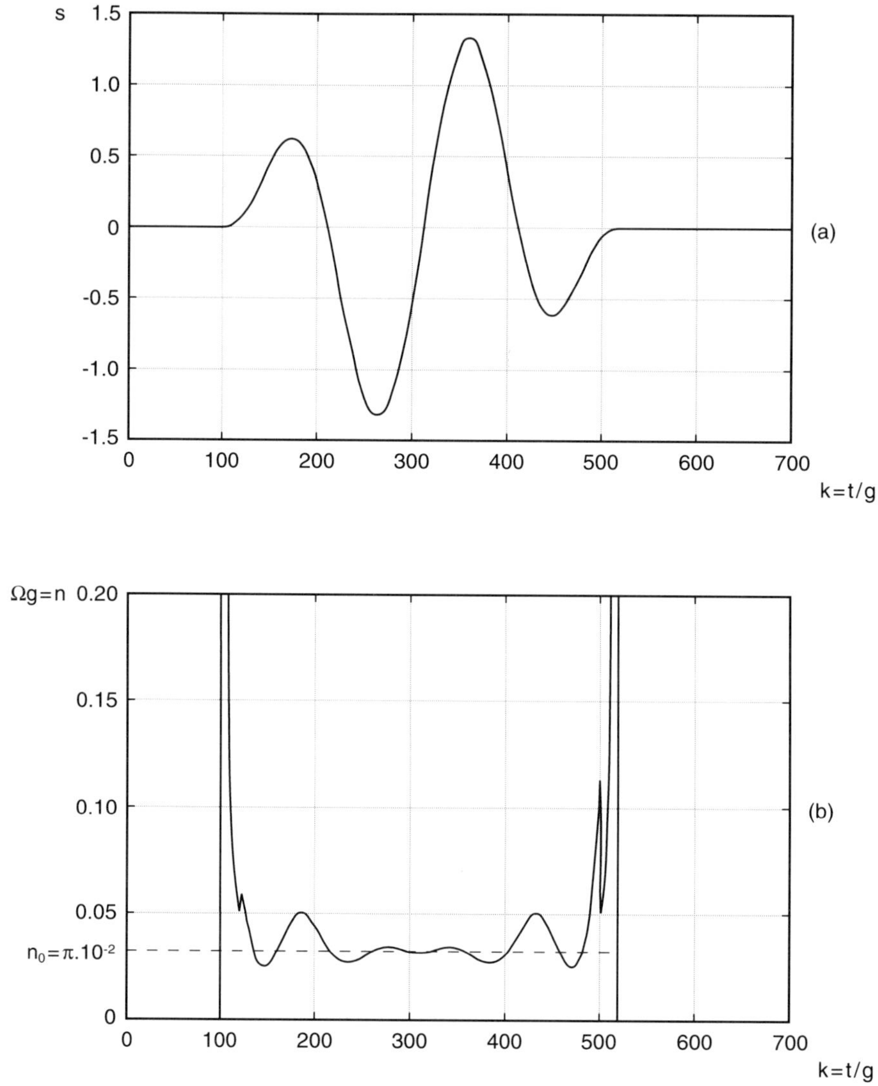

Fig. 1.28 *a* Sum of two UWB radio pulses
b Their time resolution when the second pulse amplitude is 1/2 that of the first one

because the triangular envelope is smaller than the cosine one at the moment the second pulse appears (see Fig. 1.19). Note the flat shoulder at the ω_0 level in the $\Omega(t)$ plot in the interval between the triangular envelopes' maxima.

Fig. 1.30 shows a model UWB radio signal with a triangular envelope having five time sidelobes, $N = 5$. The $s(t)$ signal has a break (discontinuity of the first derivative) at the maximum, $s_{max}(t)$, but this is practically

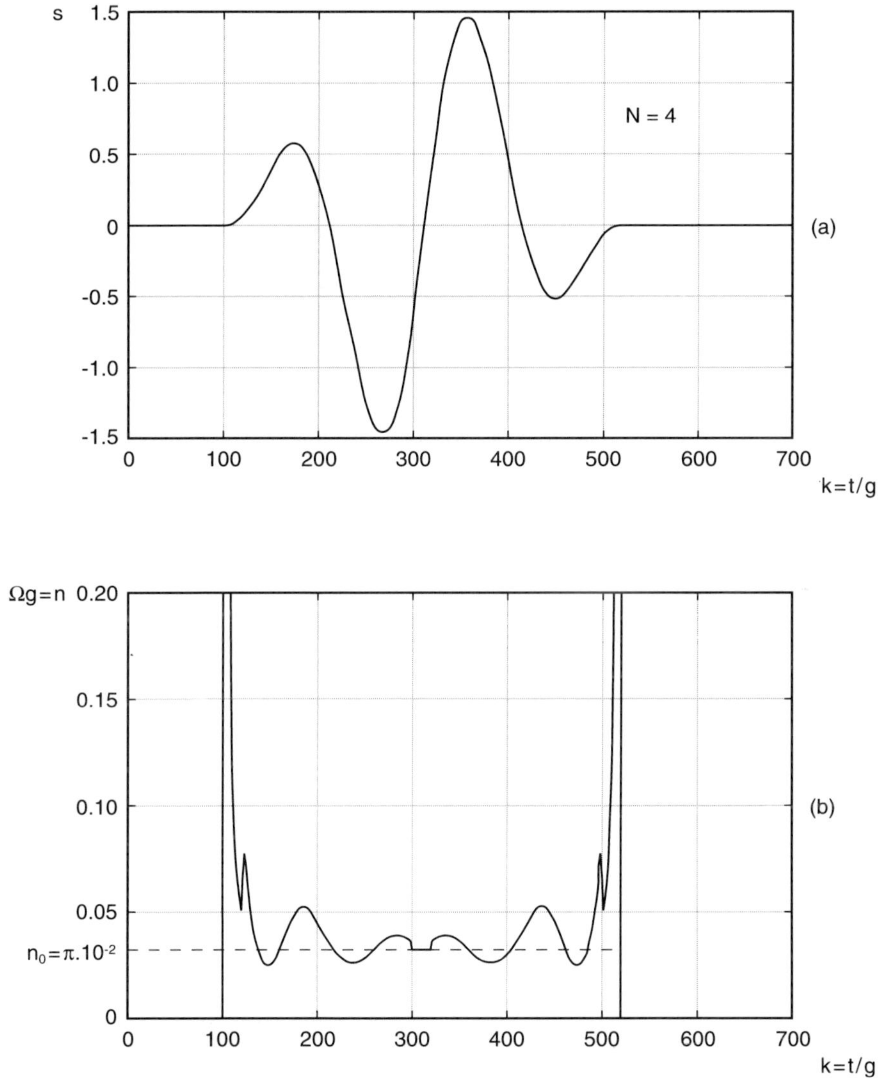

Fig. 1.29 *a* Sum of two UWB radio pulses with triangular envelope
b Its generalised instantaneous frequency

unattainable. The $\Omega(t)$ profile has an overshoot at $k = 350$. This kind of signal model may be considered as unsuitable for this problem.

The feature in $\Omega(t)$ in Figs. 1.28 and 1.29 carrying information about the time position of the second pulse depends on the type of model and, therefore, it determines the requirement on the waveform of the UWB radio signals to be used. Employing the signal models with a triangular and cosine envelope (Fig. 1.19a), we find that both signals near the pulse origin

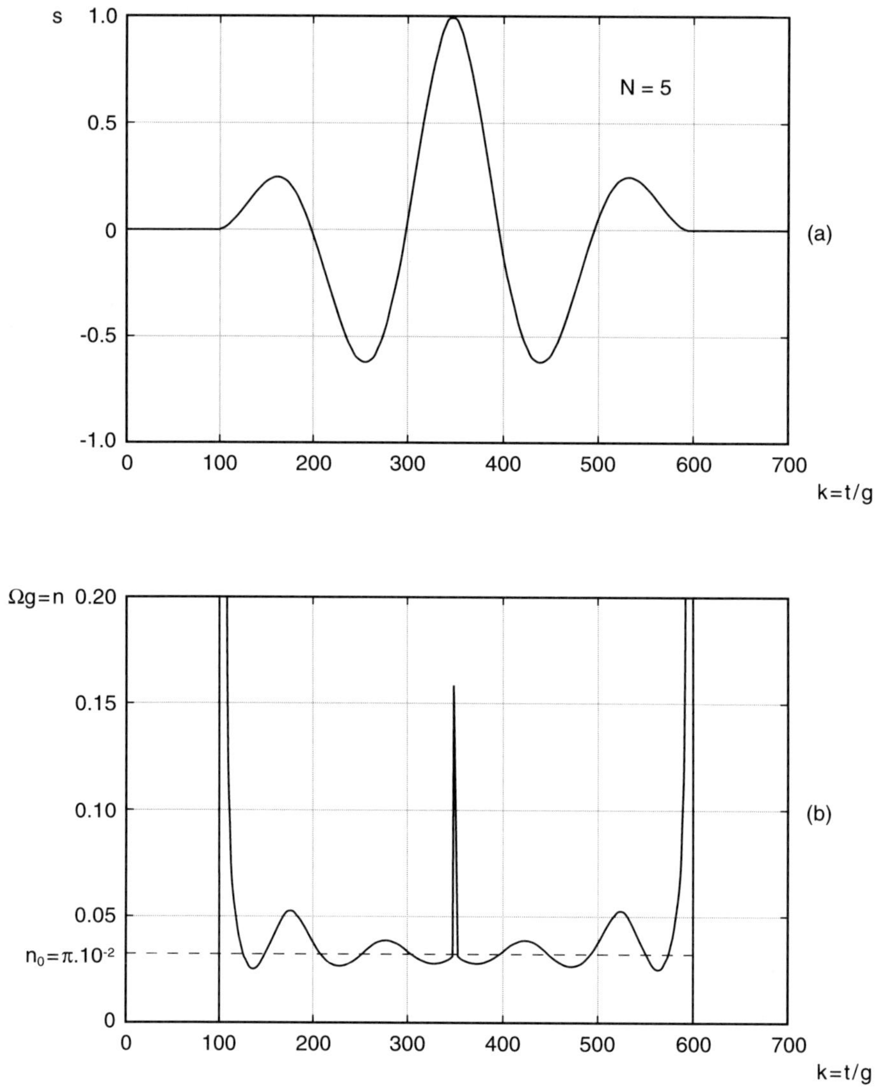

Fig. 1.30 *a* UWB signal model with triangular envelope
b Its generalised instantaneous frequency

point are described by the quadrature parabola

$$s_T(t) = \frac{8\pi}{NT_0^2} t^2$$

$$s_C(t) = \frac{4\pi^2}{NT_0^2} t^2$$

that is, at $t = 0$ both models have a break of the second derivative, d^2S/dt^2, which is larger by a factor of $\pi/2$ for $s_C(t)$. Such signals correspond to the steps in the function $\Omega(t)$ in Figs. 1.27 and 1.29.

For a signal with a smoother variation of the function $s(t)$ at the pulse origin, the step in the $\Omega(t)$ profile will be smaller when the two pulses are resolvable. Fig. 1.31 gives the time dependence, $s(t)$, of the sum of two pulses $s(t)$ at $N = 4$ and a 20 count delay, having an envelope described by the

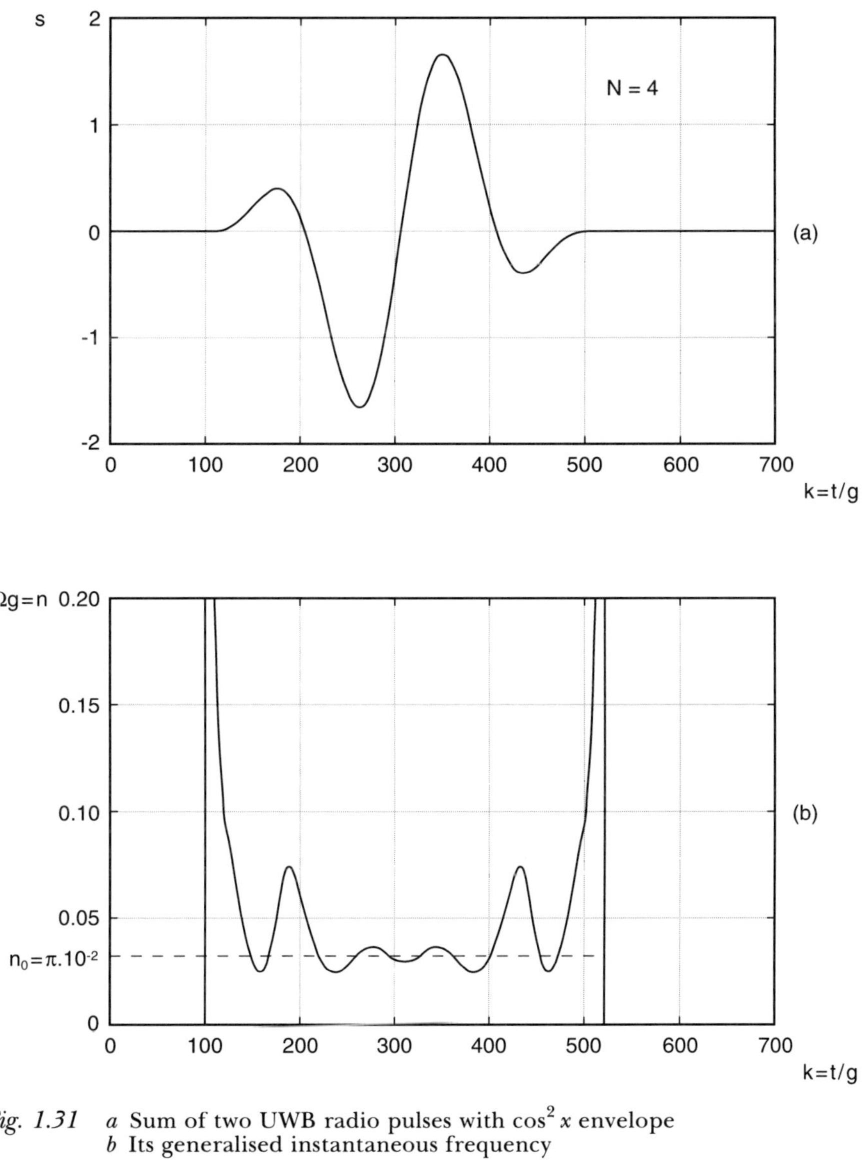

Fig. 1.31 *a* Sum of two UWB radio pulses with $\cos^2 x$ envelope
b Its generalised instantaneous frequency

function $\cos^2 x$. In this case the signal begins with a discontinuity in the third derivative, d^3s/dt^3, so that in Fig. 1.31b there are only minor steps in the $\Omega(t)$ profile (discontinuity in the first derivative, $d\Omega/dt$ at $k = 120, 500$). This complicates the problem of signal resolution.

This analysis permits the conclusion to be made that the character of $s(t)$ function growth at the pulse origin should be considered as an essential characteristic of UWB radio pulses affecting their resolvability. Therefore, this characteristic should be taken into account when setting requirements on UWB waveform shapers.

The use of other generalised characteristics does not provide tools for resolving UWB signals, as this was done with instantaneous frequency $\Omega(t)$. For example, Fig. 1.32 shows a hodograph of the sum of two UWB radio pulses, $S_\Sigma(t)$, with a triangular envelope, which is indicated by the shoulder in the M_Σ, Θ_Σ diagram. But the use of this criterion for signal resolution in the presence of noise does not seem to be fruitful. Similarly, only minor details make the generalised envelopes (Fig. 1.33a,b) and the phase diagrams

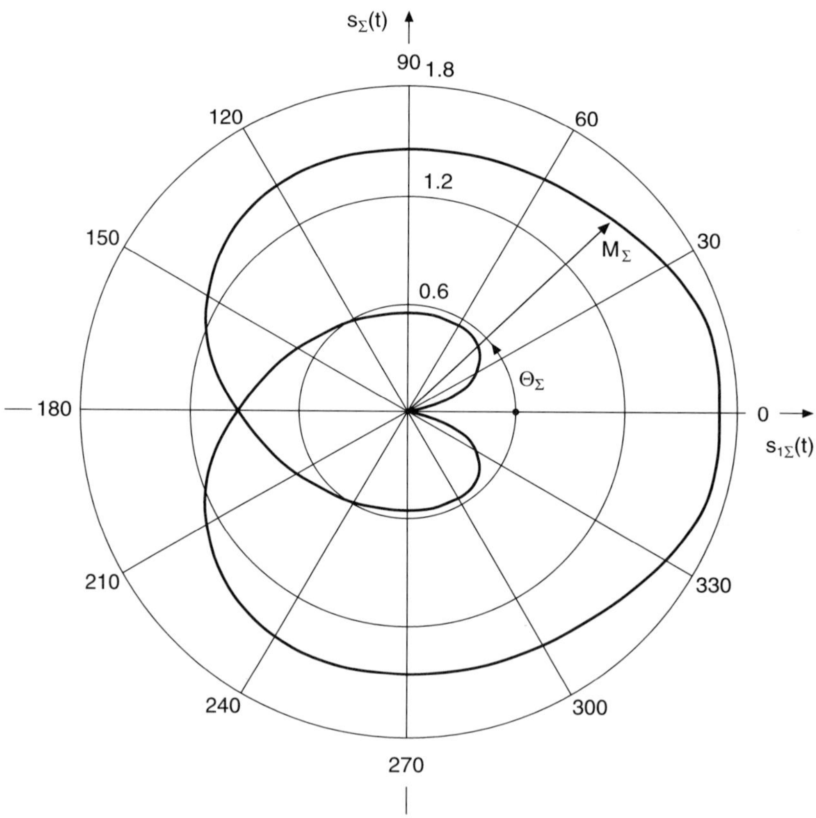

Fig. 1.32 Hodograph of the sum of two UWB pulses with triangular envelope

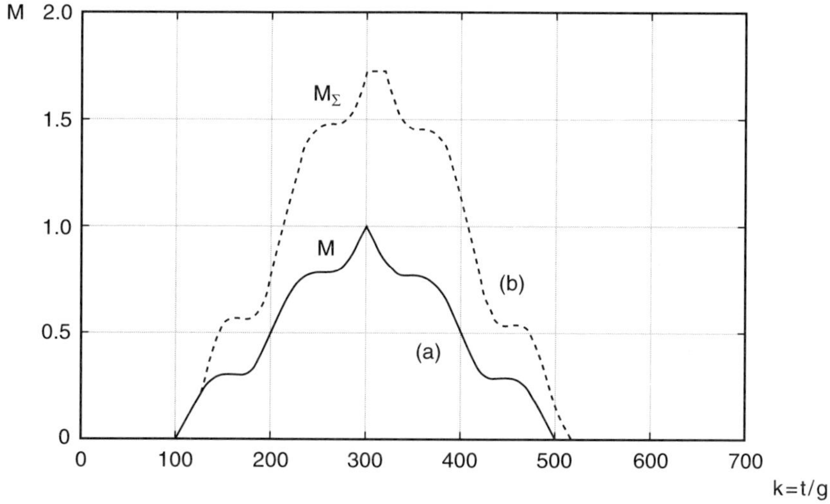

Fig. 1.33 *Generalised envelope for UWB pulse (a) and (b) for sum of two delayed pulses*

(Fig. 1.34a,b) distinguishable in the case of one or two UWB pulses with a 20 count delay.

These illustrations relate to computer-simulated UWB radio signals. The processing of real signals inevitably involves noise interference, leading to incorrect results, as in the case of the plots in Fig. 1.25. The regularisation procedures discussed can provide the problem solution in these conditions.

Examine now the computation of the generalised instantaneous frequency, $\Omega(t)$, for real UWB signals received by a ground penetrating radar (GPR).

Fig. 1.35a presents a time plot of two UWB radio pulses reflected from a ground and from a subsurface object. The pulse duration is about 4 ns, and the pulses are completely time resolvable. The measurement noise makes the problem ill-posed, which is indicated by $\Omega(t)$ overshoots prior to the origin of the first pulse, where $s(t) = 0$ (Fig. 1.35b). The application of an algorithm for count averaging over several points followed by a comparison with a threshold permits the regularisation of the result (Fig. 1.35c). The plots obtained allow one to identify the time positions of the first and second pulses. In the time interval of the pulses, the $\Omega(t)$ plot oscillates with a characteristic frequency of around $2\omega_0$, although these oscillations contain noticeable noise-induced distortions.

Fig. 1.36a presents a UWB waveform at the GPR receiver, when the pulse reflected by the subsurface object partly overlaps (in time) with a weaker pulse from the Earth's surface. The origin of the second pulse in Fig. 1.36a is indicated by the shorter sidelobe at $t \approx 12.5$ ns. The generalised frequency plot in Fig. 1.36b permits the identification of the time positions of the first

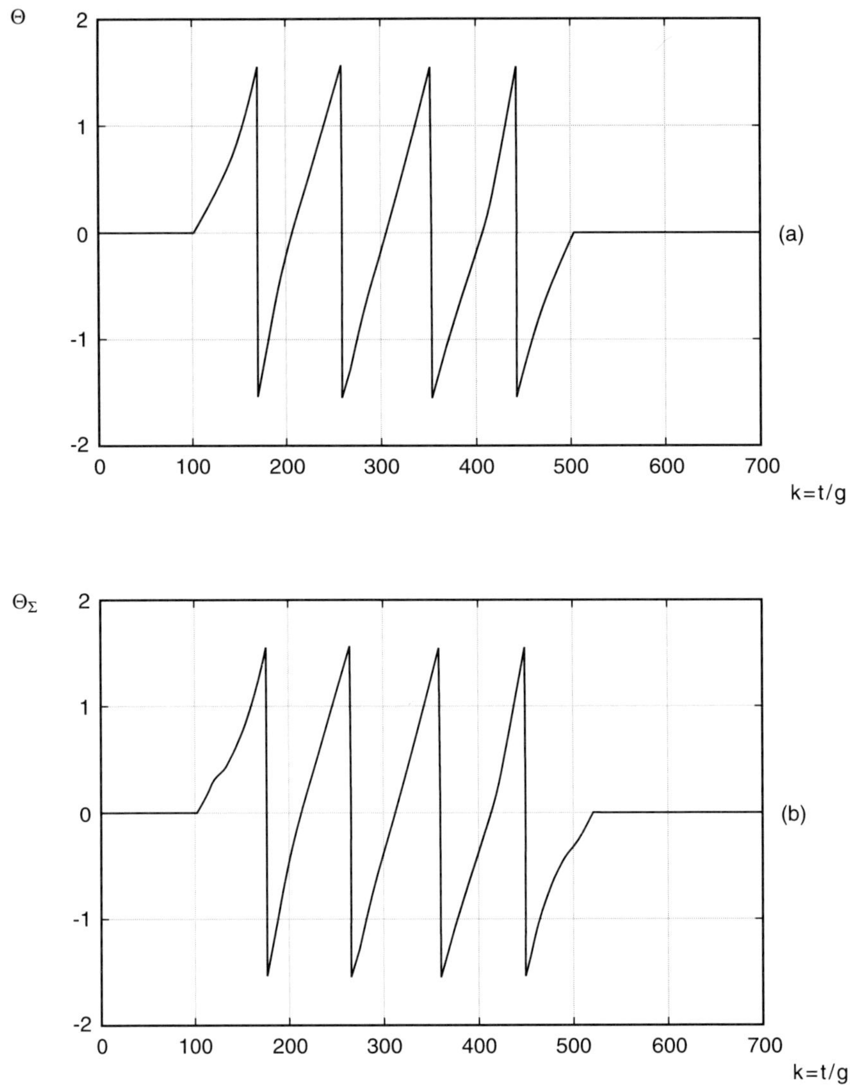

Fig. 1.34 *a* Generalised phase for single UWB pulse
b For sum of two delayed pulses

and second pulses, their time delay making up only a small fraction of the total pulse duration. Since the second pulse is observed together with the first one which has a lower amplitude, the second pulse origin is indicated by the second maximum on the $\Omega(t)$ plot. Nevertheless, the end point of the first pulse in the presence of the second one which has a larger amplitude is hardly discernible in Fig. 1.36b and can be compared with the $\Omega(t)$ minima in the time interval $t = 16$–17 ns. The third peak in this plot represents the end point of the second pulse.

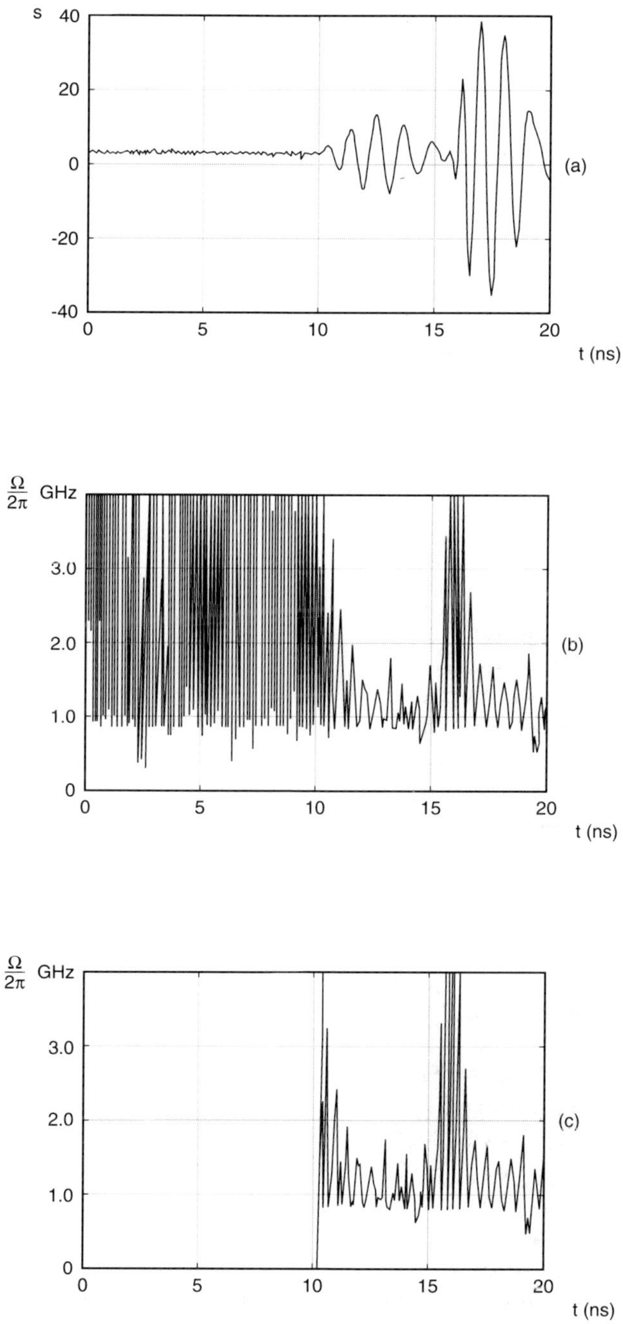

Fig. 1.35 *a* Reflected UWB radio pulses
b Their generalised instantaneous frequency
c After regularisation procedure

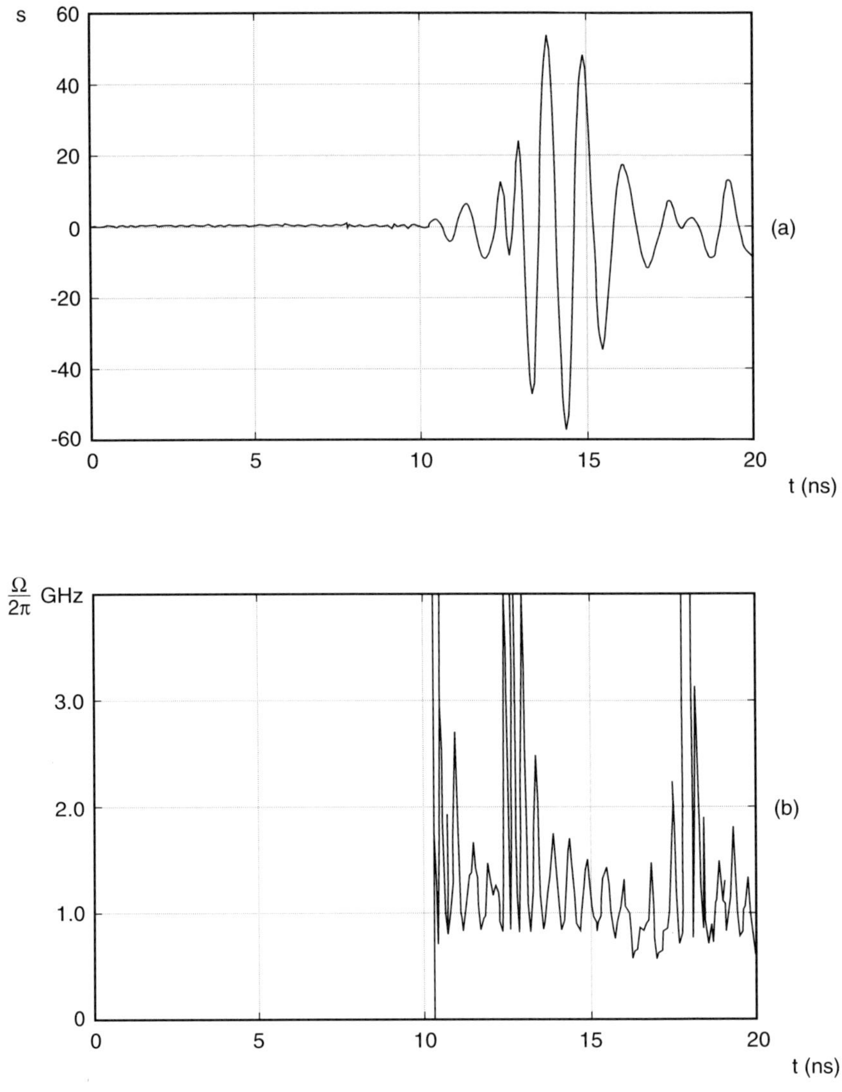

Fig. 1.36 *a* Two partly overlapped UWB radio pulses
b Possibility of their time resolution using generalised instantaneous frequency plot

Generalised modelling of UWB signals and analysis of possible applications of generalised characteristics, especially instantaneous frequency, $\Omega(t)$, permit a refined representation of signal waveform details, as compared to the Hilbert complex model. Below, we discuss other possibilities for the description of UWB waveforms.

1.5 Modelling of signal polarisation

The concept of polarisation of an electromagnetic wave has so far been applied only to narrowband, monochromatic oscillations. The polarisation structure of such a wave is defined by the strength vector, **E**, of the electric field and its direction in a plane normal to the wave propagation direction. The end of the vector, **E**, moves in this plane in an ellipse, a circle or along a line, and its direction is taken to be the angular position, Ψ, of the polarisation plane.

The vector projection, $E_x(t)$ and $E_y(t)$, of the electric field, $\mathbf{E}(t)$, in Cartesian coordinates (Fig. 1.37) may be considered as the polarisation components of an electromagnetic wave. For an amplitude and phase variable signal, these components can be described as

$$E_x(t) = A(t)\cos[\omega_0 t + \varphi(t)] \tag{1.64a}$$

$$E_y(t) = kA(t)\cos[\omega_0 t + \varphi(t) + \varphi_0] \tag{1.64b}$$

The modulation functions, $A(t)$ and $\varphi(t)$, of a narrowband signal do not change much over a frequency period, $T_0 = 2\pi/\omega_0$, and are, therefore, the same for $E_x(t)$ and $E_y(t)$. The difference between these components can only be expressed through a constant factor, k, and the value of the initial phase, φ_0. Let us define the angular position of the polarisation plane in Fig. 1.37 as a function of time

$$\Psi(t) = \arctan\frac{E_y(t)}{E_x(t)} \tag{1.65}$$

At zero initial phase, $\varphi_0 = 0$, in eqn. 1.64b, the polarisation components, $E_x(t)$ and $E_y(t)$, coincide in shape as far as the factor k, and using eqn. 1.65 we have

$$\Psi(t) = \arctan k = \text{const} \tag{1.66}$$

The angular position of the polarisation plane does not change in this case, so such an electromagnetic wave is called a linearly polarised wave. Note that the coincidence condition for the polarisation profiles, $E_x(t)$ and $E_y(t)$, is sufficient to identify a UWB vector signal as being linearly polarised because the vector, **E**, makes oscillations in the same plane, as is shown in Fig. 1.37, and its modulation rate introduces no changes.

Analysis of narrowband signals also involves the quadrature components at $\varphi_0 = -\pi/2$

$$E_x(t) = A(t)\cos[\omega_0 t + \varphi(t)]$$

$$E_y(t) = kA(t)\sin[\omega_0 t + \varphi(t)]$$

In this case the polarisation plane position is described as

$$\Psi(t) = \arctan[k\tan(\omega_0 t + \varphi(t))]$$

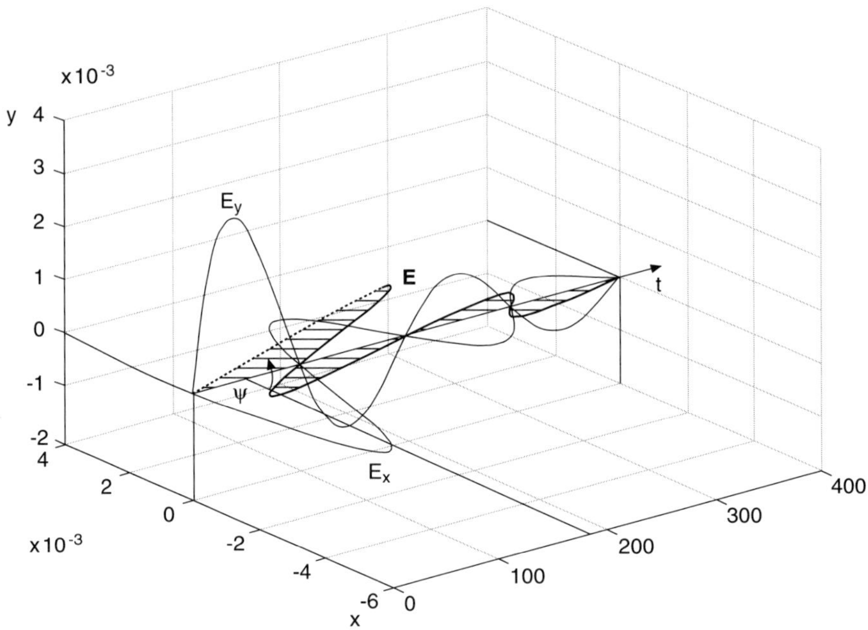

Fig. 1.37 Linearly polarised UWB electromagnetic wave

At $k = 1$, when $d\varphi/dt \ll \omega_0$, we obtain the angular velocity of the polarisation plane

$$\frac{d\Psi}{dt} = \omega_0 + \frac{d\varphi}{dt} \approx \omega_0$$

Such an electromagnetic wave is circularly polarised, while at $k \neq 1$ we have elliptical polarisation. In either case, the polarisation plane changes its position periodically, with a carrier frequency period $T_0 = 2\pi/\omega_0$. The angular velocity of the polarisation plane is of the order of ω_0 and varies with the wave ellipticity parameters.

In case of UWB signals, the nonuniform frequency response of a target (see Section 1.1) determines the waveform variation of the polarisation components of a scattered signal. The problem of signal polarisation structure description is linear; however, the signal description in the frequency domain is not particularly expressive, because the angular velocities of monochromatic oscillation vectors are different for different frequencies. Therefore, time-domain analysis with impulse response matrices (Chapter 3) is more suitable

$$[h(t)] = \begin{bmatrix} h_{xx}(t) & h_{xy}(t) \\ h_{yx}(t) & h_{yy}(t) \end{bmatrix} \tag{1.67}$$

The difference in the matrix elements determines the difference in the time-domain waveform of the polarisation components $E_x(t)$ and $E_y(t)$ of a UWB signal scattered by a target. This is what distinguishes the description of UWB signal polarisation structure from the conventional polarisation model. Indeed, whereas an incident UWB signal is linearly polarised, a scattered signal is characterised by a convolution of the incident signal with a polarisation matrix (eqn. 1.67) of a radar target.

Owing to the differences in impulse responses, $h_{ij}(t)$, which are the elements of the matrix, the polarisation components of a scattered signal acquire different time-domain waveforms, $E_y(t) \neq kE_x(t)$, so that the criterion for a wave to be linearly polarised does not hold valid. The time variation in the angular position of the polarisation plane

$$\Psi(t) = \arctan\frac{E_y(t)}{E_x(t)}$$

is then defined by the local variation in the scalar signal waveforms, $E_x(t)$ and $E_y(t)$, rather than by centre frequency, ω_0, at zero initial phase, $\varphi_0 = 0$ in eqn. 1.64. Thus, the function $\Psi(t)$ provides additional information on target properties. In Fig. 1.38 the vector $\mathbf{E}(t_i)$ position is shown when the scalar components of the vector signal, $E_x(t)$ and $E_y(t)$, do not exactly coincide.

In Section 1.4 we introduced the generalised instantaneous frequency function, $\Omega(t)$, defined by the ratio of the velocity modulus, $V(t)$, of the

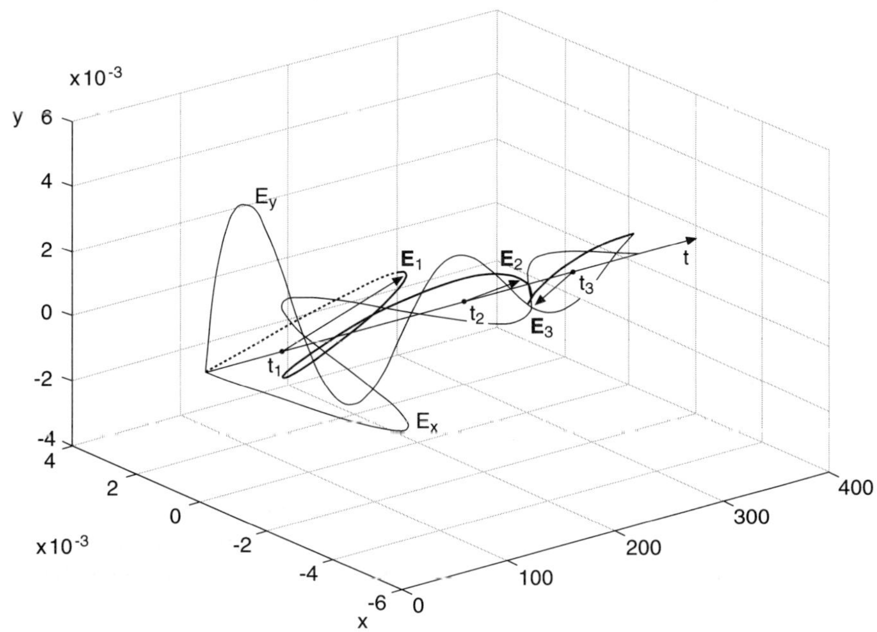

Fig. 1.38 Time variation in the angular position of the polarisation plane when scalar components $E_{x,y}$ do not exactly coincide

end of the generalised envelope vector to its length, $M(t)$. We used this function as a generalised characteristic of the UWB scalar waveform. It seems reasonable to apply a similar analysis to a polarised vector UWB signal. In this way scalar functions, $M_x(t)$ and $M_y(t)$, represent generalised envelopes of the scalar UWB signal waveform components $E_x(t)$ and $E_y(t)$. The vector, $\mathbf{M}(t)$, with the modulus,

$$M(t) = \sqrt{M_x^2(t) + M_y^2(t)}$$

oriented at an angle $\Psi(t) = \arctan M_y(t)/M_x(t)$ can be treated as a function of time in the Cartesian coordinates. If one uses components of a linearly polarised signal, $\mathbf{E}(t)$, such that $E_y(t) = kE_x(t)$ (here k is a constant factor), then the vector $\mathbf{M}(t)$ will lie in the signal polarisation plane with an angular position $\Psi = \arctan k$. The modulus of vector $\mathbf{M}(t)$ is proportional to the generalised envelopes $M_x(t)$ and $M_y(t)$

$$\begin{aligned} M(t) &= \sqrt{M_x^2(t) + M_y^2(t)} \\ &= M_x\sqrt{1 + k^2} \end{aligned}$$

Similarly, we can obtain an expression for the velocity, $V(t)$, of the end of vector $\mathbf{M}(t)$ of a linearly polarised signal

$$\begin{aligned} V(t) &= \sqrt{V_x^2(t) + V_y^2(t)} \\ &= V_x\sqrt{1 + k^2} \end{aligned}$$

The velocity vector, $\mathbf{V}(t)$, also lies in the signal polarisation plane. For this reason, the expression for the generalised instantaneous frequency of a linearly polarised UWB signal is the same for any of its polarisation components

$$\begin{aligned} \Omega(t) &= \frac{V(t)}{M(t)} \\ &= \frac{V_x\sqrt{1 + k^2}}{M_x\sqrt{1 + k^2}} \\ &= \Omega_x(t) = \Omega_y(t) \end{aligned}$$

If the component waveforms, $E_x(t)$ and $E_y(t)$, of a polarised signal have local differences, the vectors, $\mathbf{M}(t)$ and $\mathbf{V}(t)$, also show a local variation, and the condition for them to be in the same polarisation plane is violated. Then the UWB signal is no longer a linearly polarised signal, and the generalised instantaneous frequency is to be defined by the modulus ratio of the vectors

$$\Omega(t) = \frac{V(t)}{M(t)}$$

The distinction between this function and the generalised instantaneous frequencies of the polarisation components, $\Omega_x(t)$ and $\Omega_y(t)$, provides information on the local properties of the electric field vector, $\mathbf{E}(t)$, and its polarisation components, $E_x(t)$ and $E_y(t)$. In order to be able to use this information profitably, we must gain a clear understanding of the properties of the impulse response matrix, $[h]$, for various targets and obtain extensive data in test experiments.

To conclude, signal models described by eqns. 1.64a,b can be used to produce circularly ($k = 1$) or elliptically polarised UWB waveforms, keeping in mind that this can only be done for narrowband signals with a finite bandwidth (Section 1.3.2).

1.6 UWB waveform analysis with integral transforms

1.6.1 The short-time Fourier transform

We have shown (Section 1.3.1) that the UWB signal descriptions based on integral transforms in the time and frequency domains are unable to represent local elements of the signal waveform, $s(t)$, since each value of, say, spectral density, $S(\omega)$, results from a weighted integration over all instantaneous values of the function $s(t)$. The functions with discontinuity points are then described inadequately, the causality principle is violated and so on. On the other hand, the wide application of Fourier transforms in radio engineering theory has stimulated the quest for their improvement, in particular, for the analysis of transient radio signals. In Section 1.3.1 we cited a book [54], in which the concepts of instantaneous and current spectra were introduced. However, these concepts found no application at that time.

Much promise is offered by the short-time Fourier transform (STFT), which represents the integral transform of the complex envelope, $f(t)$, of a real signal, $s(t)$, as [15]

$$S(\tau, \omega) = \int_{-\infty}^{+\infty} f(t) g^*(\tau - t) \exp(-i\omega t)\, dt \tag{1.68}$$

where $g(t)$ is the time window function defining the actual integration bounds in eqn. 1.68. In this chapter, we have demonstrated some modelling errors in UWB signal analysis, using the concept of complex envelope. So eqn. 1.68 should be applied to the real signal model, $s(t)$, with a real window function, $g(t)$

$$S(\tau, \omega) = \int_{-\infty}^{+\infty} s(t) g(\tau - t) \exp(-i\omega t)\, dt \tag{1.69}$$

This equation represents the signal $s(t)$ in the time-frequency plane, $\{\tau, \omega\}$. The integration with respect to time is performed in the vicinity of the point t_i within the duration of the window function, T_w (Fig. 1.39).

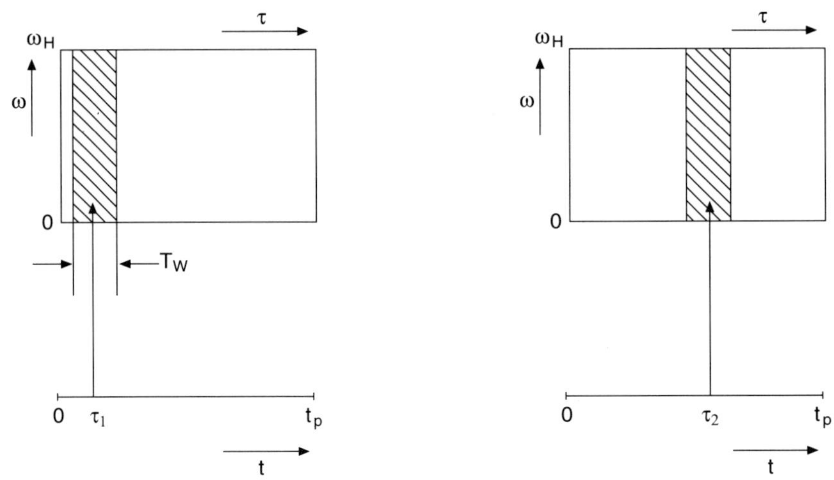

Fig. 1.39 Illustration of short time Fourier transform procedure

Of principal importance in dealing with the STFT is the choice of window duration, τ_i. If it is a small fraction of the total signal duration, t_p, the investigation of eqn. 1.69 is made with respect to the signal portion, most of which may be the fastest changing element of the signal waveform or the discontinuity region in the $s(t)$ function. In the preset dynamic measurement range it appears possible to analyse the evolution of the high frequency spectrum region, $\omega \approx \omega_H$, representing a fast variation of the waveform $s(t)$ with the window time position, τ_i. If this variation is related to the sum of UWB responses from the nearby local scatterers, the use of a short time window, T_w, permits the time resolution of these responses. As the window duration, T_w, increases, the total energy of the signal portion being analysed grows, while the fractional energy of high frequency components of the signal spectral density decreases. This makes it difficult to analyse fast changes in the signal waveform, $s(t)$. In the general case, the choice of the window duration, T_w, is said to define the compromise between the attainable time and frequency resolutions [71].

Examine as an illustration an incident UWB radio pulse, $s(t)$, of 60 count duration (Fig. 1.40a). Fig. 1.40b shows the signal $S_\Sigma(t)$ as a sum of three $s(t)$ signals with a relative delay of ten and seven counts. One can see that all time sidelobes in Fig. 1.40a have the same duration or, in the terminology of the work [38], the same sequent. But the sidelobe duration, a_i, of the summed signal, $S_\Sigma(t)$, is variable, indicating a transient character of UWB signals. This makes the signals different from monochromatic oscillations, where any summation of oscillations of identical frequency produces oscillations of the same frequency. Assuming the spacing between the adjacent zero counts of the $S_\Sigma(t)$ function to be the signal

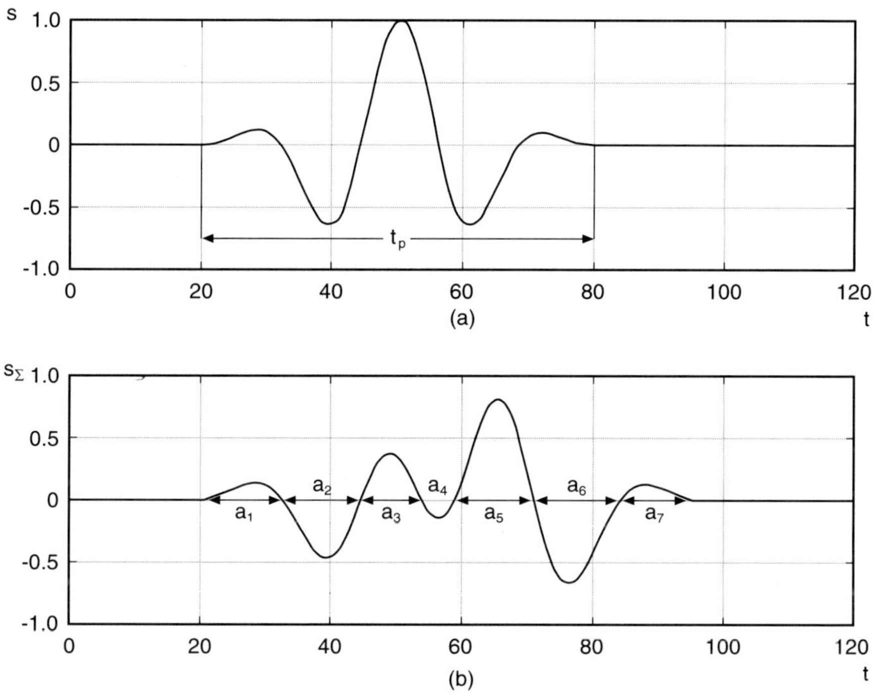

Fig. 1.40 *a* Incident UWB radio pulse
b Sum of three delayed pulses

sequent, a_i (Fig. 1.40b), we may consider the sequent, a_i, as a local discrete measure of the UWB waveform, in contrast to the averaging condition used in [38].

The three-dimensional diagram in Fig. 1.41 shows the spectral density of the summed signal, $S_\Sigma(t)$, calculated from eqn. 1.69 using a time window of duration equal to that of the signal, $T_w = t_p$. The spectral density is uniform at small τ when the window function, $g(t)$, and the signal, $S_\Sigma(t)$, are overlapped in a small time interval, t. The spectral density function is formed as τ rises. This function represents the Fourier bandwidth of the summed signal $S_\Sigma(t)$ with a bandwidth factor equal, according to eqn. 1.4, to

$$\mu = \frac{\Delta\omega}{\omega_0} = \frac{4}{N} = 0.57$$

where $N = 7$ is the number of time sidelobes of $S_\Sigma(t)$ in Fig. 1.40b. To extract information about the details of the signal waveform from Fig. 1.41 is quite difficult.

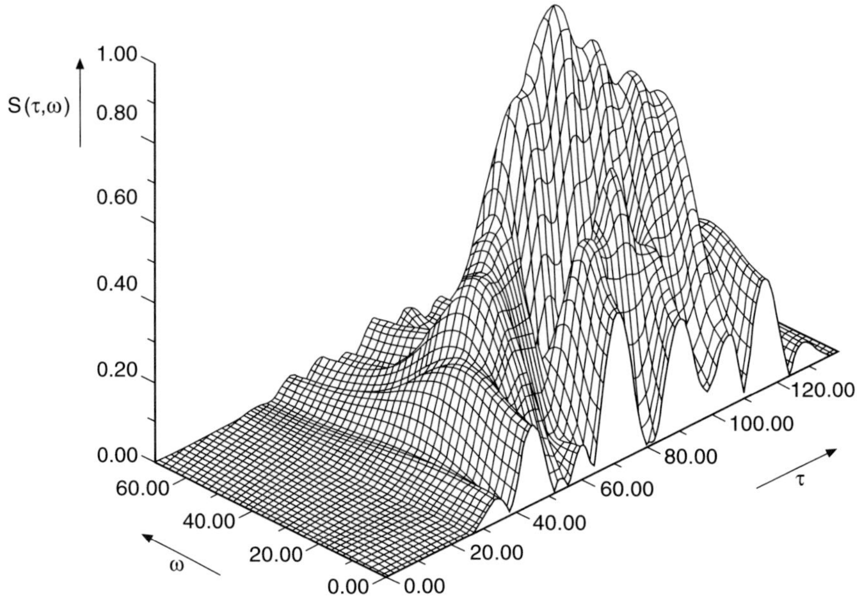

Fig. 1.41 Time-frequency diagram of the summary signal spectral density with equal duration of the signal and the time window

Fig. 1.42 presents the results of STFT calculations for the radio pulse $S_\Sigma(t)$ with the chosen time window duration, T_w, equal to the duration of one sidelobe of an incident signal, $s(t)$ (Fig. 1.40a). Small T_w values provide information on the spectral density variation with changing τ, namely, on the local details of the $S_\Sigma(t)$ waveform. It follows from this analysis that there is an energy redistribution in this signal fragment between high and low frequency regions. For example, at $\tau = 60$ in Fig. 1.42, there is a well defined peak in the high frequency region, associated with the shortest sequent, a_4, in the $S_\Sigma(t)$ signal at $t = 60$ (Fig. 1.40b). But the longest sequent, a_5, at $t = 70$, determines the energy peak in the low frequency region at $\tau = 70$. One can also see smoother frequency dependences of $S(\omega, \tau)$ at a given value of τ than in Fig. 1.41, which indicate decreasing resolution in the frequency domain.

The computational costs for the STFT procedures are quite substantial, and they reduce to the well known FFT procedures [84] for each window position in Fig. 1.39 and for each of its durations, T_w. This circumstance may interfere with STFT algorithm application in real situations.

1.6.2 The wavelet transform

Signal waveform analysis has recently come up with so-called wavelet transformation [15, 46], which consists in integrating the transient signal, $s(t)$,

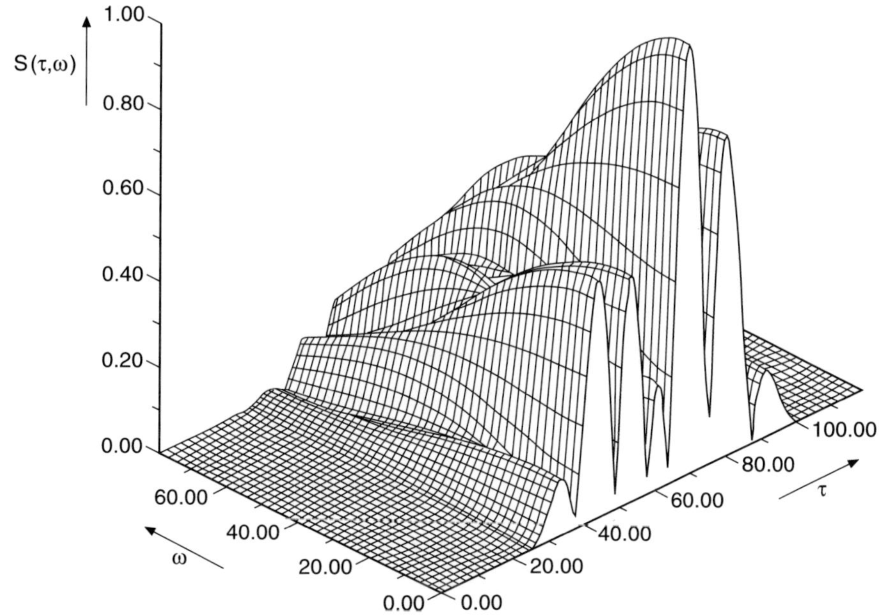

Fig. 1.42 Time-frequency diagram of the summary signal spectral density with equal duration of the signal sequent and the time window

with the transform kernel $h_{ab}(t)$ known as a wavelet

$$\mathrm{WT}(a,b) = \frac{1}{\sqrt{a}}\int_{-\infty}^{+\infty} s(t)h\left(\frac{t-b}{a}\right)dt \tag{1.70}$$

A wavelet represents a rapidly ascending function using the parameter b for the time delay and the parameter a, which prescribes the time scale transformation coefficient, i.e.

$$h_{ab}(t) = \frac{1}{\sqrt{a}}h\left(\frac{t-b}{a}\right)$$

Therefore, the wavelet transform, WT, represents the time function, $s(t)$, on the time-scale plane $\{a, b\}$. In contrast to the STFT, the WT does not use the concept of frequency and analyses the signal waveform in the time domain. The function $h_{ab}(t)$ defines the time interval, in which the analysis is made and which has a position prescribed by the parameter b. Variation in the scale coefficient, a, permits the waveform analysis to be made at different variation rates of $s(t)$. At small values of a, $a \ll 1$, the function $h_{ab}(t)$ is concentrated within a short time interval, and the computation of eqn. 1.70 allows one to analyse the rapidly varying signal fragments, as well as the $s(t)$ function discontinuity points. The factor $1/\sqrt{a}$ normalises the energies of the signal fragments being analysed. This circumstance offers an opportunity for time resolution of UWB signals reflected from nearby local scattering centres. With increasing coefficient a, WT computations provide low frequency, total information about the signal.

In order to perform the computations, the wavelet should be presented in a form having a scale coefficient and a delay changing discretely, i.e.

$$h_{mn}(t) = a_0^{-m/2} h(a_0^{-m} t - n b_0 a_0^m)$$

where m and n are integer numbers and a_0 and b_0 are the discretisation steps for the scale and delay, respectively. Then the WT takes the following form

$$\mathrm{WT}_s(m, n) = a_0^{-m/2} \int_{-\infty}^{+\infty} s(t) h(a_0^{-m} t - n b_0 a_0^m)\, dt$$

This representation can use a delay step varying with the chosen scale coefficient. Therefore, the analysis accuracy for rapidly varying signal fragments can be raised, while the body of computations to be made in the low frequency region becomes smaller.

We shall give a few examples to illustrate WT computation for UWB signals.

Fig. 1.43a shows a UWB waveform $s(t)$ resulting from truncation of the oscillating function with a Gaussian envelope over interval $t_p = 75$ counts along the t-axis. Fig. 1.43b gives WT computation with eqn. 1.70 for the wavelet $h_{1,b}(t) = s(t)$ and the time-scale transformation coefficient in the range $a \leqslant 1.5$. The results are shown as lines of a given level of $\mathrm{WT}(a, b)$ in the plane $\{a, b\}$. The regions of high line density represent steep $s(t)$ segments, while line-free regions are the segments of smooth $s(t)$ variation, for instance, extrema. The cross section of the three-dimensional $\mathrm{WT}(a, b)$ profile at $a = 1$ and $h_{1,b}(t) = s(t)$ coincides with the autocorrelation function of $s(t)$. Small coefficient a values permit an analysis of waveform details, such as discontinuities at the pulse origin and end point. Fig. 1.43c presents a cross section of a three-dimensional $\mathrm{WT}(a, b)$ profile (Fig. 1.43b) at $a = 0.05$. Note that the value of $\mathrm{WT}(0.05, b)$ is much smaller than the maximum value of $s(t)$, but $s(t)$ discontinuity points can be identified. The Gaussian envelope in Fig. 1.43c has been truncated at the discontinuity points of the derivative ds/dt. In keeping with the material in Section 1.2, the result of the convolution of $s(t)$ and this function is a double integral of the function $s(t/0.05)$. This is supported by the fact that short duration pulses in Fig. 1.43c have $N = 5$ time sidelobes, whereas the original waveform $s(t)$ has $N = 7$.

The change in the discontinuity type in the signal model, $s(t)$, by a step function changes the transformations to single integrations. This case is presented in Fig. 1.43d, showing short duration pulses at the signal, $s(t)$, origin and end point with $N = 6$ time sidelobes. Also note their increasing amplitudes relative to the pulses in Fig. 1.43c.

Fig. 1.44 presents a UWB waveform $s(t)$ as a sum of three incident signals (Fig. 1.43a) with a time delay of 20 and 14 counts. The computational results for $\mathrm{WT}(a, b)$ and its cross section are given in Fig. 1.44b,c,d illustrating

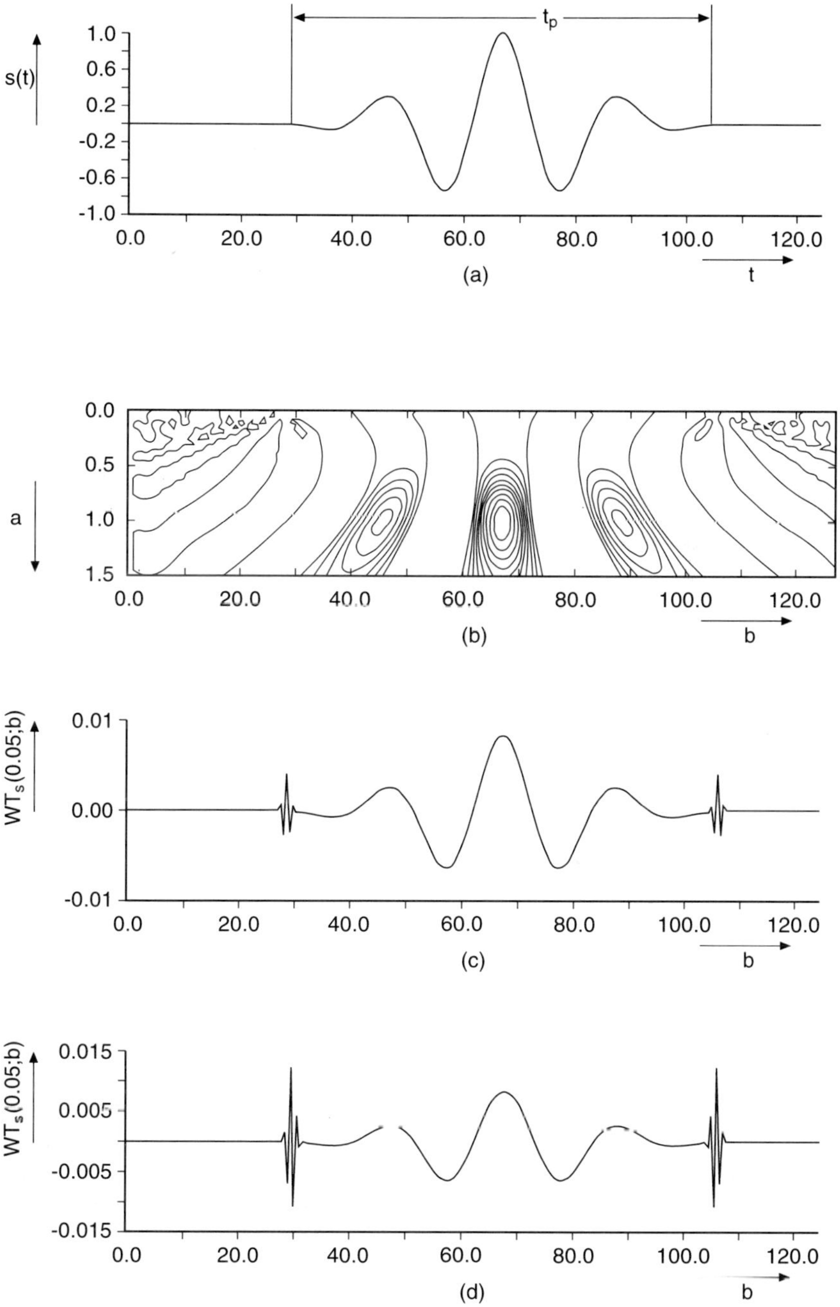

Fig. 1.43 *a* UWB signal model
b Its wavelet transform
c and *d* WT cross sections for different types of signal model discontinuities

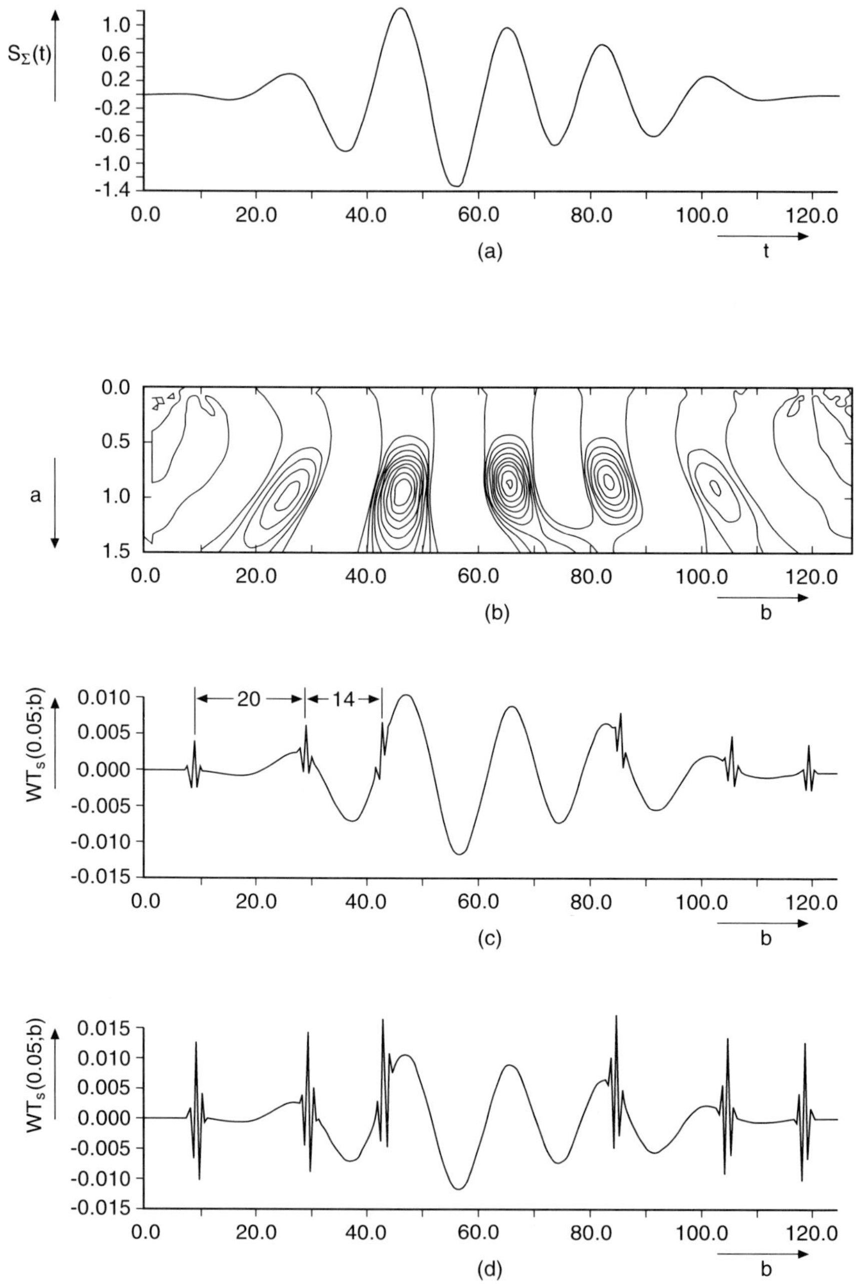

Fig. 1.44 *a* Sum of three delayed UWB signals
b Its wavelet transform
c and *d* WT cross sections for different types of signal model discontinuities

possible time resolution of partial signals, using $\mathrm{WT}(a, b)$ at $a = 0.05$. The amplitude of informative features at the partial pulse origin and end point is not large. Therefore, the radar system must have sufficient energy potential and apply special processing techniques, such as the regularisation method (see Chapter 5).

Chapter 2
Analytical description of target scattering

2.1 Methods for electrodynamic problem solution

2.1.1 *Approximate methods*

The exact solution to the problem of electromagnetic wave scattering by an object is given by Maxwell's equations with appropriate boundary conditions. However, the transition from a boundary problem in mathematical physics to applied, generalised relations in radar technology cannot be made purely mathematically. This will require simplifying approximations based on physical concepts, on signal and target parameters which can adequately disclose the physical reality [27].

There are two groups of approximate methods commonly used for the description of radar target scattering: the geometrical theory of diffraction (GTD) by Keller [51] and the physical optics methods based on Kirchhoff's approximation [14].

For monochromatic signals, the target scattering properties are described by Helmholtz' equation with the appropriate boundary conditions for the electrical and magnetic field components. General mathematical solutions for scalar, vector and more complex fields can be found by using Green's scalar, vector or tensor functions. The solution at points $\mathbf{r}$ of source-free space, V, can be represented by the sum of the scalar field components $u(\mathbf{r})$ given by Green's formula

$$u(\mathbf{r}) = \int_S \left[u(\mathbf{r}') \frac{d\mathbf{G}(k, \mathbf{r}, \mathbf{r}')}{d\mathbf{n}} - \mathbf{G}(k, \mathbf{r}, \mathbf{r}') \frac{du}{d\mathbf{n}}(\mathbf{r}') \right] d\mathbf{r}' \tag{2.1}$$

where $\mathbf{G}(k, \mathbf{r}, \mathbf{r}')$ is Green's scalar function, $u(\mathbf{r}')$ is the field on the surface S with $\mathbf{r} = \mathbf{r}'$, $\mathbf{n}$ is the normal to the surface S, $k = \omega/c$ is a wave number, ω is angular velocity and c is the speed of light.

The sources of field $u_s(\mathbf{r})$ scattered by a target are the currents induced on the target surface or in its bulk. Fig. 2.1 shows how to choose a closed

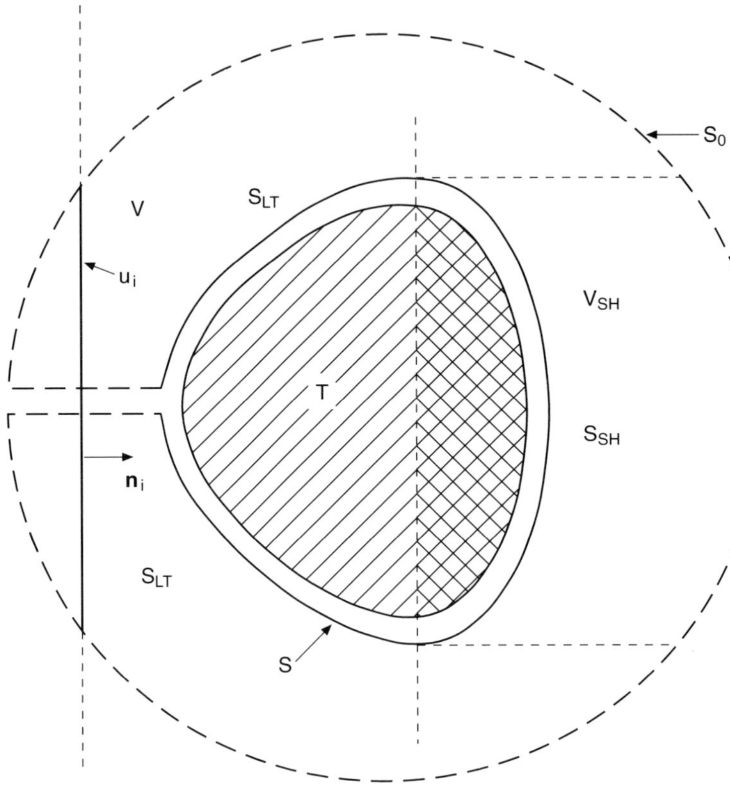

Fig. 2.1 Integration surface in Green's formula

surface, S_0, such that the space, V, around the target, T, should be free from field sources, thereby providing the conditions for applying Green's formula (eqn. 2.1). The integral over the opposite cross-cut edges and over an infinitely remote portion of S_0 is equal to zero. So the integration of eqn. 2.1 is only made over the surface, S, enclosing the target. This procedure yields the scattered field, $u_s(\mathbf{r})$, created by the target as a source located outside the space, V. At the same time, the target is excited by an incident field, $u_i(\mathbf{r})$, from a source located within the space, V. This field is represented in Fig. 2.1 as a plane wave with a normal, $\mathbf{n}_i$, to it. Therefore, space V encloses the total field, $u_T(\mathbf{r})$, which is a superposition of the incident and scattered fields

$$u_T(\mathbf{r}) = u_i(\mathbf{r}) + u_s(\mathbf{r}) \tag{2.2}$$

Under the boundary conditions, eqn. 2.2 is satisfied on the target surface coinciding with the S boundary. In order to find $u_s(\mathbf{r})$ in eqn. 2.1, one should integrate the total field (eqn. 2.2). The exact solution to the electrodynamic scattering problem is given by the following integral equation

$$u_T(\mathbf{r}) = u_i(\mathbf{r}) + \int_S L\{u_T(\mathbf{r}'), \mathbf{G}(k, \mathbf{r}, \mathbf{r}')\}\, d\mathbf{r}' \tag{2.3}$$

where $L\{\ \}$ is the linear operator from eqn. 2.1.

An approximate solution of eqn. 2.3 can be obtained by Born's iterative procedure [91]. For this, the sought-for function, $u_T(\mathbf{r})$, in the right-hand side of eqn. 2.3 is replaced by an approximate field value derived from a physically substantiated consideration. As a first step, one may use Kirchhoff's, or the physical optics, approximation, in which an approximate, geometrical optics value, $u_g(\mathbf{r})$, is introduced into eqn. 2.3. According to the boundary conditions for a perfect conductor, $u_g(S_{LT}) = u_i(S_{LT})$, then the zero approximation for the total field is $u_{T0}(S_{LT}) = 2u_i(S_{LT})$ in the light region and $u_{T0}(S_{SH}) = 0$ in the shadow region. Then eqn. 2.3 is written as

$$u_{T1}(\mathbf{r}) = u_i(\mathbf{r}) + u_{s1}(\mathbf{r}) = u_i(\mathbf{r}) + \int_{S_{LT}} L\{u_{T0}(\mathbf{r}'), \mathbf{G}(k, \mathbf{r}, \mathbf{r}')\}\, d\mathbf{r}' \tag{2.4}$$

where S_{LT} and S_{SH} are, respectively, the field illuminated and the shadowed regions of the target surface (see Fig. 2.1).

This solution is inconsistent: $u_{T1}(\mathbf{r}) \neq u_{T0}(\mathbf{r})$, since the geometrical optics approximation does not exactly meet the boundary conditions of the problem. As a matter of fact, the incident field partly penetrates, due to diffraction, into the geometrical shadow. Therefore, the difference $u_{T1}(\mathbf{r}) - u_{T0}(\mathbf{r}) = u_{d1}(\mathbf{r})$ is a first approximation to the estimation of the diffraction component, $u_d(\mathbf{r})$, by solving eqn. 2.3, and this component distinguishes this solution from that derived from geometrical optics. Note that the solution to eqn. 2.4, taking into account the approximation used, is valid for any point, $\mathbf{r}$, in the space region, V, including the geometrical shadow, V_{SH}.

2.1.2 Physical optics approximation

The physical optics approximation is widely applied in engineering design practice to solving problems of electromagnetic scattering by radar targets. Its popularity is due to the universal nature of the algorithm (eqn. 2.4) used for solving the problems, and to the high accuracy of the results for the quasioptical frequency region (see Section 1.1) and for the directions close to mirror reflections, in which geometrical components predominate. The integral in eqn. 2.4 is calculated by the stationary phase method equivalent to finding the asymptotic far-scattered field (Fraunhoffer diffraction). For finite-dimensional objects, the scattered field, $u_s(\mathbf{r})$, in the physical optics approximation is represented as a sum of diffractional components of waves formed at the object's edges (edge waves). An important advantage of the physical optics approach is an adequate description of the scattered field in the vicinity of the light–shadow boundary and in the more complex regions of focusing, light–shadow overlap and caustic zones [14]. The solution derived from Kirchhoff's approximation is continuous in the neighbourhood of these regions, although edge waves have poles at their boundaries (these compensate one another, when summed). This advantage results from integral representation of the scattered field in eqn. 2.4.

The possibility of a high frequency approximation in the framework of the physical optics approach depends on whether the following inequality is fulfilled

$$ka \gg 1 \tag{2.5}$$

where $k = 2\pi/\lambda = \omega/c$ and a is a characteristic size of the target (Section 1.1).

In the case of UWB signals providing the time resolution of local scattering centres, one should include their dimensions in expr. 2.5, which are a fraction of the total target size. This will weaken expr. 2.5 and impart to the problem some properties of the resonance and Rayleigh regions. This feature of UWB signals was characterised as frequency dualism in Chapter 1. Besides, at the bandwidth factor $\mu = \Delta\omega/\omega_0 \geqslant 1$, expr. 2.5 is fulfilled quite differently for the upper and lower spectral components, complicating the error estimation in the solutions.

The physical optics approximation can be improved by using P. Ufimtsev's physical theory of diffraction or the edge wave method [108]. In this theory, the contribution of edge wave components to the scattered field is found from the difference between the exactly known solution to the model problem (e.g. a conducting halfplane) and its physical optics approximation. This mathematical quantity refines the diffractional component, which was already partly accounted for by eqn. 2.4, and therefore has no physical sense of its own, so that the respective edge wave cannot be separated and measured. However, inclusion of this component can provide a method applicable to objects of smaller electrical size, *ka*, for the near-resonance frequency region and for larger deviations from the mirror reflection and the light–shadow interfaces.

Incorporation into the low frequency range is also possible due to the account of secondary diffraction components in Ufimtsev's method. This is done by iterations which result in cumbersome expressions, and so this method has only been applied to simple problems with no error estimation of the solution [108]. Nevertheless, it represents a convenient technique for the estimation of scattering characteristics of complex objects.

2.1.3 The geometrical theory of diffraction

The geometrical theory of diffraction (GTD) is an effective tool for the analysis and calculation of wave field radiation and scattering. This theory is based on the visual perceptions and simple concepts of geometrical optics, such as rays, wave fronts, reflection, refraction and shadowing. These concepts have been extended to diffraction phenomena, where an incident ray generates a great many new rays at the points where the object's surface is curved. The diffraction localisation principle suggested by Fock [29] is the basic idea of GTD. In this method, geometrical optics is combined with solutions of diffraction problems for model objects, known as canonical problems, with properties which reflect only local characteristics of an object of complex geometry. An important point in this approach is that the rays produced by diffraction propagate by geometrical optics' laws, so that the secondary scattering can be treated in terms of the same or a similar model problem, without complicating its description. For this reason, GTD possesses a strong localisation property. This property is revealed to the greatest extent when applied to UWB signal analysis, and it is the basis for representation of a complex object as a sum of simpler objects.

The solution of model problems in GTD can be obtained with Helmholtz' equation as an exact (in a few cases) or as an asymptotic, high frequency expansion. The expansion yields the principal term ($k \to \infty$), sometimes followed by other terms. In contrast to physical optics, the solution accuracy of GTD is determined exclusively by convergence of the mathematical methods used; it is unrelated to the ill-formulation of the initial model problem. In line with the localisation principle, diffraction rays are produced by local scattering centres, that is, by target surface regions resolvable by UWB signals. The sizes of such regions by signal bandwidth factor $\mu \approx 1$ are comparable with the incident signal duration and the centre signal wavelengths ($k_0 a \sim 1$). These facts ensure an effective application of numerical methods, and therefore GTD provides a sure foundation for error estimation.

In a typical GTD model diffraction problem for a halfplane edge (Fig. 2.2), the monochromatic diffraction wave is represented as

$$u_{d1} = \frac{1}{\sqrt{r}} \exp(ikr) D(\varphi, \varphi_0, k) u_i \tag{2.6}$$

where φ and r are the polar coordinates of the observation point, P, φ_0 is the angle position of the incident plane wave, u_i, $k = \omega/c$ is the wave number and D is a diffraction coefficient.

The diffraction coefficient, $D(\varphi, \varphi_0, k)$, in eqn. 2.6 defines the angular dependence of the amplitude of a cylindrical diffraction wave, u_{d1}. It can be derived from an exact or approximate solution of the model problem. For a perfectly conducting halfplane (Pauli's problem), the exact solution is expressed by the Fresnel integral. The first term in its asymptotic expansion is used in the GTD as the diffraction coefficient (Keller's coefficient)

$$D(\varphi, \varphi_0, k) = -\frac{\exp(i\pi/4)}{2\sqrt{2\pi k}} \left(\frac{1}{\cos[(\varphi - \varphi_0)/2]} \pm \frac{1}{\cos[(\varphi + \varphi_0)/2]} \right) \tag{2.7}$$

where the plus and minus signs denote the boundary conditions for the H and E polarisations of the incident field [14].

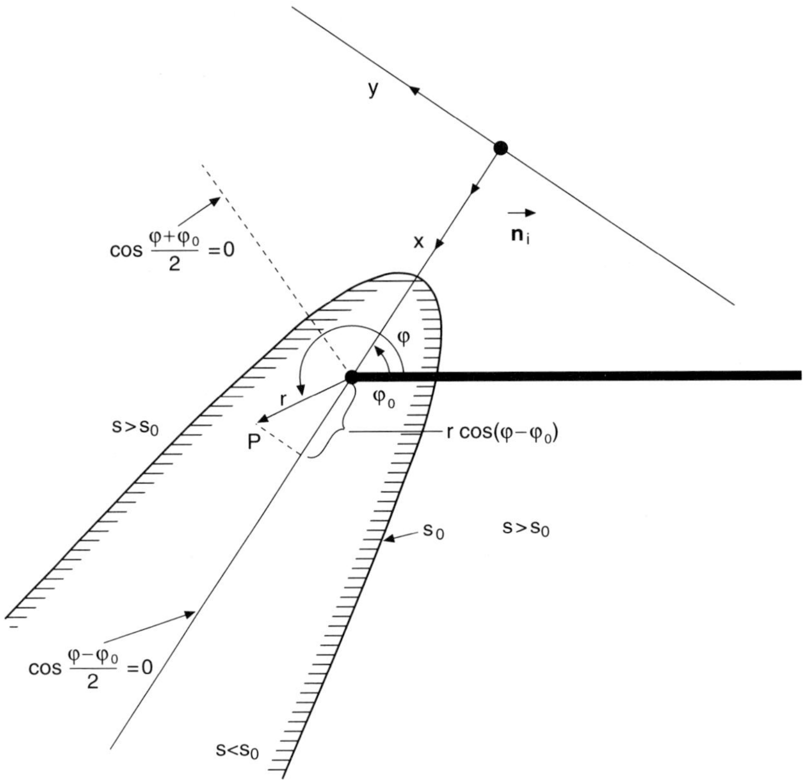

Fig. 2.2 Space-frequency domain of GTD applicability

In this approximation, eqn. 2.6 for the diffraction wave, u_{d1}, has discontinuities, which means that it has poles at $\cos[(\varphi \pm \varphi_0)/2]$ and is, naturally, inapplicable. It follows from Fig. 2.2 that the poles in eqn. 2.7 correspond to the lines of vision coinciding with the light–shadow boundaries for the incident and reflected plane waves (these are usually treated separately). An applicability criterion for eqn. 2.7 is the minimum denominator value in each summand, which is proportional to the space-frequency parameter

$$s = \sqrt{2kr}\cos\frac{\varphi \pm \varphi_0}{2} \tag{2.8}$$

The squared parameter

$$s^2 = kd = kr[1 + \cos(\varphi \pm \varphi_0)]$$

defines the phase shift of plane and cylindrical waves, depending on the wavenumber, k, and the quantity, d. The latter designates the ray path difference between a plane and a cylindrical wave, or the eikonal [14] (Fig. 2.2). The quantity s may, therefore, serve as a measure of distance between the observation point and the light–shadow boundary.

For a monochromatic wave, one can find, at minimum value of s_0, the boundary of a region, $s \geqslant s_0$, in the plane (Fig. 2.2), in which the approximation (eqn. 2.7) would be satisfactory. This boundary can then be conveniently represented by a quadratic parabola in the Cartesian coordinate system, $\{x, y\}$, with the x-axis directed along $\mathbf{n}_i$. The region inside the parabola adjoining the light–shadow boundary is known as the transition zone, or semi-shadow. In order to obtain solutions for the semi-shadow, we need more than one asymptotic term for the diffraction coefficient or to have strict solutions to the problem. For high frequency problems ($k \to \infty$), however, the semi-shadow zone width, according to eqn. 2.8, becomes smaller, and eqn. 2.7 applies to most of the plane (Fig. 2.2), as in the geometrical optics approximation.

The localisation and visualisation we have discussed are very important merits of the GTD method. In addition, its solution algorithm is well adapted to the description of repeated diffractions. On the other hand, the existence of space regions, like the semi-shadow, for which exact solutions cannot be obtained, reduces the value of the method. With UWB signals, the situation becomes more complicated due to the dependence of these regions' boundaries on the wavelength of signal frequency components. This dependence introduces errors in estimations of the scattering characteristics of wideband radar signals. Limitations on the method applicability grow with increasing bandwidth factor because of the increasing low frequency components in the spectral density. It is impossible to find a point outside the semi-shadow region in the limit of the δ function with an unlimited bandwidth. Here, high frequency GTD methods become inapplicable. The use of exact solutions to model problems in the framework of the GTD could remove many difficulties. However, attempts to find such solutions for all space-frequency regions of a complex object involve numerous complications. Some of these are due to the need to consider multiple diffraction. On the other hand, the scattering problem for a monochromatic wave presents a complicated interference pattern of locally scattered waves, which are to be described by complex oscillating time–space functions.

Some of the difficulties can, in principle, be removed or at least reduced by using UWB signals with a high time resolution. Then, in addition to the scattered field localisation would make it possible to register its time-resolvable components, thus considerably simplifying the problem solution. This supports the fairly general conclusion of the essential advantages of the time-domain approach to the study of target-scattered UWB signals [9, 54]. Many researchers have pointed out the greater simplicity of this approach and the larger scope of problems which may be solved.

We shall analyse a possible solution to the target scattering problem by a hybrid method combining the merits of the high-frequency GTD approach with exact model solutions for high time resolution UWB signals.

2.2 A hybrid solution for conducting edge scattering

2.2.1 Space-frequency description of the halfplane response

Consider the possibility of time resolution of field components in the diffraction problem for a plane wave scattered by a screen edge. Interest in this problem is due to the available exact solution for a halfplane problem (the Sommerfeld problem) [108]. It defines, for monochromatic incident field $u_i = A_0 \exp(-\mathrm{i}\omega t)$, the complex-valued amplitude, u, of scattered field as

$$u = u_1 \pm u_2 = A_0 \exp[-\mathrm{i}kr\cos(\varphi - \varphi_0)]F\left(\sqrt{2kr}\cos\frac{\varphi - \varphi_0}{2}\right)$$
$$\pm A_0 \exp[-\mathrm{i}kr\cos(\varphi + \varphi_0)]F\left(\sqrt{2kr}\cos\frac{\varphi + \varphi_0}{2}\right)$$

where $k = \omega/c$ is the wave number and the signs in the sum $u_1 \pm u_2$ are chosen depending on the boundary conditions used. The Fresnel integral, $F(\xi)$, is written in terms of the tabular functions, $C(\xi)$ and $S(\xi)$

$$F(\xi) = \frac{1}{\sqrt{2\mathrm{i}}}\left\{\frac{1}{2} + C\left(\sqrt{\frac{2}{\pi}}\xi\right) + \mathrm{i}\left[\frac{1}{2} + S\left(\sqrt{\frac{2}{\pi}}\xi\right)\right]\right\} \tag{2.9}$$

and in this form is usually used to find u in the semi-shadow region. In the designations of Fig. 2.2, the boundaries for the plane incident, u_i, and the plane mirror image waves correspond to the conditions $\cos[(\varphi \pm \varphi_0)/2] = 0$ with the vanishing argument, $F(\xi)$. The scattered field intensity, $u(\xi)$, in the shadow region is characterised by a smooth curve (Fig. 2.3) and has oscillations in the light region because of the interference of plane waves and

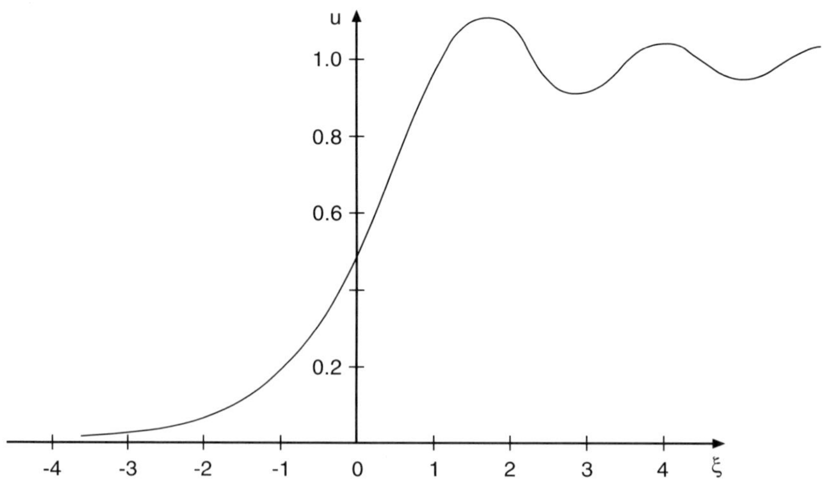

Fig. 2.3 Halfplane scattered monochromatic field intensity

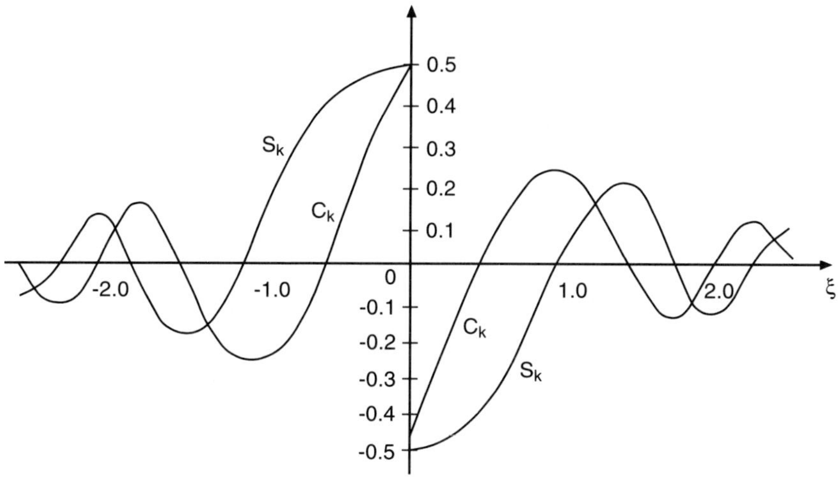

Fig. 2.4 Discontinuous components of halfplane scattered field

waves reflected by the halfplane edge. At an arbitrary point, P (Fig. 2.2), these waves differ by the time delay and can be registered individually in case of short duration UWB radio pulses. In order to separate the geometrical optics and diffraction components in the solution, we represent the functions $C(\xi)$ and $S(\xi)$ as the sum of discontinuous functions (Fig. 2.4) and the sign functions $\operatorname{sgn}\xi$

$$C(\xi) = (1/2)\operatorname{sgn}\xi + C_k(\xi)$$
$$S(\xi) = (1/2)\operatorname{sgn}\xi + S_k(\xi)$$

where $\operatorname{sgn}\xi = \pm 1$ in the light and shadow regions, respectively.

The total field then has the form

$$\begin{aligned} u = {} & A_0 \exp[-ikr\cos(\varphi - \varphi_0)]\left[\frac{1}{2} + \frac{1}{2}\operatorname{sgn}\left(\cos\frac{\varphi - \varphi_0}{2}\right)\right] \\ & + A_0 \exp[-ikr\cos(\varphi - \varphi_0)]\frac{1}{\sqrt{2\mathrm{i}}}\left[C_k\left(\sqrt{\frac{4kr}{\pi}}\cos\frac{\varphi - \varphi_0}{2}\right)\right. \\ & \left. + \mathrm{i}S_k\left(\sqrt{\frac{4kr}{\pi}}\cos\frac{\varphi - \varphi_0}{2}\right)\right] \pm \left\{A_0 \exp[-ikr\cos(\varphi + \varphi_0)]\right. \\ & \times \left[\frac{1}{2} + \frac{1}{2}\operatorname{sgn}\left(\cos\frac{\varphi + \varphi_0}{2}\right)\right] \pm A_0 \exp[-ikr\cos(\varphi + \varphi_0)] \\ & \left. \times \frac{1}{\sqrt{2\mathrm{i}}}\left[C_k\left(\sqrt{\frac{4kr}{\pi}}\cos\frac{\varphi + \varphi_0}{2}\right) + \mathrm{i}S_k\left(\sqrt{\frac{4kr}{\pi}}\cos\frac{\varphi + \varphi_0}{2}\right)\right]\right\} \end{aligned} \tag{2.10}$$

The first and third terms in eqn. 2.10 are the plane incident and reflected fields in the geometrical optics approximation: the field amplitude is A_0 in the light region and zero in the shadow. The second and fourth terms result from the plane wave diffraction by the screen edge and must be expressed by functions with a delay proportional to the distance, r, from the screen edge (see Fig. 2.2).

The diffraction-related change of the incident field will be treated as the effect of a linear filter formed by two-dimensional space, $\{r, \varphi\}$, and a screen edge. The frequency response of the filter can be defined as the complex-valued amplitude ratio of the field components at a certain point, P, and the incident field at $r = 0$: $H(\omega) = u/A_0$. For a plane reflected wave in eqn. 2.10, with the first and the third terms we have

$$H(\omega) = \exp[-ikr\cos(\varphi + \varphi_0)]\left[\frac{1}{2} + \frac{1}{2}\,\mathrm{sgn}\left(\cos\frac{\varphi + \varphi_0}{2}\right)\right]$$

Inverse Fourier transformation gives the half-plane impulse response for plane waves, which has the form of the δ function

$$h(t) = \delta\left[t + \frac{r}{c}\cos(\varphi + \varphi_0)\right]\left[\frac{1}{2} + \frac{1}{2}\,\mathrm{sgn}\left(\cos\frac{\varphi + \varphi_0}{2}\right)\right]$$

Therefore, reflected plane waves propagate without changes in the pulse waveform, irrespective of the bandwidth, and the frequency and impulse responses only determine the time delay.

From the analysis of edge waves (the second and fourth terms in eqn. 2.10), we can distinguish the delay factor, $\exp(ikr)$

$$u_{d1,2} = A_0\exp(ikr)\exp\left(-\mathrm{i}2kr\cos\frac{\varphi \pm \varphi_0}{2}\right)\frac{1}{\sqrt{2\mathrm{i}}}\left[C_k\left(\sqrt{\frac{2}{\pi}}\sqrt{2kr}\cos\frac{\varphi \pm \varphi_0}{2}\right)\right.$$
$$\left. + \mathrm{i}S_k\left(\sqrt{\frac{2}{\pi}}\sqrt{2kr}\cos\frac{\varphi \pm \varphi_0}{2}\right)\right]$$

Then, the halfplane frequency response for an edge wave is given by the function of the only space-frequency argument, s, which is the path length difference, Δ, between a plane and an edge wave at the point (r, φ) in Fig. 2.2, normalised to the wavelength, λ, i.e.

$$\begin{aligned} 2kr\cos^2\frac{\varphi \pm \varphi_0}{2} &= kr + kr\cos(\varphi \pm \varphi_0) \\ &= 2\pi\frac{\Delta}{\lambda} \\ &= s^2 \end{aligned} \tag{2.11}$$

In the expression for the halfplane frequency response to an edge wave

$$\begin{aligned} H_d(\omega) &= \frac{u_{d1,2}}{u_i} \\ &= \exp(-\mathrm{i}s^2)\frac{1}{\sqrt{2\mathrm{i}}}\left[C_k\left(\sqrt{\frac{2}{\pi}}s\right) + \mathrm{i}S_k\left(\sqrt{\frac{2}{\pi}}s\right)\right] \end{aligned} \tag{2.12}$$

the arguments of the functions C_k and S_k have different signs in the light and shadow regions in the geometrical optics approach. The sign of argument s is determined by the sign of $\cos[(\varphi \pm \varphi_0)/2]$. Complex-value function, $H_d(\omega)$ (eqn. 2.12), may be presented as the functions U and Φ of the space-frequency parameter s

$$U(s) = \frac{1}{\sqrt{2}}\sqrt{C_k^2\left(\sqrt{\frac{2}{\pi}}s\right)+S_k^2\left(\sqrt{\frac{2}{\pi}}s\right)}$$

$$\Phi(s) = \arctan\frac{S_k\sqrt{\frac{2}{\pi}}s}{C_k\sqrt{\frac{2}{\pi}}s} - s^2 - \frac{\pi}{4}$$

These are plotted in Fig. 2.5 with full lines and may be treated as amplitude-frequency and phase-frequency responses of the space-frequency filter formed by the halfplane edge. At different points of space (r, φ) in Fig. 2.6 the magnitude of argument s determines a different amplitude-frequency response $U(\sqrt{\omega})$ and phase-frequency response $\Phi(\sqrt{\omega})$. The functions $U(s)$ and $\Phi(s)$ in Fig. 2.5 are monotonic despite the presence in eqn. 2.12 of the oscillating factor $\exp(-is^2)$, whose influence is compensated for by oscillations of the functions $C_k(\xi)$ and $S_k(\xi)$ (see Fig. 2.4). Physically, the

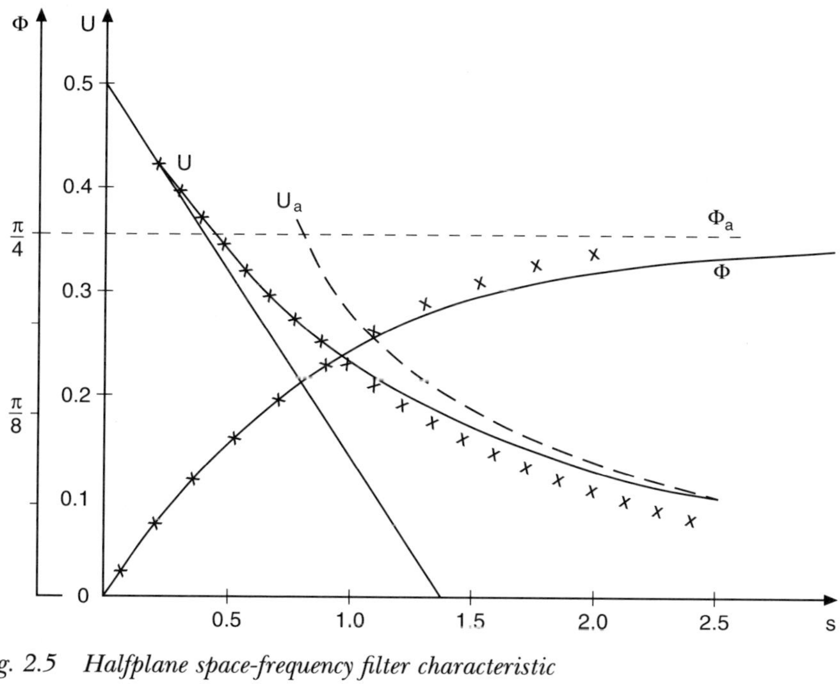

Fig. 2.5 Halfplane space-frequency filter characteristic

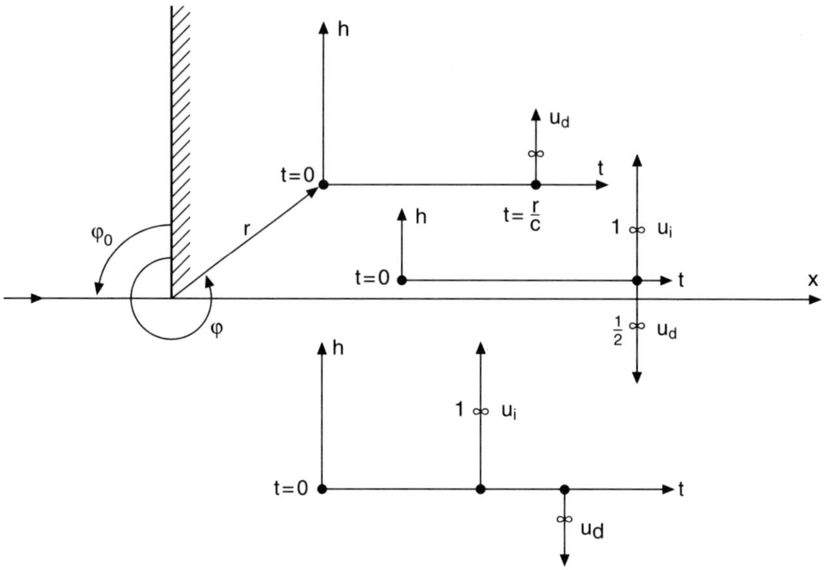

Fig. 2.6 Impulse responses for the halfplane edge wave

absence of oscillations is due to the absence of interference between the edge and plane waves. As a result, we can achieve a simple and fairly exact approximation of the halfplane space-frequency response to an edge wave for small values of s

$$U(s) = \frac{1}{2}\exp\left(-\sqrt{\frac{2}{\pi}}s\right) \tag{2.13a}$$

$$\Phi(s) = \frac{\pi}{4}\left[1 - \exp\left(-\sqrt{\frac{2}{\pi}}s\right)\right] \tag{2.13b}$$

These functions are marked by crosses in Fig. 2.5; they describe satisfactorily the halfplane space-frequency response as an edge wave filter up to $s \leqslant \sqrt{\pi/2}$. For larger s the response can be obtained by an asymptotic representation of the Fresnel integrals

$$H_a(s) = \pm\frac{1}{2\sqrt{\pi}\,s}\exp\left(\frac{i\pi}{4}\right) \tag{2.14}$$

It is clear from Fig. 2.5 that the asymptotic representation (dashed lines) is fairly good starting with $s \geqslant \pi/2$. Note that the $U_a(s)$ function is similar to an integrator response.

According to eqn. 2.11, the argument s of the halfplane frequency response to an edge wave can be expressed as a product of two parameters: that of frequency, ω, and of spatial position, $p(r, \varphi)$, of the point P, at which the field is examined, i.e.

$$s = \sqrt{2kr}\cos\frac{\varphi \pm \varphi_0}{2}$$
$$= \sqrt{\omega}\sqrt{\frac{2r}{c}}\cos\frac{\varphi \pm \varphi_0}{2}$$
$$= \sqrt{\omega}\,\sqrt{p}$$

This expression shows that the parameter s and, hence, the halfplane frequency response, does not meet the invariance criterion with respect to the spatial shift: the response varies with the position, p. By substituting the variables in the formula for the inverse Fourier transform, we find the halfplane impulse response to a halfplane edge

$$h_d(t) = \frac{1}{2\pi}\int H(\sqrt{\omega p})\exp(\mathrm{i}\omega t)\,d\omega$$
$$= \frac{1}{2\pi\sqrt{p}}h\left(\frac{t}{\sqrt{p}}\right) \qquad (2.15)$$

where the argument, $t/\sqrt{p}$, shows the variation in the time scale, $h(t)$, with the spatial position of the observation point (r, φ) as a function of p. It is known that the time-scale transformation of a signal, $s(t)$, in lumped circuits requires the use of nonstationary elements which do not meet the delay invariance criterion [119]. Eqn. 2.15 indicates that this occurs in electrodynamic problems which are not invariant to a spatial shift.

The method of finding the diffraction component of a scattered field by separating the geometrical portion from the total field can also be employed to solve other problems (wedge or tip diffraction problems and others), if the exact or approximate solutions are known for all values of the space-frequency parameter, s.

The above expression for the frequency response of a halfplane edge allows one to describe UWB waveform distortions due to diffraction. These are determined by the nonuniform distribution of the response over the signal bandwidth, $\Delta\omega$. Let us estimate the relative variation in the frequency response, $(1/U(\omega))\,dU/d\omega$, at the centre frequency, $\omega = \omega_0$. From eqn. 2.13a we have

$$\frac{1}{U_0}\frac{dU}{d\omega}\bigg|_{\omega=\omega_0} = -\frac{1}{2}\sqrt{\frac{2}{\pi}}s_0\omega_0 \qquad (2.16)$$

where $s_0 = \sqrt{\omega_0 p}$ is the value of the space-frequency parameter (eqn. 2.11) at ω_0 and $U_0 = U(\omega_0)$.

The absolute variation of the frequency response magnitude is

$$\Delta U = \frac{dU}{dw}\Delta\omega$$

and using eqn. 2.16 we find that

$$\frac{\Delta U}{U_0} = \frac{\Delta\omega}{\omega_0}\left(-\frac{1}{2}\sqrt{\frac{2}{\pi}}s_0\right)$$
$$= -\frac{\mu}{2}\sqrt{\frac{2}{\pi}}s_0 \qquad (2.17)$$

This result indicates that the wide bandwidth character of a signal described as $\mu = \Delta\omega/\omega_0$ reveals itself as a nonuniform frequency response proportional to the parameter s_0. In particular, at the light–shadow boundary of the geometrical optics solution $s = 0$, the halfplane frequency response to an edge wave does not distort the UWB waveform at any bandwidth

$$\frac{\Delta U}{U_0} = 0$$
$$H(\omega) = U(0)\exp[-\mathrm{i}\Phi(0)] = \tfrac{1}{2} \qquad (2.18)$$

Plane and edge waves are time-unresolvable at the light–shadow boundary, and the condition of eqn. 2.18 for undistorted edge waveforms corresponds to the physically necessary condition of continuity for instantaneous values of the electromagnetic field components along the boundary at any distance, r. With distance from the boundary, the edge wave frequency response becomes, according to eqn. 2.17, nonuniform, the nonuniformity increasing with s. This means (see Fig. 2.5) that the relative parameter variation, $\Delta s/s_0$, at a given signal frequency variation, $\Delta\omega/\omega_0$, grows with s. It is the necessity of matching the incident and reflected plane and edge waves that leads in GTD to a nonuniform convergence of solutions expressed in the semi-shadow region by poorly converging series with discontinuous terms [14].

Our analysis is applicable to a wide frequency range for all points in the plane of Fig. 2.2. It provides continuous solutions and an understanding of the waveform changes in terms of its time-scale variation.

To derive eqn. 2.16 we made use of the frequency response approximation (eqn. 2.13) valid at $s < \sqrt{\pi/2}$. At larger s, eqn. 2.14 similarly gives

$$\frac{\Delta U}{U_0} = -\frac{1}{2}\frac{\Delta\omega}{\omega_0} \qquad (2.19)$$

At the semi-shadow value, $s = \sqrt{\pi/2}$, the calculations with eqns. 2.17 and 2.19 coincide, thus confirming the approximation validity.

It follows from eqn. 2.19 that the frequency distortions of the edge wave at $s > \sqrt{\pi/2}$ no longer depend on the spatial parameter. This also follows directly from the asymptotic representation of the frequency response (eqn. 2.14)

$$H_a(\omega) = \pm\frac{1}{2\sqrt{\pi}\sqrt{\omega}\sqrt{2r/c}\cos[(\varphi \pm \varphi_0)/2]}\exp\left(\mathrm{i}\frac{\pi}{4}\right) \qquad (2.20)$$

which is now the direct product of the frequency and space functions (rather than their product function), that is, a function with separable variables. The spatial factor in the general expression (eqn. 2.15) is taken out of the integral, and the impulse response waveform becomes independent of the position of the reception point in space.

At a given centre frequency, ω_0, the lines of constant values of

$$s_0 = \sqrt{\omega_0}\sqrt{2r/c}\cos[(\varphi \pm \varphi_0)/2]$$

in the plane $\{r, \varphi\}$ are parabolas aligned with the light–shadow boundary (see Fig. 2.2). The GTD considers such parabolas as boundaries of the semi-shadow region. The above analysis of the frequency properties of edge waves has shown that frequency and impulse responses change within the semi-shadow region, producing changes in the edge-scattered UWB waveform: $H = H(\omega, p)$; $h = h(t, p)$. Therefore, the parabolas with constant s are lines of identical waveforms in the transient scattering problem. Outside the semi-shadow region, the spatial parameter has no influence on the frequency and impulse response shapes and only determines the edge wave amplitude: $H = H(\omega)/\sqrt{p}$, $h = h(t)/\sqrt{p}$. Hence, outside the semi-shadow region, identical scattered waveforms can be described in terms of the usual directivity pattern by defining it with eqn. 2.20 as

$$u_d(\varphi, \varphi_0) = \frac{1}{\cos(\varphi - \varphi_0)/2} \pm \frac{1}{\cos(\varphi - \varphi_0)/2}$$

The dependences on the distance, $1/\sqrt{r}$, for both types of edge wave are identical.

We shall now use the halfplane frequency response to define the waveform and parameters of the edge-scattered UWB pulse. The absence of oscillations in the expression for the frequency response (eqn. 2.12) gives the opportunity to use relations derived in Section 1.3.3 from a Taylor-series expansion of the UWB signal bandwidth around a centre frequency value, ω_0.

Consider a radio pulse of duration $t_p = 2/\beta$ with a Gaussian envelope, $f_1(t) = \exp(-\beta^2 t^2)$. Let us estimate the shape of the complex envelope of a scattered signal from the first two terms of the series (eqn. 1.44)

$$\begin{aligned} f_2(t) &= a_0 f_1(t)\left(1 - \mathrm{i}\frac{a_1}{a_0}\frac{df_1/dt}{f_1}\right) \\ &= a_0 \exp(-\beta^2 t^2)\left(1 + \mathrm{i}\beta^2 t\frac{a_1}{a_0}\right) \end{aligned}$$

where a_0 and a_1 are the expansion coefficients of eqn. 1.44. With the approximation of eqn. 2.13 and the conditions $s < \sqrt{\pi/2}$, $\mu \leqslant 1$, $t \leqslant 1/\beta$, we shall define the scattered signal waveform as

$$f_2(t)/U(s_0) = \exp(-\beta^2 t^2)\exp(\mathrm{i}\{\Phi(s_0) + \omega_0[1 + (\mu/4)^2\sqrt{2/\pi}\, s_0]t\})$$

where $\mu = \Delta\omega/\omega_0$ is the bandwidth factor.

Outside the semi-shadow region, we set $\Phi(s_0) = \pi/4$, $s_0 = \sqrt{\pi/2}$ in the exponent and rewrite

$$f_2(t)/U(s_0) = \exp(-\beta^2 t^2)\exp(-\mathrm{i}\{\omega_0[1 - (\mu/4)^2]t + \pi/4\}) \qquad (2.21)$$

Thus, the edge UWB waveform distortion reveals itself as a change in the pulse phase structure: the high frequency fill-in is shifted by $\pi/4$ relative to the envelope and acquires a linear time-varying term, $\omega_0(\mu/4)^2 t$, which is the same as a $(1 - (\mu/4)^2)$-fold reduction in centre frequency ω_0. The distortion in the semi-shadow region decreases as the parameter s_0 decreases. The accuracy of this result depends on the accuracy of the analytical approximations discussed in Section 1.3. For example, the time-scale change of an edge wave, which follows from the general analysis of the half-plane frequency response (see eqn. 2.15), is manifested as a time-scale change in the high frequency fill-in and is unrelated to the complex model envelope, valid for a medium bandwidth factor. The fill-in frequency shift is similar to a Doppler effect, and its value at $\mu \approx 1$ is as large as 6%.

2.2.2 *Impulse response of model targets*

We have shown that a solution to the diffraction problem can be represented as a superposition of impulse responses for the incident, reflected and diffracted edge waves. For highly resolvable UWB signals, these components can be registered individually.

Fig. 2.6 shows a family of impulse responses $h(t)$ of the halfplane ($\varphi = 0$) for several points $\{r, \varphi\}$ in the Cartesian coordinates (h, t) with the reference at points $\{r, \varphi\}$. Impulse responses for an edge wave are, by convention, depicted by the δ functions. The actual difference from the δ function, corresponding to eqns. 2.12 and 2.15, will be discussed below. One can see in Fig. 2.6 that there is a single u_d edge impulse in the shadow region. In the light region, there is an incident impulse u_i of unit amplitude and an edge impulse u_d. On the light-shadow region boundary, these impulses are unresolvable in time and are strictly described by the δ function. With distance from the boundary, u_d shows a delay relative to u_i, and this produces interference of the waves in case of a monochromatic signal (see Fig. 2.3).

We should like to emphasise again the absence of any discontinuous field components in this solution. This is an essential advantage over the more commonly used GTD methods owing to the rigorous solution of the model problem and to the representation of the Fresnel integrals, $C(\xi)$ and $S(\xi)$, by the discontinuous functions, $C_k(\xi)$ and $S_k(\xi)$ (see Fig. 2.4).

The results obtained by solving model electrodynamic problems prove to be especially valuable if they can be used for obtaining solutions for more complex objects. In particular, the diffraction problem for a slit in a conducting screen ($x = 0$, see Fig. 2.7) can be solved using the components of a field scattered by the edges of halfplanes A and B. The solution is found as a superposition of the incident field, u_i, the primary diffraction components

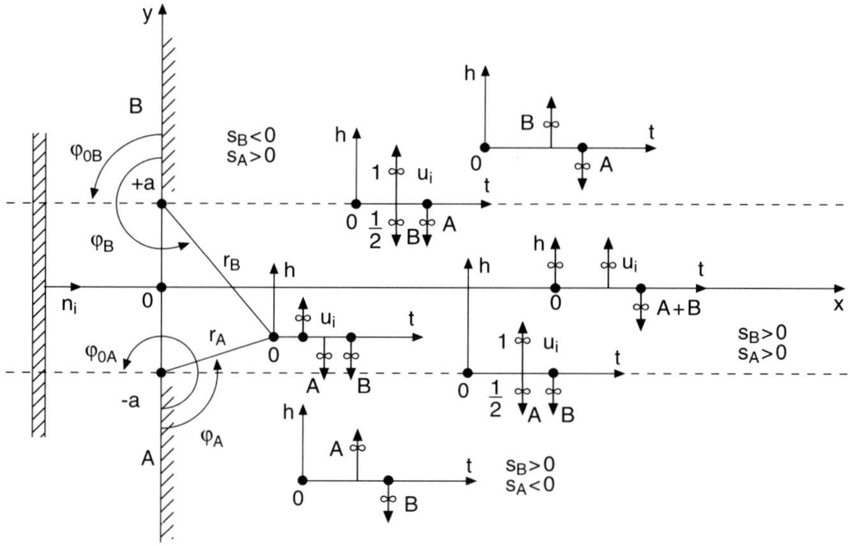

Fig. 2.7 Spatial-frequency representation of the slit diffraction problem

at the edges of planes A and B ($x = \pm a$), the secondary diffraction components produced by re-reflections between the edges of A and B with the appropriate time delay, and so on. A comparison of these components offers good grounds for the use of approximate solutions.

Fig. 2.7 shows a family of impulse responses $h(t)$ in the problem of slit diffraction in the Cartesian coordinates (h, t) with the reference at points $\{r, \varphi\}$. In Fig. 2.7 are shown components of incident field u_i and primary diffractions A and B, conventionally represented with δ functions.

Consider the primary diffraction field incident along the x-axis of a plane pulse wave (Fig. 2.7). The geometrical optics solution for the halfplanes is prescribed by Heaviside's functions, $1(y)$

$$G(y) = G_A(y) - G_B(-y)$$

where

$$G_A(y) = 1(y + a)$$
$$G_B(y) = 1[-(y - a)]$$

By replacing the functions $1(y)$ by the Fresnel integrals with the arguments

$$r_B - x = 2r_B \cos^2[(\varphi_B - \varphi_{0B})/2]$$
$$= s_B^2/k$$

and

$$r_A - x = 2r_A \cos^2[(\varphi_A - \varphi_{0A})/2]$$
$$= s_A^2/k$$

we obtain a strict solution for the primary diffraction field

$$F(s) = F(s_A) - F(-s_B)$$
$$= F(s_B) - F(-s_A)$$

where s_A and s_B are the space-frequency arguments of the fields from the A and B edges.

This solution is valid over the whole plane in Fig. 2.7. With eqn. 2.10, we obtain

$$F(s_A) - F(-s_B) = \exp(ikx)\,\tfrac{1}{2}[\mathrm{sgn}(s_A) - \mathrm{sgn}(-s_B)] + u_d(s_A) + u_d(-s_B)$$

The bracketed term denotes the incident wave field with an impulse response $h_i(t) = \delta(t - x/c)$.

The impulse responses from a slit defining the overall field are shown in Fig. 2.7 with the edge wave components conventionally represented by the δ function. In the shadow regions, the signs of s_A and s_B are different, and the impulse response is described by two time-delayed discontinuous functions with opposite signs. This is consistent with the concept of a primary diffraction field formed by edge waves. If the incident pulse is short enough, the edge wave components become time resolvable.

Now we shall use the resultant solution to examine the structure of the field radiated into the half space, $x > 0$, for an arbitrary incident signal. We shall first consider the radiation along the x-axis which is seen in Fig. 2.7 to be described by the impulse response consisting of the incident wave component and the sum of edge waves. The latter is, by convention, represented as a sum of the δ functions with opposite signs but identical time delay

$$\Delta t = a^2/(2xc) \tag{2.22}$$

The factor in the total edge response will be denoted as $2U(s)$, in accordance with eqn. 2.13. Then the impulse response is written as

$$h(t) = \delta(t - x/c) - 2U(s)\delta(t - x/c - \Delta t) \tag{2.23}$$

The incident signal, $u(t) = A_0 \sin(\omega_0 t) \cdot 1(t)$, at a distance x in the limits $-a \leqslant y \leqslant a$ and over the time interval $0 < t - x/c < \Delta t$ acts as a signal, u_0, which may be called a precursor

$$u_0(t, x) = A_0 \sin(\omega_0 t) \cdot [1(t - x/c) - 1(t - x/c - \Delta t)]$$

The precursor is the incident signal segment of duration Δt which is created only by the incident field and contains no diffractional components. The precursor duration decreases with distance as in eqn. 2.22.

An incident field in the form of a short high frequency pulse with a sinusoidal fill-in propagates without reduction in the first time sidelobe amplitude until the edge wave lag (eqn. 2.22) has decreased to a quarter period of the sinusoid. This type of radiation is characteristic of coherent light generators, which have oscillations in a high frequency problem propagating in a straight line for large distances, until the path length difference, Δt, becomes as small as a fraction of the oscillation period (see eqn. 2.22).

It was pointed out in Section 2.1 that the first approximation in the physical theory of diffraction is found by calculating the Kirchhoff integral from the aperture of the transmitter or scatterer. In this case the contributions of individual aperture elements to the total field become obscured, and only the overall monochromatic field can be defined. This field is then represented mathematically as a sum of edge waves, so the precursor is unpredicted.

After a time interval, Δt, the diffracted field already consists of both impulse response components (eqn. 2.23), so the overall signal can be written as

$$u = A_0 \sin \omega_0 (t' + \Delta t) - 2U(s) \sin \omega_0 t' \tag{2.24}$$

where $t' = t - \Delta t$. The prime will further be omitted in time designations, and there is no need to use the switching functions. The calculations from eqn. 2.24 must be equivalent to the classical solutions of the diffraction problem for a monochromatic signal. Let us express eqn. 2.24 as

$$u = [A_0 \cos \omega_0 \Delta t - 2U(s)] \sin \omega_0 t + A_0 \sin \omega_0 \Delta t \cos \omega_0 t \tag{2.25}$$

With a time delay, Δt, comprising a small fraction of the period of frequency ω_0, we can put $\cos \omega_0 \Delta t \approx 1$ and $\sin \omega_0 \Delta t \approx \omega_0 \Delta t = \pi a^2/\lambda x$. The approximation of eqn. 2.12 for the diffractional component amplitude yields

$$\begin{aligned} 2U(s) &= \exp\left(-\sqrt{\frac{2}{\pi}} s\right) \\ &\approx 1 - \sqrt{\frac{2}{\pi}} s \\ &= 1 - \sqrt{\frac{2}{\pi}} \sqrt{\frac{2\pi \Delta t c}{\lambda_0}} \\ &= 1 - 2a \frac{1}{\sqrt{2x\lambda_0}} \end{aligned}$$

where $\lambda_0 = 2\pi c/\omega_0$. Then we find from eqn. 2.25

$$u = A_0 \frac{2a}{\sqrt{2x\lambda_0}} \sin \omega_0 t + A_0 \frac{\pi a^2}{\lambda_0 x} \cos \omega_0 t \tag{2.26}$$

The classical solution

$$u = A_0 \frac{2a}{\sqrt{2x\lambda_0}} \sin \omega_0 t$$

derived within GTD completely coincides with the first term of eqn. 2.26 and describes a wave propagating with an amplitude decaying as $1/\sqrt{x}$. The second term in eqn. 2.26 describes a wave decaying as $1/x$. This means that such a wave is meaningful only in the near zone of the radiator.

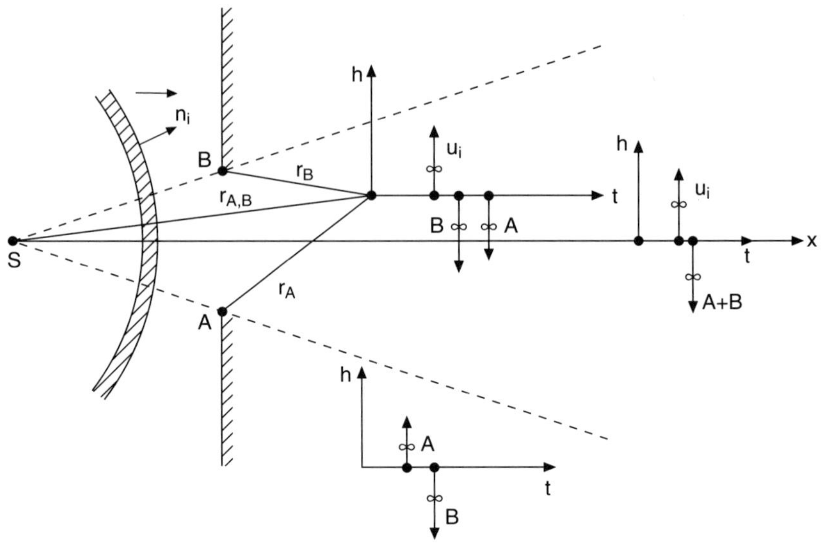

Fig. 2.8 Slit diffraction problem of a nonplane wave

The impulse response comprising the far and near-zone components has been found for such small elements as a Hertz dipole and a point source [119]. The transient solution to eqn. 2.23 refers to a finite-dimensional object, describes the field in the time domain and is of a general nature, since it is applicable to a nonplane incident wave, to the changed geometry of model objects, etc. Fig. 2.8 shows, in particular, the possibility of constructing a solution for a cylindrical incident wave, which only differs from the above solution by a different algorithm for the computation of r_A, r_B or r_i, depending on the problem geometry.

The total diffraction field can be found by supplementing the primary diffraction field with the components of repeated diffractions of wave A by edge B, of wave B by edge A, and so on. In time-domain analysis, the impulse response components are also time-resolvable relative to the primary diffraction impulses. A comparison of the repeated diffraction component amplitudes provides the necessary number of components in the general solution for the required accuracy.

Let us consider a conducting strip (see Fig. 2.9) for the case of a monostatic radar observation: $\varphi_{0A} = \varphi_A = \Theta$, $\varphi_{0B} = \varphi_B = \Theta + \pi$. The impulse response comprises the primary diffraction components conventionally represented by the δ functions and delayed by $\pm 2(a/c)\cos\Theta$ (Fig. 2.10). Therefore, their time-domain position does not permit a separate estimation of the strip size, a, and its orientation, Θ.

The secondary diffraction consists in the scattering of wave A by edge B and *vice-versa*. Both components have the same delay relative to the wave arrival at the point $x = 0$. The tertiary diffraction components are delayed by $(1/c)(4a \pm 2a\cos\Theta)$.

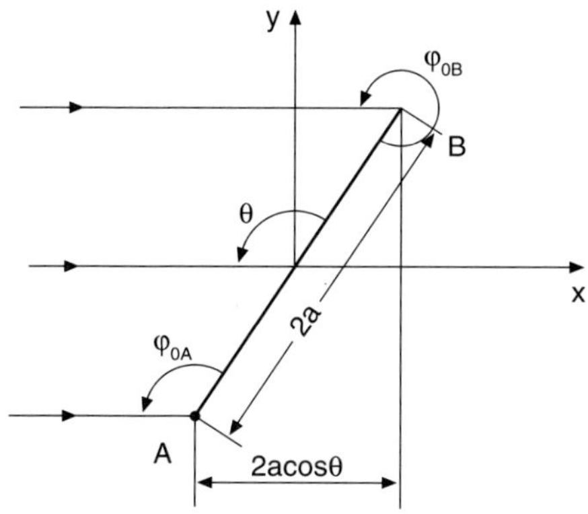

Fig. 2.9 Conducting strip diffraction geometry

With the asymptotic representation of the Fresnel integrals, we can write the expressions for the primary, secondary and tertiary components for a strip

$$u_I = \frac{\exp(-\mathrm{i}\pi/4)}{\sqrt{2\pi kr}}\left[\frac{1}{2}\exp(-\mathrm{i}2ka\cos\Theta)\left(1 \pm \frac{1}{\cos\Theta}\right) + \frac{1}{2}\exp(\mathrm{i}2ka\cos\Theta)\left(1 \pm \frac{1}{\cos\Theta}\right)\right]$$

$$u_{II} = \left(\frac{\exp(-\mathrm{i}\pi/4)}{\sqrt{2\pi kr}}\right)^2 \frac{\exp(\mathrm{i}2ka)4}{\sqrt{2a}\sin\theta}$$

$$u_{III} = \left(\frac{\exp(\mathrm{i}\pi/4)}{\sqrt{2\pi kr}}\right)^3\left(-\frac{1}{2a}\right) \times \left[\frac{\exp(\mathrm{i}k(4a - 2a\cos\Theta))}{\cos^2(\Theta/2)} + \frac{\exp(\mathrm{i}k(4a + 2a\cos\Theta))}{\sin^2(\Theta/2)}\right]$$

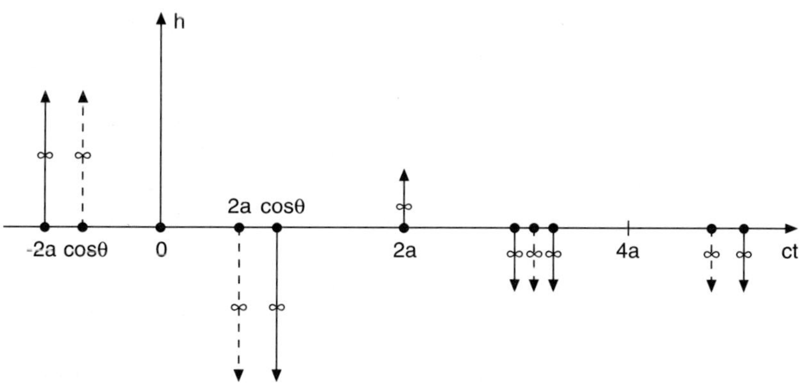

Fig. 2.10 Spatial invariance components of strip diffraction problem

Evidently, the difference of the impulse response components from the δ function due to the nonuniformity of edge frequency responses increases with repeated reflections as if they had passed n consecutive filters, $(\exp(\mathrm{i}\pi/4)/\sqrt{2\pi kr})^n$. One may assume that the UWB waveform distortions described in this Section grow additively. For instance, a Gaussian radio pulse shows a 12% centre frequency shift in the secondary diffraction. The amplitude of the nth component of repeated diffraction is proportional to $(1/\sqrt{k_0 a})^n(1/\sqrt{k_0})$. For high frequency problems, $k_0 a \gg 1$ and the impulse response can only be represented by primary diffraction. Closer to the resonance region, $(k_0 a \approx 1)$, it is possible to detect repeated diffraction components, and this raises the information value of measurements in this region.

For instance, separate measurements of strip size $2a$ and its orientation, Θ, in the monostatic observation require discrimination between the primary and secondary diffraction responses (Fig. 2.10). This is possible because the UWB signal waveform, multiply scattered by an object's edge, differs from that of the primary component. This difference can be registered for UWB signals.

If the bandwidth factor is small, there is an alternative way, namely, the analysis of the field spatial distribution in, say, a bistatic radar observation. The impulse response variation with the angle Θ is shown by a dashed line in Fig. 2.10. It can be seen that the secondary response has, in contrast to the primary response, a constant position in the time domain at $ct = 2a$. This invariance relative to the aspect angle corresponds to the invariance of the target transfer function poles in a complex plane [7]. Therefore, separate registration of the primary and secondary diffraction responses imparts to UWB signals a new quality – in our example, this is the ability to determine the object's size, $2a$, and its orientation, Θ, even in a monostatic transmission.

2.2.3 Wedge-scattered waveforms

In constructing the overall solution to the diffraction problem (Figs. 2.6–2.10), we conventionally represented diffraction components as δ functions. But this is impossible in actual reality, since all components of the total field in the semi-shadow region must be physically continuous. The incident field in the light region has a constant amplitude while, in the shadow, the diffraction component decays as $1/\sqrt{r}$ with distance from the halfplane edge. Clearly, two differently varying quantities cannot be equated at their boundary, so the transient region must also be considered because here the field is not a sum of plane and cylindrical (or spherical) waves.

Moreover, the diffraction component in the shadow region is the only element of the impulse response; it must meet Rayleigh's postulates (Section 1.1), that is, it must necessarily be a bipolar function. Therefore, the diffraction component of the impulse response cannot be described by

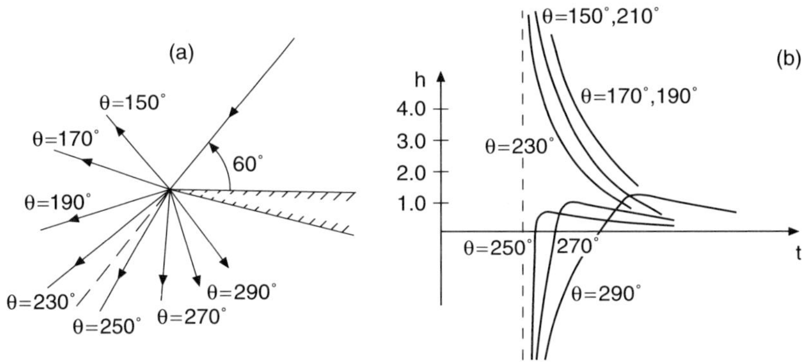

Fig. 2.11 *a* Wedge diffraction problem
b Impulse responses from a wedge

the δ function, and the diffraction component of the UWB signal scattered by a halfplane will have a waveform different from that of the incident signal.

The diffracted impulse responses obtained from the solution of a transient problem for a wedge at various angles Θ [105] are shown in Fig. 2.11. In the light region ($\Theta < 240°$) the response is unipolar, so Rayleigh's postulates can only be met for a combination of response components for the incident and reflected waves. In the shadow ($\Theta > 240°$) the impulse response is bipolar, as was suggested, and meets Rayleigh's criteria.

In order to define the diffraction component waveform, we need to know the nature of the impulse response discontinuity at the wave front and its parameters must be compared with the incident signal parameters. For example, for a low frequency signal of long duration, all the impulse responses in Fig. 2.11 may be approximately taken to be δ functions.

The limiting behaviour of the diffraction component discontinuity can be defined as $1/\tau$ from Green's function of the wave equation in cylindrical coordinates [14]. Similarly, it was shown in [105] that the diffraction problem solution had no terms decreasing faster than $t^{-1+\gamma}$, where γ is a small quantity. Thus, the discontinuity at the front of the diffraction component varies in the strict solution from $1/\tau$ at the geometrical solution boundary to $1/\sqrt{\tau}$ outside the transition region. In the latter situation, the asymptotic relationships hold true, so the response waveform does not change with distance from the light–shadow boundary.

Analytical description of the convolution of radio pulses with an impulse response of arbitrary discontinuity is impossible in the general case. There have only been a few attempts to reduce this operation to special functions. For this reason, the application of the analytical results presented here requires the use of numerical methods specific to discontinuous functions [45]. A generalised method for the computation of the diffraction component waveform is based on the integrability of discontinuous functions of the type

$$h(t) = t^p 1(t) \tag{2.27}$$

where $p \neq -1$ and $1(t)$ is a step function.

In our problem, the impulse response is given by eqn. 2.27 with

$$-1 < p < -\tfrac{1}{2} \tag{2.28}$$

A single integration of eqn. 2.27 removes the response discontinuity, i.e.

$$\begin{aligned} a(t) &= \int_0^t h(t)\, dt\, 1(t) \\ &= \frac{1}{p+1} t^{p+1} 1(t) \end{aligned} \tag{2.29}$$

The convolution integral of a step response, $a(t)$, and the incident signal, $s_i(t)$, gives the function $y(t)$ as an integral of the desired function, $s_d(t)$, describing the diffraction component waveform

$$\begin{aligned} y(t) &= s_i(t) * \int_0^t h(t)\, dt \\ &= s_i(t) * a(t) \\ &= \int_0^t s_d(t)\, dt \end{aligned}$$

where $s_i(t)$ is an incident radio pulse and $*$ denotes the integral convolution. Then the desired function is found by differentiation

$$s_d(t) = \frac{d}{dt} y(t) \tag{2.30}$$

The step responses (eqn. 2.29) are plotted in Fig. 2.12, in which the hatched region corresponds to responses satisfying eqn. 2.28. The asymptotic form of the step response for $p \to -1$ is expressed by a step function, corresponding

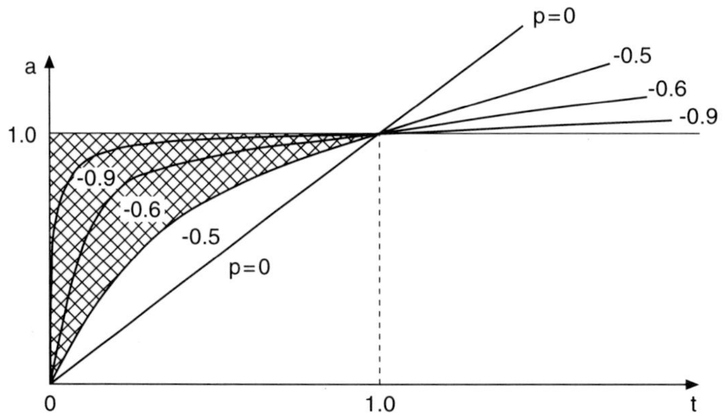

Fig. 2.12 Step responses of an edge wave with p-parameter

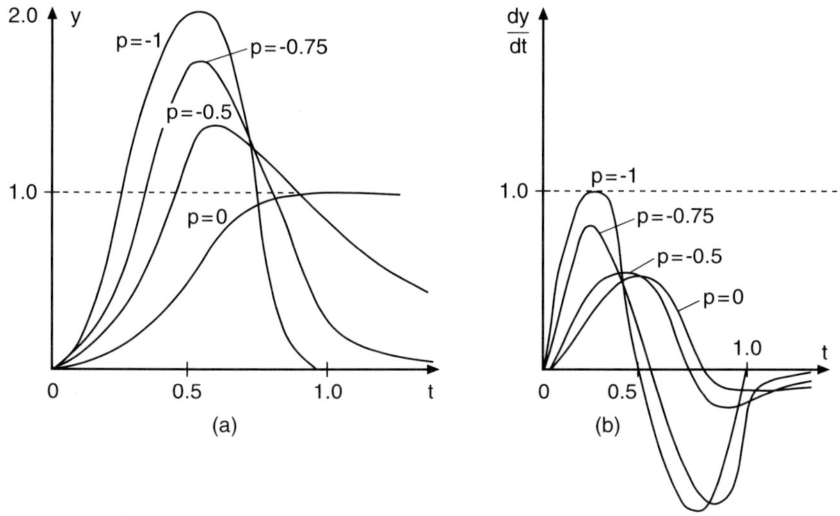

Fig. 2.13 *Edge scattered UWB radio pulse waveform*

a After convolution with step response
b After differentiation

in the convolution procedure to the integration of the incident signal, whose waveform is completely restored by further differentiation (eqn. 2.30). Thus, the effect of the diffraction response on the incident signal on the light–shadow boundary at $p \to -1$ is equivalent to the effect of the δ function, as was expected.

Fig. 2.13 shows the convolution computations for a UWB pulse consisting of one sinusoid period with a step response, $a(t)$, and the functions obtained by differentiation (eqn. 2.3) to describe the diffraction component waveform. At the light–shadow boundary with $p \to -1$ the diffraction waveform reproduces the incident waveform. Within the semi-shadow region ($p \to -1/2$), the diffraction waveform is extended in time. This time-scale change of UWB diffracted pulses was predicted in the above analysis. For the signal being examined, the bandwidth factor is $\mu \approx 2$, so its period must change by 25%, according to eqn. 2.21. The plot in Fig. 2.13b shows a 50% increase in the first oscillation half period at $p \to -1/2$. Taking into account the general tendency for variation in the UWB waveform, the transformation at $p \to -1/2$ may be termed as semi-integration.

2.3 Decomposition of the target impulse response

2.3.1 The aim of impulse response decomposition

Target impulse response is widely used as a radar characteristic in the analysis of UWB signal changes in the time domain (see Chapter 3). We have

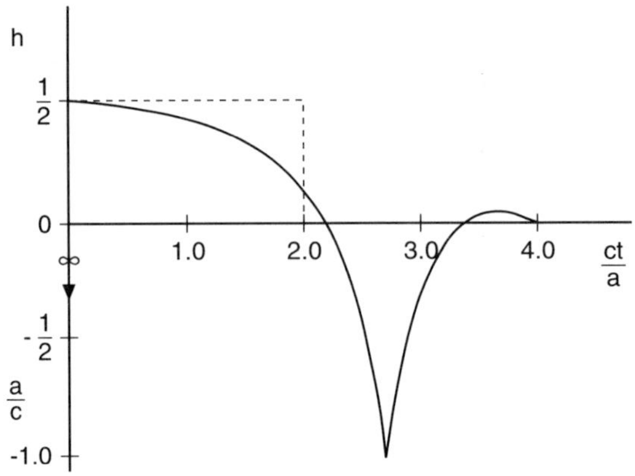

Fig. 2.14 Impulse response from a conducting sphere

shown that in the model scattering problem the changes in the incident waveform are determined by the nature of the response discontinuity. The impulse response generally includes discontinuities of different orders. Fig. 2.14 presents the response of a conducting sphere of radius a, comprising a δ function and a step function at $ct/a = 0$ and discontinuity of the first derivative, dh/dt, at $ct/a = 2.6$. In either case, the waveform of a scattered signal, $s_s(t)$, can be found by calculating the integral convolution of the incident signal, $s_i(t)$, and the impulse response, $h(t)$. The waveform variation for $s_i(t)$ exhibits a certain specificity largely related to pulse scattering by target local elements. The responses from local elements by use of UWB signals can be registered separately. For instance, for an $s_i(t)$ signal belonging to the class of radio pulses when $\int_{-\infty}^{+\infty} s_i(t)\,dt = 0$ (Section 1.2), the calculation of convolution for a sphere response gives a value close to zero at the sites of a smooth variation of $h(t)$. The main contribution to the response is made by its discontinuity points (Fig. 2.14). This may serve as a basis for response decomposition into elements responsible for the scattered signal parameters and waveforms.

As an illustration, we shall analyse the decomposition of a rectangular impulse response, $h(t)$ (Fig. 2.15a), having a Fourier transform, $H(\omega)$ (Fig. 2.15b). To describe the change in signals with a spectral density lying within ω_L–ω_H bandwidth (Fig. 2.15b), it is necessary to find an impulse response in the form of the inverse Fourier transform of the function $H(\omega)$ in the given frequency range. Such a response is known as the smooth, or truncated, impulse response

$$h_a(t) \approx \frac{h_0}{\pi}\frac{\Delta\omega}{\omega_0}\left\{\frac{\sin[\Delta\omega/2(t+t_p/2)]}{\Delta\omega/2(t+t_p/2)}\sin\left[\omega_0\left(t+\frac{t_p}{2}\right)\right]\right.$$

$$\left.-\frac{\sin[\Delta\omega/2(t-t_p/2)]}{\Delta\omega/2(t-t_p/2)}\sin\left[\omega_0\left(t_0-\frac{t_p}{2}\right)\right]\right\}$$

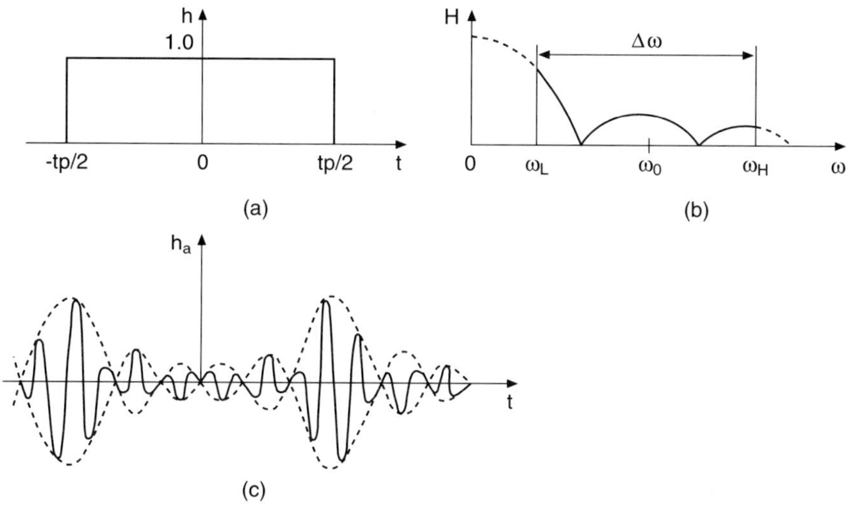

Fig. 2.15 Truncated impulse response for finite-spectrum signal

a Rectangular impulse response
b Truncated Fourier transform
c Truncated impulse response

where $\Delta\omega = \omega_H - \omega_L$ and $\omega_0 = (\omega_H + \omega_L)/2$, and this represents a response to a rectangular signal spectral density uniformly distributed within the range $\omega_H - \omega_L$. Two local radio pulses are produced in this case, with time-domain positions and amplitudes which correspond to the parameters of $h(t)$ in Fig. 2.15a. Under the above conditions, the response can be decomposed into two local scattering centres.

As ω_L tends to zero, the local responses in Fig. 2.15c begin to interact and, in the limit $\omega_L = 0$, give one common response with characteristic Gibbs rise times determined by the limiting value of ω_H. This means that the low frequency properties of the signal do not permit response decomposition into identifiable local scatterers.

Another aspect of this problem is the specific application of impulse responses to define the scattered signal waveforms. When transient processes are investigated in aperiodic radioelectronic circuits, it is generally assumed that the output response waveform of a system excited by a rather short video pulse, $s(t)$, coincides with the waveform of the system impulse response, $h(t)$. Fig. 2.16 shows a rectangular impulse response of duration t_0 and an exciting radio pulse, $s_i(t)$, as a sinusoid of one period duration, $t_p \ll t_0$. The responses, $s_{1,2}^{(s)}(t)$, to short video pulses, $s_{1,2}^{(i)}(t)$, practically coincide with $h(t)$, while the total response, $s^{(s)}(t)$, is quite different from both the incident signal, $s(t)$, and the impulse response. A shorter duration of the input excitation does not bring the response closer to $h(t)$. This example is another illustration of a possible decomposition: the properties of an

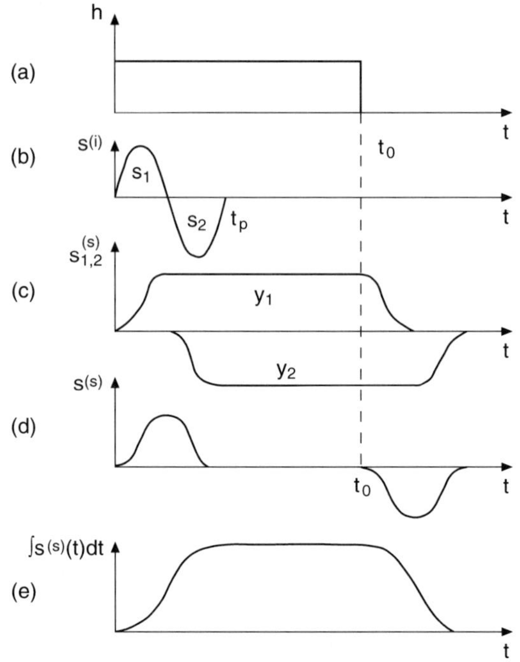

Fig. 2.16 UWB radio pulse convolution with rectangular impulse response

a Impulse response
b Exciting pulse
c Scattered signal decomposition
d Scattered signal
e Restored impulse response

impulse response with respect to a radio pulse can be fully represented by the local points $t = 0$ and $t = t_p$.

This analysis allows us to choose a general decomposition method consisting in decomposition by the discontinuity points of the impulse response and its derivatives. Let us describe the response, $y(t)$, of a target with an impulse response, $h(t)$, to an input signal, $x(t)$, with spectral density $X(\mathrm{i}\omega)$. For this, we shall take advantage of the concept of current spectrum for a segment of $h(t)$ over the time interval $(0, t)$

$$H(\mathrm{i}\omega, t) = \int_0^t h(t)\exp(-\mathrm{i}\omega t)\,dt$$

Then

$$y(t) = \frac{1}{2\pi}\int_{-\infty}^{+\infty} X(\mathrm{i}\omega)H(\mathrm{i}\omega, t)\exp(\mathrm{i}\omega t)\,d\omega$$

After N-fold integration by parts, we have

$$y(t) = \frac{1}{2\pi}\int_{-\infty}^{+\infty} X(\mathrm{i}\omega)\left(\frac{\mathrm{i}}{\omega}\right)^N \exp(\mathrm{i}\omega t)\int_0^t \frac{dh^{(N)}}{dt^N}\exp(-\mathrm{i}\omega t)\,dt\,d\omega$$

Keeping in mind that the product of spectra corresponds to the convolution of time-domain functions, we can write

$$y(t) = \iint\int_0^t x(t)\, dt^N * \frac{dh^{(N)}}{dt^N} \tag{2.31}$$

where $*$ denotes the integral convolution. The contribution of the high order derivatives, $dh^{(N)}/dt^N$, to the target-scattered signal is lower, the higher the frequency of the incident signal, $x(t)$, since its spectrum, $X(\mathrm{i}\omega)$, is divided by ω^N when integrated. Therefore, the problem we have presented can be solved by decomposing $h(t)$ over the time derivatives. For this, we must approximate $h(t)$ by the sum of piecewise smooth, nonoverlapping time functions, $h_k(t)$, and single out the δ functions

$$h(t) = \sum_{k=1}^{L} h_k(t) + \sum_{k=0}^{L} a_k \delta(t - t_k)$$

where $k = 0, 1, 2$ are the numbers of segments of the functions, $h_k(t)$, and of their bounds, t_k.

Let us now represent the piecewise smooth functions, $h_k(t)$, as power series

$$h_k(t) = \sum_{p=0}^{\infty} m_{pk}(t - t_{k-1})^p[1(t - t_{k-1}) - 1(t - t_k)]$$

By superposition we obtain

$$y(t) = \sum_{j=-1}^{\infty} y_j(t) \tag{2.32}$$

where

$$y_{-1}(t) = x(t) * \sum_{k=0}^{L} a_k \delta(t - t_k)$$

$$= \sum_{k=0}^{L} a_k x(t - t_k)$$

$$y_0(t) = x(t) * \sum_{k=1}^{L} m_{0k}[1(t - t_{k-1}) - 1(t - t_k)] \ldots$$

Only the δ functions can be retained in the sums of the series (eqn. 2.32) for a narrowband high frequency signal. Every point $a_k \neq 0$ where the response smoothness is broken represents its local segment.

When dealing with UWB signals that contain low frequency components, one should take into account the existence of high order discontinuities in the series (eqn. 2.32). The transformation of the signal $x(t)$ with such discontinuities is given by eqn. 2.31 as the integration of $x(t)$ with respect to time in order to raise the low frequency components of the scattered signal, $y(t)$. This necessitates taking account of the interaction among neighbouring points, a_k, of $h(t)$.

We must now parametrise the impulse responses at the discontinuity points of $h(t)$ and its derivatives by differentiating each term in eqn. 2.32 as many times as is necessary for obtaining the δ functions. As a result, we have

$$\frac{dy_0}{dt} = x(t) * \sum_{k=1}^{L} m_{0k}[\delta(t - t_{k-1}) - \delta(t - t_k)]$$

$$\frac{d^2y_1}{dt^2} = x(t) * \sum_{k=1}^{L} m_{1k}[\delta(t - t_{k-1}) - \delta(t - t_k)] \ldots$$

Having regrouped and integrated the terms, we get the desired expression

$$y_j(t) = \sum_{k=0}^{L} M_{jk} \int\int_{t_k}^{t} \cdots \int x(t - t_k)\, dt^{(j+1)} \tag{2.33}$$

where

$$M_{jk} = (j!)(m_{j(k+1)} - m_{jk})$$

$$m_{j0} = 0$$

$$j = 0, 1, 2, \ldots$$

Eqn. 2.33 proves the possibility of decomposing an impulse response, because target scattering properties are only defined by discontinuity points of the response and its derivatives. This result expresses the localised character of a radar response: the scattered signal is defined by M_{jk} coefficients allowing for disturbances in the smoothness of $h(t)$ at some points. Target segments corresponding to a smooth response function do not scatter a radar signal. The causality criterion for $h(t)$ of a radar target rules out, in principle, designing a nonreflecting object.

2.3.2 *UWB waveform definition by a smooth target*

We shall analyse some specific features of transient scattering by objects with smooth surfaces. As an illustration, consider the scattering of UWB radio pulses by a conducting sphere. We shall employ the impulse response decomposition method for finding the parameters of scattered signals. With the exact impulse response of a sphere (see Fig. 2.14) and the δ function properties for the variable $z = ct/a$ and the incident pulse $s(t) = A(t) \cos \omega_0 t$, we shall find with eqn. 2.33

$$y_p(z) = M_{pk} \left(\frac{c}{a\omega_0}\right)^p \int\int \cdots \int A\left(\frac{z}{\omega_0}\right) \cos z\, dz^p \tag{2.34}$$

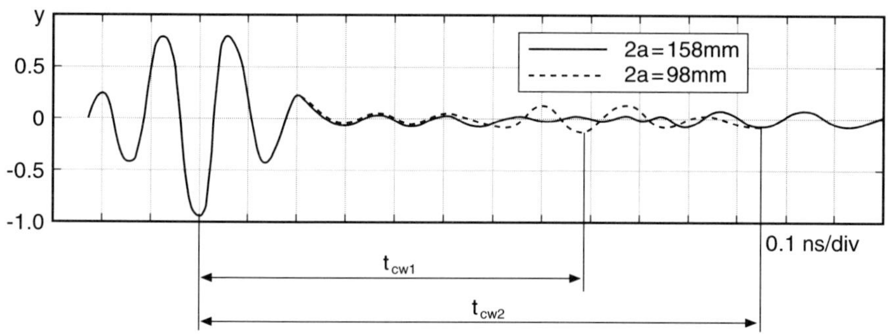

Fig. 2.17 UWB radio signal scattered from a conducting sphere

Here, the coefficients M_{pk} are dimensionless values. The sphere response (Fig. 2.14) has three kinds of singularity – the δ function, the discontinuity of $h(z)$ at $z = 0$ and the discontinuity of the derivative dh/dz at $z = 2.6$. The scattered signal components will be designated as y_0, y_1 and y_2, respectively.

The multiple integrals in eqn. 2.34 define the waveform of the scattered signal, $y_p(z)$. The peak values of the pulses integrated in dimensionless time, z, are approximately equal to the peak value of the incident signal. The scattered signal components, $y_0(z)$ and $y_1(z)$, coexist. The waveform of $y_1(z)$ is obtained by a single integration of the incident signal, and the amplitude ratio is expressed as

$$\frac{y_{1m}}{y_{0m}} = \frac{M_{10}}{M_{00}} \frac{c}{a\omega_0} = \frac{M_{10}}{M_{00}} \frac{\lambda_0}{2\pi a}$$

For the impulse response of a sphere $M_{00} = M_{10} = 1/2$, and the ratio $y_1(z)/y_0(z)$ is small in the high frequency region and is 0.08 at $\lambda_0 = 2\pi c/\omega_0 = a/2$. The influence of $y_1(z)$ on the total signal scattered by a point, $z = 0$, is seen as a small shift of zeros in the total function, $y_0(z) + y_1(z)$.

The discontinuity of dh/dz at $z = 2.6$ determines the component $y_2(z)$ known as a creeping wave. With eqn. 2.34, the waveform of $y_2(z)$ is found by double integration of the incident signal, and its amplitude is

$$y_{2m} = M_{21}\left(\frac{c}{a\omega_0}\right)^2 = M_{21}\left(\frac{\lambda_0}{2\pi a}\right)^2$$

The discontinuity of the derivative dh/dz at the point $z = 2.6$ can be estimated from Fig. 2.14 as $M_{21} = 10$. Then we have

$$\frac{y_{2m}}{y_{0m}} = \frac{1}{2}\left(\frac{\lambda_0}{a}\right)^2 \tag{2.35}$$

One can see that the formation of a creeping wave is associated with the low frequency components in the UWB signal spectral density.

Fig. 2.17 presents experimental time diagrams for pulses scattered by conducting spheres of 158 and 98 mm diameter. One can clearly identify the creeping wave pulses. In both cases, $f_0 = \omega_0/2\pi = 5\,\mathrm{GHz}$ and the creeping pulse amplitudes calculated using eqn. 2.35 fit fairly well the experimental values. Eqn. 2.34 also gives a correct estimate of the waveform fine structure: the creeping wave pulses obtained by double integration of the incident signal have a larger centre fill-in period and a smaller number of time sidelobes. Therefore, the results we have presented confirm the validity of the methods used for the description of scattered UWB waveforms.

Chapter 3

Ultrawideband radar target responses

3.1 The radar target as a linear dynamic system

3.1.1 Target description with system characteristics

Ultra-wideband (UWB) signal analysis and synthesis requires a clear understanding of the changes which radar signals undergo during the processes of transmission, reflection and reception. In this respect, the relationship between UWB radar science and transient electrodynamics is more intimate than that between classical radar and electrodynamics. An important trend today is the development of mathematical target models capable of providing an adequate analysis of UWB signals reflected from a target of complex geometry. The progress in this field has been primarily due to the use of concepts and methods of the general dynamic systems theory and to the physical interpretation of the models in terms of solutions to inverse transient electrodynamic problems. This line of research was initiated in the 1960s [74, 77]. The first studies dealt with frequency responses (FR), impulse responses (IR), and the related step (SR) and ramp (RR) responses. Later, other system characteristics were introduced into the linear dynamic target model in the form of linear differential and difference equations, FR, transfer functions (TF) and series expansion of their natural components with respect to singular points (poles) – complex-valued scattering resonance frequencies – and others. The efficiency of the systems approach has been proved in a number of investigations [24, 58, 85]. It has been shown that the model of a linear scatterer excited by a plane wave can be represented by two linear dynamic elements connected in series. The first element represents the radar response of the scatterer proper in the direction φ, when illuminated from the direction Θ; it is described by a family of impulse-scattering responses $h(t, \Theta, \varphi)$ which are independent of the distance to the observation point (Fig. 3.1). For simplicity, our treatment will be confined to a point radiator and receiver, a bistatic measuring system, and plane geometry. The second dynamic element characterises

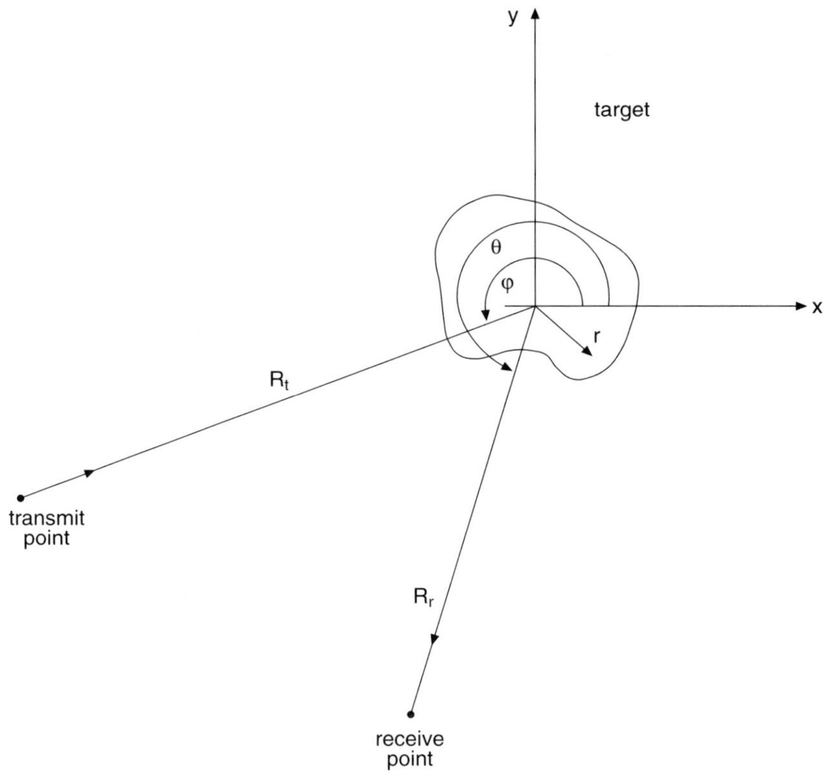

Fig. 3.1 Geometry of RTR measurements

the attenuation and time delay of the scattered signal in free space, introducing no error in the time-domain UWB waveform. A radar target can be described by other characteristics commonly used in treatments of linear systems, such as IR, FR, TF, etc.

As a linear dynamic system, the radar exhibits some special properties that should be taken into account in theoretical and experimental investigations. First, a target response depends on the aspect angle relative to the points of signal transmission and reception, Secondly, the dynamic system chosen as a target model is multi-dimensional, as are the incident and scattered electromagnetic pulses. Target surveillance may be performed by multistatic registration systems from different points in space. Finally, target responses contain two qualitatively different components – the early-time (forced) and the late-time (free, or natural) components. A forced response is formed while the incident wave front moves along the target ($0 \leqslant t \leqslant 2l_t/c$, where l_t is the target length). It has a finite duration and is associated with signals from local scattering centres characteristic of the target geometry (see below). A natural response with dominant damped sinusoids reveals itself after the wave front has passed the target ($t > 2l_t/c$).

The energies of the two components vary with the target geometry and the frequency range of the incident signal [85].

With the singularity expansion method (SEM) suggested by C. Baum [7], one can derive the expression for the scattered field and transfer function, assuming for simplicity that the transmission and reception points coincide

$$\mathbf{E}^{(s)}(s,\Theta,\varphi) = \mathbf{E}^{(s)'}(s,\Theta,\varphi) + \mathbf{E}^{(s)''}(s,\Theta,\varphi) \tag{3.1}$$

$$\mathbf{H}(s,\Theta,\varphi) = \mathbf{H}'(s,\Theta,\varphi) + \mathbf{H}''(s,\Theta,\varphi) \tag{3.2}$$

The first terms in eqns. 3.1 and 3.2 are the integral functions of the complex-valued frequency, s, having no singular point in the entire complex-valued plane, except, perhaps, for the point $s \to \infty$. The second terms represent rational fractional functions (theoretically, of an infinite order) which can be decomposed into simple fractions

$$\mathbf{E}^{(s)''}(s,\Theta,\varphi,R_n) = \xi \sum_{m=1}^{\infty} B_m(\Theta,\varphi,R_r)(s - s_m)^{-1} \tag{3.3}$$

$$\mathbf{H}''(s,\Theta,\varphi) = \xi \sum_{m=1}^{\infty} A_m(\Theta,\varphi)(s - s_m)^{-1} \tag{3.4}$$

where $B_m(\Theta,\varphi,R_r)$ and $A_m(\Theta,\varphi)$ are complex-valued coefficients (residues of the expansion in singular points) and ξ is the signal polarisation vector. Note that since the transmission bandwidth of any radar system is finite, we actually deal with a finite number of terms in the expansions in eqns. 3.3 and 3.4 (see below).

The inverse Laplace transforms of eqns. 3.1 and 3.2 give time responses: the scattered signal $\mathbf{e}^{(s)}(t,\Theta,\varphi,R_r)$ and the target impulse response $\mathbf{h}(t,\Theta,\varphi)$, which also contain a forced and a natural component

$$\mathbf{e}^{(s)}(t,\Theta,\varphi,R_r) = \mathbf{e}^{(s)'}(t,\Theta,\varphi,R_r) + \mathbf{e}^{(s)''}(t,\Theta,\varphi,R_r) \tag{3.5}$$

$$\mathbf{h}(t,\Theta,\varphi) = \mathbf{h}'(t,\Theta,\varphi) + \mathbf{h}''(t,\Theta,\varphi) \tag{3.6}$$

The first terms here have a finite duration, being the inverse Laplace transforms of integral functions. They represent early-time components containing local information on the target scattering behaviour and vary with the target aspect angle and polarisations of the incident and received signals.

The second terms are theoretically infinite. The inverse Laplace transforms of eqns. 3.3 and 3.4 represent them as sums of complex-valued exponents, namely

$$\mathbf{e}^{(s)''}(t,\Theta,\varphi,R_n) = \xi \sum_{m=1}^{\infty} B_m(\Theta,\varphi,R_r)\exp(s_m t) \tag{3.7}$$

$$\mathbf{h}''(t,\Theta,\varphi) = \xi \sum_{m=1}^{\infty} A_m(\Theta,\varphi)\exp(s_m t) \tag{3.8}$$

Since the residues, B_m and A_m, and the complex-valued resonance frequencies, s_m, are either real or exist as complex conjugate pairs, these components are sums of real, damped sinusoids. The amplitudes and phases of the sinusoids depend on the target aspect angle and signal polarisation, while the damped decrements and frequencies γ_m and ω_m, do not, being the real and imaginary parts of the poles, s_m. The natural component and its parameters contain global information on the target.

3.1.2 Forms of target description: total and local responses

We have shown that a target model permitting the analysis of target-related changes in incident UWB signals represents a linear dynamic system. The input and output signals are generally vector signals, which is due to the vector structure of the electromagnetic field and, possibly, to the multistatic nature of the measuring system. There are various ways of describing this equivalent dynamic system, which may be grouped into two major classes – parametric and nonparametric descriptions. A parametric description of the scattering model is based on a finite set of parameters, the number of which in the discrete case is much less than the number of samples of time signals and their spectral density functions. Otherwise, the description of target scattering behaviour is termed as nonparametric.

A parametric description involves the use of differential equations with respect to state variables or to input–output variables, models with parametrised IR, FR and transfer functions. The parametrisation is performed by expanding the functions in a set of orthogonal functions or by their rational fractional representation, etc. Nonparametric target descriptions may involve a convolution integral with nonparametrised impulse (step) responses, models with nonparametrised FR or transfer functions, etc.

The parameters of linear dynamic scattering models are generalisations of classical radar characteristics, such as radar cross section (RCS), polarisation scattering matrix and others. These may be united under a common term – UWB radar responses. Such responses are similar to systems or hardware characteristics widely used for the description of transient processes in lumped circuits. They describe radar target behaviour in a wide frequency range and contain information on the target geometry to be used for target discrimination, imaging and orientation detection [10, 58]. The parameters and responses in target scattering models can be calculated or estimated experimentally using radar sensors with various UWB signal waveforms. The one-to-one correspondence between the responses allows one to choose an appropriate approach to the problem in question and to compare the results obtained by different methods.

Consider some types of UWB radar response in more detail. A radar target can be represented as a dynamic system with two inputs and outputs and an internal coupling defined by scalar dynamic elements which transform the polarisation components of the transmitted signal (Fig. 3.2).

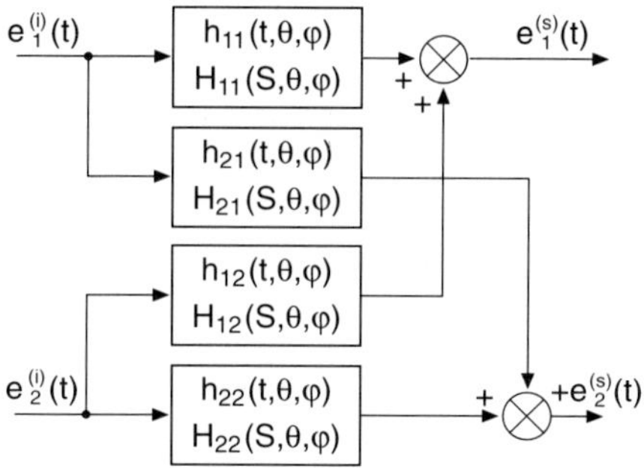

Fig. 3.2 Radar target model

UWB radar responses can be subdivided into total and local responses, sometimes termed as integral and differential responses, respectively. A total response represents the response of an object as a whole to test UWB signals. Such responses characterise target detectability and the capability of the radar to recognise and discriminate a target. Local responses describe the scattering properties of individual target elements and surface fragments. Their measurement allows identification of the main scattering sources, in particular at the stage of designing and testing various means of reducing target detectability.

3.2 Nonparametric target scattering models

3.2.1 Nonparametric description of targets

For UWB signals, the description of scalar dynamic elements of a target model is made in terms of impulse responses, $h(t, \Theta, \varphi)$, frequency responses, $H(\omega, \Theta, \varphi)$, and transfer functions, $H(s, \Theta, \varphi)$. A radar target is, respectively, described in terms of the polarisation matrices (PM) of impulse responses

$$\mathbf{h}(t, \Theta, \varphi) = [h_{ij}(t, \Theta, \varphi)], \qquad i, j = 1, 2 \tag{3.9}$$

transfer functions

$$\mathbf{H}(s, \Theta, \varphi) = [H_{ij}(s, \Theta, \varphi)] \tag{3.10}$$

or, at $s = \mathrm{i}\omega$, of frequency responses

$$\mathbf{H}(\mathrm{i}\omega, \Theta, \varphi) = [H_{ij}(\mathrm{i}\omega, \Theta, \varphi)] \tag{3.11}$$

where

$$h_{ij}(t, \Theta, \varphi) = 2\sqrt{\pi}R_r e_{ij}^{(s)}(t + (R_r/c), \Theta, \varphi, R_r)/E_{0j}^{(i)}(\varphi) \tag{3.12}$$

$$H_{ij}(\mathrm{i}\omega, \Theta, \varphi) = 2\sqrt{\pi}R_r E_{ij}^{(s)}(\mathrm{i}\omega, \Theta, \varphi, R_r) \exp(\mathrm{i}\omega R_r/c)/E_j^{(i)}(\mathrm{i}\omega, \varphi) \tag{3.13}$$

Here $e_{ij}^{(s)}(t, \Theta, \varphi, R_r)$ is the ith polarisation signal at the reception point, scattered when the jth polarisation signal in the form of the delta function $e_j^{(i)}(t, \varphi) = E_{0j}^{(i)}(\varphi)\delta(t)$ is incident on the target; $E_{ij}^{(s)}(\mathrm{i}\omega, \Theta, \varphi)$ and $E_j^{(i)}(\mathrm{i}\omega, \varphi)$ are the Fourier or Laplace transforms of the respective time signals; $i, j = 1, 2$. Such normalisation of the UWB radar response is widely used because it simplifies the change from FR polarisation matrices to RCS polarisation matrices [1]

$$\sigma(\omega, \Theta, \varphi) = \mathbf{H}(\mathrm{i}\omega, \Theta, \varphi)\mathbf{H}^{T^*}(\mathrm{i}\omega, \Theta, \varphi) \tag{3.14}$$

The IR and FR matrices (TR matrices) are interrelated through the Fourier (or Laplace) transforms. Other popular tools are the step response (SR) matrix $\mathbf{g}(t, \Theta, \varphi) \equiv [g_{ij}(t, \Theta, \varphi)]$ and the ramp response (RR) matrix $\mathbf{r}(t, \Theta, \varphi) \equiv [r_{ij}(t, \Theta, \varphi)]$ which are related to the IR matrices as

$$\begin{aligned} \mathbf{g}(t, \Theta, \varphi) &= \int_0^t \mathbf{h}(t', \Theta, \varphi)\, dt' \\ \mathbf{r}(t, \Theta, \varphi) &= \int_0^t \mathbf{g}(t', \Theta, \varphi)\, dt' \end{aligned} \tag{3.15}$$

They are also related to the FR polarisation matrices as

$$\frac{\mathbf{H}(\mathrm{i}\omega, \Theta, \varphi)}{\mathrm{i}\omega} = \int_0^\infty \mathbf{g}(t, \Theta, \varphi) \exp(-\mathrm{i}\omega t)\, dt \tag{3.16}$$

$$\frac{\mathbf{H}(\mathrm{i}\omega, \Theta, \varphi)}{(\mathrm{i}\omega)^2} = \int_0^\infty \mathbf{r}(t, \Theta, \varphi) \exp(-\mathrm{i}\omega t)\, dt \tag{3.17}$$

The scattered signals reduced to the point of target location can be calculated using UWB radar responses,

$$\mathbf{e}_0^{(s)}(t, \Theta, \varphi) = \int_0^\infty \mathbf{h}(t - \mathrm{i}, \Theta, \varphi)\mathbf{e}^{(i)}(\tau)\, d\tau \tag{3.18}$$

$$\mathbf{E}_0^{(s)}(\mathrm{i}\omega, \Theta, \varphi) = \mathbf{H}(\mathrm{i}\omega, \Theta, \varphi)\mathbf{E}^{(i)}(\mathrm{i}\omega) \tag{3.19}$$

where $\mathbf{e}_0^{(s)}(t)$, $\mathbf{e}^{(i)}(t)$ are the incident and scattered signals at the point of target location.

The scattered signal at the reception point has the form

$$\begin{aligned} e^{(s)}(t, R_r) &= e^{(s)}(t) \\ &= e_0^{(s)}(t - R_r/c)/(2\sqrt{\pi}R_r) \end{aligned} \tag{3.20}$$

and its spectral density function is

$$\begin{aligned} E^{(s)}(\mathrm{i}\omega, R_r) &\equiv E^{(s)}(\mathrm{i}\omega) \\ &= E_0^{(s)}(\mathrm{i}\omega, \Theta, \varphi) \exp(\mathrm{i}\omega R_r/c)/(2\sqrt{\pi}R_r) \end{aligned} \tag{3.21}$$

These forms of target description are nonparametric and in practice do not differ from those used in the linear lumped systems theory. The nonparametric UWB response is found under certain generalised conditions (see below), and this is a merit which allows experimental estimation of the response. Target modelling can be performed by using the classical methods of systems theory. This approach is free from methodological errors caused by the improper choice of the parametric model structure. On the other hand, ignoring the *a priori* information on the nature of electromagnetic signal scattering, one decreases the standards of target model testing and makes it difficult to interpret and apply the model itself.

3.2.2 The relationship between UWB radar response and target geometry

The physical interpretation of a target model tested experimentally is an essential factor for model applicability, and here we shall present some results (for details, see [77]).

Suppose that a plane monochromatic wave with the wave number $k = \omega/c$ is incident on a target along the z-axis (Fig. 3.3). For the backscatter ($\Theta = \varphi$), one can obtain the following expressions for the UWB response in the physical optics approximation [58, 115]

$$H'(\mathrm{i}\omega, \Theta) = \frac{\mathrm{i}\omega}{\sqrt{\pi}} \int_0^{l_t} \frac{dA(z)}{dz} \exp(-\mathrm{i}2z\omega/c)\, dz \tag{3.22}$$

$$h'(t, \Theta) = -\frac{c}{4\sqrt{\pi}} \frac{d^2A(z)}{dz^2} \qquad z = ct/2 \tag{3.23}$$

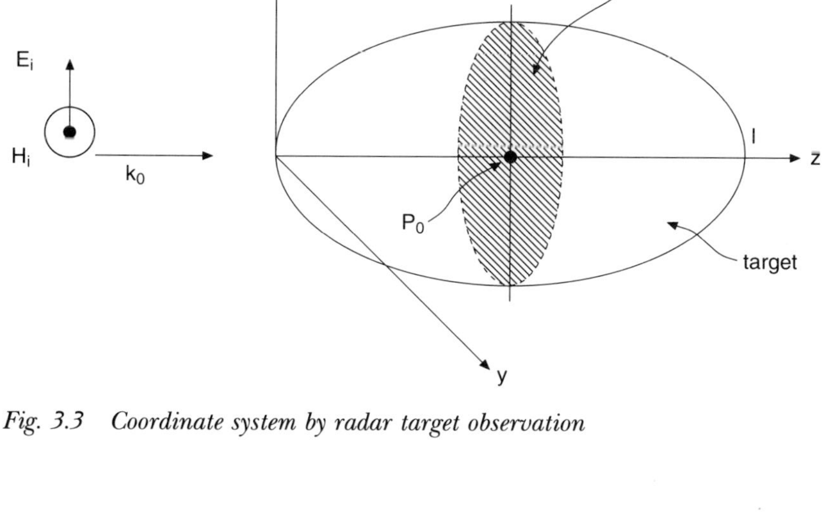

Fig. 3.3 Coordinate system by radar target observation

where $A(z)$ is a function of target area projection onto the plane normal to the z-axis and intercepting it at the point z, l_t is the total extent of the target in the z direction. It can be seen in Fig. 3.3 that the function $A(z)$ increases monotonically with z

$$A(z) = \begin{cases} 0, & z < 0 \\ A(l_0), & z \geqslant l_0 \end{cases} \tag{3.24}$$

These expressions have been obtained by a standard physical optics method. They offer a physical interpretation of the UWB radar response, applicable only for analysis of high frequency scattering, when the incident signal bandwidth lies above the resonance scattering region. In the low frequency and resonance regions, one makes use of the generalised physical optics method which allows for the fact that induced currents penetrate the geometrical shadow region [77]. With the help of this method, Adachi, Kennaugh, Moffatt, Young and others have shown [77] that eqn. 3.23 can be modified into

$$h'(t, \Theta) = -\frac{c}{4\pi}\frac{d^2S(z)}{dz}\bigg|_{z=ct/2'} \tag{3.25}$$

where $S(z)$ is a profile function describing the dependence of the target cross section on the z-coordinate (Fig. 3.3). For convex targets, the functions $A(z)$ and $S(z)$ coincide in the interval $0 \leqslant z \leqslant l_0$ but differ in the geometrical shadow.

Using eqn. 3.15, we can find the relationship between the target geometry and other time-dependent UWB responses, namely

$$g'(t, \Theta) = -\frac{1}{2\sqrt{\pi}}\frac{dS(z)}{dz}\bigg|_{z=ct/2'} \tag{3.26}$$

$$r'(t, \Theta) = -\frac{1}{\sqrt{\pi}c}S(z)\bigg|_{z=ct/2} \tag{3.27}$$

Eqns. 3.25 to 3.27 are variants of the Kennaugh–Moffatt identity which gives an estimate of the target profile function from UWB response measurements. For example, the profile function can be estimated from the forced component of an element of the ramp response PM, i.e.

$$S(z) = \sqrt{\pi}cr'\left(\frac{2z}{c}\right)\left[1\left(\frac{2z}{c}\right) - 1\left(\frac{2z - l_t}{c}\right)\right] \tag{3.28}$$

where $1(x)$ is a unit function (step function).

The experimental results cited, for instance in [115], show good agreement between the estimates of eqn. 3.28 and real profile functions for both the illuminated target surface and the shadow region. The function $S(z)$ is a highly informative geometrical characteristic of a target. It is related to the target FR through

$$H(\mathrm{i}\omega, \Theta)/(\mathrm{i}\omega)^2 \sim \int_0^{l_t} S(z) \exp(-\mathrm{i}2z\omega/c)\, dz \tag{3.29}$$

We infer from this that the principal information on target geometry is obtained from the low frequency components of the scattered signal, while the high frequency components are defined by small geometric target details [58, 77]. It is desirable, therefore, that the transmitted signal bandwidth of a radar system, designed for target recognition and imaging from target geometry, should lie in the Rayleigh scattering and resonance regions.

The characteristic target function

$$\gamma(\mathbf{r}) = \begin{cases} 1, & \mathbf{r} \in \mathbf{V} \\ 0, & \mathbf{r} \notin \mathbf{V} \end{cases} \tag{3.30}$$

where $\mathbf{V}$ is the space region bounded by the target surface, gives a more detailed description of target geometry than the profile function. Obviously, the profile function can be obtained by double integration of the characteristic function, i.e.

$$S(z) = \iint_{-\infty}^{\infty} \gamma(x, y, z)\, dx\, dy \tag{3.31}$$

The target frequency response can be represented as a Taylor series (see Section 3.3.4)

$$H'(\mathrm{i}\omega, \Theta) = a_0 + (\mathrm{i}\omega)a_1 + (\mathrm{i}\omega)^2 a_2 + \ldots \tag{3.32}$$

where a_j $(j = 1, 2, \ldots)$ are coefficients.

In the low frequency region, the terms of the series with the power index $j \geqslant 3$ can be neglected. Besides, it has been shown (Rayleigh's axiom) that

$$a_0 = a_1 = 0 \tag{3.33}$$

Therefore, in the Rayleigh scattering region

$$\begin{aligned} H'(\mathrm{i}\omega, \Theta) &= H'(\mathrm{i}\omega) \\ &= (\mathrm{i}\omega)^2 a_2 \end{aligned} \tag{3.34}$$

It is clear that in this region the target is equivalent to a double differentiating element. The factor a_2 is known as the Rayleigh coefficient which is relatively weakly dependent on the signal polarisation, the target shape and aspect angle. It follows from eqn. 3.34 that

$$a_2 = H'(\mathrm{i}\omega)/(\mathrm{i}\omega)^2 \tag{3.35}$$

The Fourier transform of eqn. 3.35 yields the expression

$$\frac{1}{2\pi}\int_{-\infty}^{\infty} a_2 \exp(\mathrm{i}\omega t)\, d\omega = \frac{1}{2\pi}\int_{-\infty}^{\infty} \frac{H'(\mathrm{i}\omega)}{(\mathrm{i}\omega)^2}\, d\omega \tag{3.36}$$

in which the left-hand side represents the weighted delta function and the right-hand side the function $r'(t, \Theta)$, as follows from eqn. 3.17, i.e.

$$a_2\delta(t) = r'(t) \tag{3.37}$$

The integration in time of both sides of eqn. 3.37 (taking into account the filtering property of the delta function) gives

$$a_2 = \int_0^\infty r'(t)\, dt \tag{3.38}$$

As was pointed out above

$$r'(t) \sim -S(z)|_{z=ct/2}$$

and

$$\int_0^\infty S(z)\, dz = V$$

therefore

$$a_2 \sim -V \tag{3.39}$$

where V is the target volume.

Expansion of eqn. 3.32 permits the formulation of conditions to be met by the moments of the target impulse response. For this we shall take advantage of the results known from Fourier transformation theory [79]

$$F(i\omega) = \sum_{j=0}^{\infty} \frac{m_j(-1)^j}{j!}(i\omega)^j \tag{3.40}$$

where $m_j = \int^\infty t^j f(t)\, dt$.

Comparison of eqns. 3.40 and 3.32, taking into account eqn. 3.37, shows that

$$\begin{aligned}\int_0^\infty h'(t)\, dt &= 0\\ \int_0^\infty th'(t)\, dt &= 0\\ \int_0^\infty t^2 h'(t)\, dt &= 2a_2\end{aligned} \tag{3.41}$$

3.3 Parametric scattering models for radar targets

3.3.1 Types of parametric model

There are two approaches commonly used in target modelling – physical and phenomenological.

In the physical approach, the analysis of target-related changes in the incident signal is based on the solutions of Maxwell's equations, or equivalent integral equations, as applied to a particular radar target or to a class of targets. The parameters employed in the mathematical model have a definite physical sense, facilitating interpretation of the model.

The other approach implies no analysis of the real, physical mechanism of signal waveform variation but is aimed at developing model signals which would be, in a sense, close to real, target-scattered signals. Phenomenological models are based on the concepts of a target either as a black box, with a structure that is unknown, or as a grey box, with a structure that is known but with parameters which have not been defined. In this case, the model and its parameters are sometimes difficult to interpret [62].

Unfortunately, in many cases physical target models are hard to build up because of the absence of exact solutions to the electrodynamic problems for objects of complex geometry and because most targets have special surface coatings. For this reason, phenomenological models still play an important role in theoretical and experimental investigations.

One's choice of a particular parametric description depends on the aim of target modelling. If a mathematical model is to simulate UWB signals scattered by a target, a phenomenological model is quite suitable, because the main criterion is the model applicability and, in a sense, consistency with experimental data. This approach is widely used in automatic control theory and in systems identification.

If a model is aimed at investigating the process of signal scattering by a given target, the phenomenological approach, even with close agreement with experimental data, may give physically absurd results. A physical model would be more appropriate in this case, because the main requirement of a model of this type is its validity.

The principal forms of parametric description of targets, although not the only ones, are differential equations in input–output variables (and the closely related rational fractional transfer functions) and the recently introduced differential equations in state variables. An advantage of such a description is that one directly obtains the scattered signal waveforms in time representation from digital or analogue processors (target simulators) [85].

3.3.2 Models based on differential equations

The set of differential equations in input–output variables (DEIOV) describing the scattering of a vector signal has the form

$$y_{ij}^{(n)}(t) + \sum_{m=1}^{n-1} a_{ijm}(\Theta, \varphi) y_{ij}^{m}(t) = \sum_{m=1}^{n-1} b_{ijm}(\Theta, \varphi) u_j^{m}(t) \tag{3.42}$$

$$y_i(t) = \sum_{j=1}^{2} y_{ij}(t) \tag{3.43}$$

Here, $u_j(t) = e_j^{(j)}(t, \varphi)$ for the incident signal and $y_i(t) = e_{0i}^{(s)}(t)$ for the reduced reflected signal.

This set of equations must be supplemented by the initial conditions associated with residual target excitation by the preceding incident signal. If the signal incidence frequency is such that the target response to the preceding pulse, by the moment the next pulse arrives, can be neglected, then we shall have $y^{(n-1)}(t_0) = \ldots = y(t_0) = 0$.

In contrast to the models based on the DEIOV used in automatic control theory, some radar targets have $m = n$ and even $m > n$. This is typical, for example, of the description in the Rayleigh or high frequency regions, since the asymptotic FR of some targets represents a nondecreasing function [85]. Nevertheless, we shall consider, for simplicity, the case of $m < n$ permitting the scattering approximation for a signal with a finite bandwidth.

Eqn. 3.42 is closely related to target description (under the zero initial conditions) by a rational fractional TF of the form

$$H_{ij}(s, \Theta, \varphi) = \frac{b_{ijm}(\Theta, \varphi)s^m + \ldots + b_{ij0}(\Theta, \varphi)}{s^n + a_{ij(n-1)}(\Theta, \varphi)s^{n-1} + \ldots + a_{ij0}(\Theta, \varphi)} \tag{3.44}$$

The frequency response

$$H_{ij}(\mathrm{i}\omega, \Theta, \varphi) = H_{ij}(s, \Theta, \varphi)|_{s=\mathrm{i}\omega}$$

has a similar structure.

It is known that a rational fractional function can be decomposed into simple fractions [56] and represented as

$$H_{ij}(s, \Theta, \varphi) = \sum_{m=1}^{n} \frac{A_{ijm}(\Theta, \varphi)}{s - s_{ijm}(\Theta, \varphi)} \tag{3.45}$$

where $s_{ijm}(\Theta, \varphi)$ are the transfer function poles (characteristic equation roots)

$$s^n + \sum_{m=1}^{n-1} a_{ijm}(\Theta, \varphi)s^m = 0 \tag{3.46}$$

$A_{ijm}(\Theta, \varphi)$ are the transfer function residues that can be found by the indeterminate coefficients method or by other standard procedures [56].

Note that the above expression is valid for the case of simple poles. A similar expression can be written for the case when the polynomial denominator of eqn. 3.46 has multiple roots.

With the inverse Laplace transform for the TF in eqn. 3.45, one can represent the target IR as

$$h_{ij}(t, \Theta, \varphi) = \sum_{m=1}^{n} A_{ijm}(\Theta, \varphi) \exp(s_{ijm}(\Theta, \varphi)t) \tag{3.47}$$

and this representation is valid over the whole time interval $t \geqslant 0$.

The relationship between the target description by the rational fractional TR or FR and the input–output differential equations follows from a direct comparison of eqns. 3.42 and 3.44. It should be emphasised that transfer functions and frequency response can only be used for simulating nontransient radar targets, with parameters remaining constant over the transmitted pulse duration, whereas the input–output differential equations can, in principle, simulate a target with an aspect angle which varies with an arbitrary velocity.

For the moments of time $t \geqslant \tau_p$, where τ_p is the incident pulse duration, eqn. 3.42 becomes homogeneous

$$y_{ij}^{(n)}(t) + \sum_{m=1}^{n-1} a_{ijm}(\Theta, \varphi) y_{ij}^{(m)}(t) = 0 \tag{3.48}$$

The initial conditions $\{y_{ij}(0) = y_{ij0}, \ldots, y_{ij}^{(n-1)}(0) = y_{ij0}^{(n-1)}\}$ are determined by the type of incident signal and the parameters of eqn. 3.42. In the stationary or quasistationary case, when the time dependence of the parameters can be neglected, each polarisation component of the scattered signal can be represented as

$$y_{ij}(t) = \sum_{m=1}^{n} B_{ijm}(\Theta, \varphi) \exp(s_{ijm}(\Theta, \varphi) t) \tag{3.49}$$

where $B_{ijm}(\Theta, \varphi)$ are the expansion coefficients defined by the initial conditions and $s_{ijm}(\Theta, \varphi)$ are thee poles of the TF (eqn. 3.44) or the roots of the characteristic eqn. 3.46.

Radar targets can also be described in terms of differential equations in state variables. This approach has recently been used for the investigation of dynamic systems in automatic control theory [1, 24]. The equations describe the target model structure, which may be simulated using an analogue computer. With a digital computer, the method of state variables yields effective algorithms for solving the differential equations. Target representation in state variables simplifies the synthesis of algorithms for optimal signal processing in designing multipurpose UWB radar systems.

Let us show that the differential equations in state variables are able to describe the dynamic elements modelling the target properties under parallel and cross polarisations of the transmitted and received signals (Fig. 3.2). Any of the differential equations (eqn. 3.42) can be reduced to a set of n first-order equations for additional variables (state variables or state coordinates) and the so-called output equation relating them to the output signal. By uniting the additional variables $\{x_{ij1}(t), \ldots, x_{ijn}(t)\}$ into the state vector $\mathbf{x}_{ij}(t) \equiv [x_{ij1}(t), \ldots, x_{ijn}(t)]^T$ and using the vector matrix form for a linear problem, we can write the differential equations in state variables and the

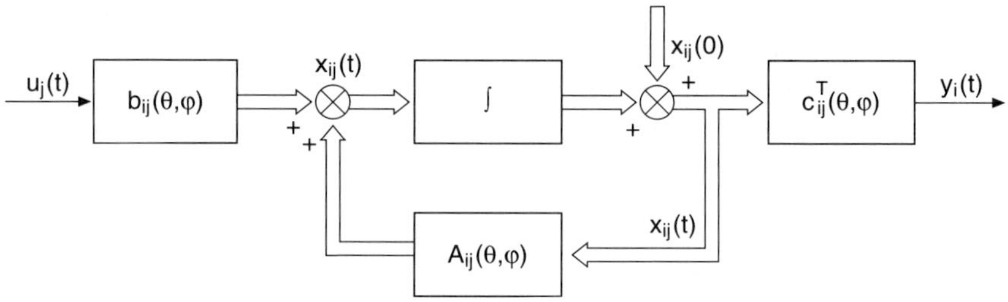

Fig. 3.4 Block diagram for target parametric model in state-space

output equation as

$$\dot{\mathbf{x}}_{ij}(t) = \mathbf{A}_{ij}(\Theta, \varphi)\mathbf{x}_{ij}(t) + \mathbf{b}_{ij}(\Theta, \varphi)u_j(t) \tag{3.50}$$

$$y_{ij}(t) = \mathbf{c}_{ij}^T(\Theta, \varphi)\mathbf{x}_{ij}(t) \tag{3.51}$$

where $\mathbf{A}_{ij}(\Theta, \varphi) \equiv [\alpha_{ijkl}(\Theta, \varphi)]$ is a dynamic $n \times n$ matrix, $\mathbf{b}_{ij}(\Theta, \varphi) \equiv [\beta_{ij1}(\Theta, \varphi), \ldots, \beta_{ijn}(\Theta, \varphi)]^T$ is the input column matrix and $\mathbf{c}_{ij}^T(\Theta, \varphi) \equiv [c_{ij1}(\Theta, \varphi), \ldots, c_{ijn}(\Theta, \varphi)]$ is the output row matrix.

The elements of the matrices $\mathbf{A}_{ij}(\Theta, \varphi)$, $\mathbf{b}_{ij}(\Theta, \varphi)$ and $\mathbf{c}_{ij}^T(\Theta, \varphi)$ are also defined by the coefficient of eqn. 3.42. The initial conditions vector $\mathbf{x}_{ij}(0) \equiv \mathbf{x}_{ij0}$ is related to the initial conditions of eqn. 3.42. After the preceding signal scattering has damped down, we have $\mathbf{x}_{ij0} \equiv 0$. Thus, the elements of the matrices $\mathbf{A}_{ij}(\Theta, \varphi)$, $\mathbf{b}_{ij}(\Theta, \varphi)$ and $\mathbf{c}_{ij}^T(\Theta, \varphi)$ $(i, j = 1, 2)$ play the role of the target UWB response.

Eqns. 3.50 and 3.51 can be schematically represented in a convenient form (Fig. 3.4). It is known that the representation of a dynamic system (target, in our case) through eqns. 3.50 and 3.51 is not unique, because any nonsingular linear transformation $\mathbf{x}'(t) = \mathbf{T}\mathbf{x}(t)$ $(\mathbf{T} \equiv [t_{kl}])$ of state variables produces a new set of differential equations in state variables having equal status with the initial set of equations. These issues are discussed in more detail elsewhere [1, 37].

The description of a radar target in terms of the state variables (eqns. 3.50 and 3.51) allows one to obtain the UWB radar response, impulse and frequency responses, and the coefficients of the input–output differential equations. In order to pass from the state differential equations to a parametrised transfer function in the form of eqns. 3.44 and 3.42, we shall perform Laplace transformations for eqns. 3.50 and 3.51, i.e.

$$p\mathbf{X}_{ij}(p) = \mathbf{A}_{ij}(\Theta, \varphi)\mathbf{X}_{ij}(p) + \mathbf{b}_{ij}(\Theta, \varphi)U_j(p) \tag{3.52}$$

$$Y_{ij}(p) = \mathbf{c}_{ij}^T(\Theta, \varphi)\mathbf{X}_{ij}(p) \tag{3.53}$$

Then it is easy to find

$$\begin{aligned} H_{ij}(\Theta, \varphi, p) &\triangleq \frac{Y_{ij}(p)}{U_j(p)} \\ &= \mathbf{c}_{ij}^T(\Theta, \varphi)(p\mathbf{I} - \mathbf{A}_{ij}(\Theta, \varphi))^{-1}\mathbf{b}_{ij}(\Theta, \varphi) \end{aligned} \tag{3.54}$$

By reducing eqn. 3.54 to the rational fractional form (eqn. 3.44), we obtain the input–output differential equation (eqn. 3.42).

The ith polarisation signal scattered by a target which has been illuminated by the jth polarisation signal can be found using the transition matrix of the target state. Under the steady-state conditions, the scattered signal can be represented as

$$\begin{aligned} y_{ij}(t) &= \mathbf{c}_{ij}^T(\Theta,\varphi)\Phi(t-t_0,\Theta,\varphi)\mathbf{x}_{ij}(0) \\ &\quad + \mathbf{c}_{ij}^T(\Theta,\varphi)\int_0^t \Phi_{ij}(t-\tau,\Theta,\varphi)\mathbf{b}_{ij}(\Theta,\varphi)u_j(\tau)\,d\tau \end{aligned} \tag{3.55}$$

where $\Phi(t,\Theta,\varphi)$ is the state transition matrix for the given polarisation component.

For the quasisteady-state case [85], we have

$$\begin{aligned} \Phi_{ij}(t,\Theta,\varphi) &= \exp(\mathbf{A}_{ij}(\Theta,\varphi)) \\ &= \sum_{k=0}^{\infty}[(k!)^{-1}(\mathbf{A}_{ij}(\Theta,\varphi)t)^k] \end{aligned} \tag{3.56}$$

Other ways of describing the transition matrix have been discussed elsewhere [85].

The signal reflected by an excited target only contains the target response and the zero initial conditions, i.e.

$$y_{ij}(t) = \mathbf{c}_{ij}^T(\Theta,\varphi)\int_0^t \Phi_{ij}(t-\tau,\Theta,\varphi)\mathbf{b}_{ij}(\Theta,\varphi)U_j(\tau)\,d\tau \tag{3.57}$$

After the incident signal has passed along the whole target length, it contains the free component only

$$y_{ij}(t) = \mathbf{c}_{ij}^T(\Theta,\varphi)\Phi_{ij}(t-t',\Theta,\varphi)\mathbf{x}_{ij}(t') \tag{3.58}$$

where $x_{ij}(t')$ is the state vector at the moment of time $t' \geqslant 2l_{max}/c$.

The state transition matrix allows us to define the respective polarisation component of the scattering impulse response

$$\mathbf{h}_j(t,\Theta,\varphi) = \begin{cases} \mathbf{c}_{ij}^T(\Theta,\varphi)\Phi_{ij}(t,\Theta,\varphi)\mathbf{b}_{ij}(\Theta,\varphi) & t \geqslant 0 \\ 0 & t < 0 \end{cases} \tag{3.59}$$

The method of state variables admits a similar multi-dimensional model for both components of the incident and reflected signals, including all polarisation state vectors.

With the state transition matrix, we can find the scattered vector signals from the known incident signals

$$\dot{\mathbf{x}}(t) = \mathbf{A}(\Theta,\varphi)\mathbf{x}(t) + \mathbf{B}(\Theta,\varphi)\mathbf{u}(t) \tag{3.60}$$

$$\mathbf{y}(t) = \mathbf{C}(\Theta,\varphi)\mathbf{x}(t) \tag{3.61}$$

where

$$\mathbf{x}(t) = [\mathbf{x}_{11}^T(t) \mid \mathbf{x}_{12}^T(t) \mid \mathbf{x}_{21}^T(t) \mid \mathbf{x}_{22}^T(t)]^T$$

$$\mathbf{u}(t) = [u_1(t) \quad u_2(t)]^T$$

$$\mathbf{y}(t) = [y_1(t) \quad y_2(t)]^T$$

$$\mathbf{A}(\Theta, \varphi) = \operatorname{diag}[\mathbf{A}_{11}(\Theta, \varphi) \quad \mathbf{A}_{12}(\Theta, \varphi) \quad \mathbf{A}_{21}(\Theta, \varphi) \quad \mathbf{A}_{22}(\Theta, \varphi)]$$

$$\mathbf{B}(\Theta, \varphi) = \left[\begin{array}{c|c} \mathbf{b}_{11}(\Theta, \varphi) & 0 \\ 0 & \mathbf{b}_{12}(\Theta, \varphi) \\ \mathbf{b}_{21}(\Theta, \varphi) & 0 \\ 0 & \mathbf{b}_{22}(\Theta, \varphi) \end{array}\right]$$

$$\mathbf{C}(\Theta, \varphi) = \begin{bmatrix} \mathbf{c}_{11}^T(\Theta, \varphi) & \mathbf{c}_{12}^T(\Theta, \varphi) & 0 & 0 \\ 0 & 0 & \mathbf{c}_{21}^T(\Theta, \varphi) & \mathbf{c}_{22}^T(\Theta, \varphi) \end{bmatrix}$$

Just as in the one-dimensional model, the representations of eqns. 3.60 and 3.61 can be obtained with matrices of different structure, which would define the UWB target response in the state space. By applying formulas similar to eqns. 3.54 to 3.59 to the multi-dimensional model of the state space, we can bypass IR TF polarisation matrices and find the scattered vector signals from the known transmitted signals.

3.3.3 Description of the natural component

We have shown that the apparatus of both differential equations and TF are natural tools for the description of free components of a scattered signal, since the models and their parameters have a clear physical interpretation. Strictly speaking, the equations and transfer functions are of infinite order. In practice, however, the frequency bandwidth, to which the model is applicable, is finite, and so is the order.

In the parametric model of natural target scattering, the parameters in the left-hand side of eqn. 3.42, the polynomial denominator in the transfer function (eqn. 3.44), the dynamic matrix in eqn. 3.50 and the state transition matrix in eqn. 3.55 are independent of the incident signal polarisation and the target aspect. Therefore, the vector model of eqns. 3.60 and 3.61 can be written in a simpler form: $\mathbf{x}(t) = [x_1(t), \ldots, x_n(t)]^T$ is an n-dimensional vector, $\mathbf{A}(\Theta, \varphi) \equiv \mathbf{A}$ is an $n \times n$ matrix, with elements which are independent of the target aspect and signal polarisation, $\mathbf{B}(\Theta, \varphi) = [b_1(\Theta, \varphi) \quad b_2(\Theta, \varphi)]$ is an $n \times 2$ matrix and

$$\mathbf{C}(\Theta, \varphi) = \begin{bmatrix} \mathbf{c}_1^T(\Theta, \varphi) \\ \mathbf{c}_2^T(\Theta, \varphi) \end{bmatrix}$$

is a $2 \times n$ matrix.

The elements $\mathbf{b}_1(\Theta, \varphi)$ and $\mathbf{c}_i^T(\Theta, \varphi)$ of the matrices $\mathbf{B}(\Theta, \varphi)$ and $\mathbf{C}(\Theta, \varphi)$ are determined as above.

If we represent the natural component of the impulse response as a sum of damped exponents (eqn. 3.47), we shall find the late-time target response to an arbitrary incident signal, namely

$$\begin{aligned} e_{0ij}^{(s)''}(t) &= h_{ij}''(t, \Theta, \varphi) \times e_j^{(i)}(t) \\ &= \sum_{m=1}^{n} A_{ijm}(\Theta, \varphi) \int_0^\infty e_j^{(i)}(\tau) \exp(s_m(t - \tau))\, d\tau \\ &= \sum_{m=1}^{n} B_{ijm}(\Theta, \varphi) \exp(s_m t) \end{aligned} \tag{3.62}$$

where $B_{ijm}(\Theta, \varphi) = A_{ijm}(\Theta, \varphi) \int_0^\infty e_j^{(i)}(\tau) \exp(-s_m \tau)\, d\tau$ are complex-valued expansion coefficients depending on the target aspect as well as on the incident waveform and polarisation.

Eqn. 3.62 means that the target response to an arbitrary signal contains, as the impulse response, a late-time component ($t \geqslant 2l_t/c$) which is a sum of complex-valued exponents with the same resonance frequencies. This is consistent with the results of Section 3.1.2.

It should be noted that the models discussed above can define the scattered signal for the late-time interval only, $t \geqslant 2l_t/c$. The state variables in the canonical diagonal equations of state represent the scattered signal modes (natural oscillations).

For the majority of radar targets that can be used as target standards in experimental investigations, the resonance frequencies have already been calculated and tabulated (see, for example, [33]). In particular, the late-time transfer function for a perfectly conducting sphere is found to be [85, 107]

$$H(s) = \sum_{m=1}^{\infty} \frac{(2m+1)(s')^{2n}}{f_m(s')g_m(s')} \tag{3.63}$$

where

$$f_m(s') = \sum_{l=1}^{m} \frac{(m+l)!}{l!\,(m-l)!} \frac{1}{2^l} (s')^{m-1}$$

$$g_m(s') = \sum_{l=0}^{m} \frac{(m+l)!}{l!\,(m-l)!} \frac{1}{2^l} (l + s')(s')^{m-1}$$

$$s' = (a/c)s$$

The poles of the transfer function (eqn. 3.63) are defined by finding the zeros of the functions $f_m(s')$ and $g_m(s')$ in the operating frequency range.

3.3.4 Description of the forced component

3.3.4.1 Expansion in power series

It was shown in Section 3.1 that the forced component of the transfer function $H_{ij}(s, \Theta, \varphi)$ is an integral function of the variable s having no singular points (or poles) in a finite complex plane. Therefore, the rational fractional description (approximation) in the plane s and the application of the appropriate differential equations are possible and expedient only within the phenomenological approach, since the poles that arise have no physical sense and do not exist in reality.

Theoretically, a parametric description of transient scattering is possible if the forced component of the transfer function is expanded in a Laurent series

$$H_{ij}(s, \Theta, \varphi) = \sum_{k=-\infty}^{\infty} c_{ijk}(\Theta, \varphi) s^k \tag{3.64}$$

$$c_{ijk}(\Theta, \varphi) = \frac{1}{2\pi \mathrm{i}} \int_S \frac{H_{ij}(s, \Theta, \varphi)}{s^{k+1}} ds \tag{3.65}$$

where S is an arbitrary circle with a centre at the point $s = 0$. This expansion corresponds to the integral differential equation for the scattered signal. The scattering is often described by using partial sums of the series (eqn. 3.64) with respect to positive and negative powers, i.e.

$$H_{ij}(s, \Theta, \varphi) = \sum_{k=0}^{\infty} c_{ijk}(\Theta, \varphi) s^k \tag{3.66}$$

$$H_{ij}(s, \Theta, \varphi) = \sum_{k=0}^{\infty} c_{ij(-k)}(\Theta, \varphi) s^{-k} \tag{3.67}$$

A disadvantage of this kind of model is that we have to retain a large number of series terms in order to describe the target scattering properties in a wide frequency range [85]. Moreover, the models described by eqns. 3.64 and 3.66 are difficult to use in practice, because they involve differentiation procedures extremely sensitive to input noises and rounding errors.

3.3.4.2 Rational fractional approximation

When a model is used to simulate a signal scattered by a real target, it is expedient to employ the above apparatus of differential equations, rational fractional transfer functions and frequency responses. Note that the problem of rational fractional approximation of functions for numerical analyses is fairly well understood and is called the Padé expansion problem [65].

The rational fractional approximation is more cost effective than the power approximation. However, its application can be shown to be meaningless when the target size greatly exceeds the range resolution, because the model dimensionality then rises abruptly. To simplify the model, one may allow for the local properties of the UWB radar response.

3.3.4.3 Glittering point models

When the target geometry is such that its profile function (or cross section function) tolerates a physically valid piecewise linear approximation, the forced component of the impulse response is the weighted sum of the delta functions, i.e.

$$h'_{ij}(t,\Theta) = \sum_{l=1}^{L} a_{ijl}(\Theta)\delta(t - T_{ijl}(\Theta)) \tag{3.68}$$

and the reduced target response, $e_{0ij}^{(s)}(t)$ to an arbitrary incident signal, $e_j^{(i)}(t)$, is equal to the weighted sum of delayed signals, namely

$$e_{0ij}^{(s)}(t) = \sum_{l=1}^{L} a_{ijl}(\Theta)e_j^{(i)}(t - T_{ijl}(\Theta)) \tag{3.69}$$

This model is often termed the glittering points model. Here, $a_{ijl}(\Theta)$ are the intensities of glittering points (local scattering centres) and $T_{ijl}(\Theta)$ are their delay times. Although this model is quite simple and convenient for application, it is not always applicable even within the phenomenological approach, since signal waveforms reflected by real local scatterers may considerably differ from incident radar waveforms. It is useful to introduce a model of complex glittering points, which would be a hybrid of the model described by eqn. 3.68 and the scatterer description in terms of differential equations and rational fractional transfer functions, namely

$$h'_{ij}(t,\Theta) = \sum_{l=1}^{L} h'_{ijl}(t - T_{ijl}(\Theta)) \tag{3.70}$$

where $h'_{ijl}(t,\Theta)$ is the impulse response of the lth glittering point, whose Laplace transform (the point transfer function) is a rational fractional function. Such models have been shown to have a more extensive application than those presented by eqns. 3.51 and 3.68. They can be used for target simulation and interpretation of experimental data.

3.4 Polarisation properties of UWB radar response

We have shown that the polarisation matrix elements for UWB responses depend on the polarisation basis which is used for their description and measurement. This considerably complicates the process of target recognition because the polarisation basis varies with target orientation in the picture plane. Also, one has to use signals of only one polarisation type in recognition radar systems and in radar meters for obtaining response standards.

Two ways may be suggested for resolving this difficulty: using radar responses invariant with the polarisation structure of the reflected signals (so called polarisation invariants) or applying algorithms for the response recalculation from one polarisation basis to another.

UWB polarisation invariants can be obtained by generalising the invariants of the polarisation scattering matrix which completely describes the target scattering properties at the given frequency, ω_i. The polarisation invariants of the matrix $\mathbf{S} = [s_{ij}]$ $(i, j = 1, 2)$ are the determinant and the sum of squared modules of its elements [49]

$$\begin{aligned} \det(\mathbf{S}) &= s_{11}s_{22} - s_{12}s_{21} \\ &= \mathrm{inv} \end{aligned} \tag{3.71}$$

$$\begin{aligned} \mathrm{span}(\mathbf{S}) &= |s_{11}|^2 + |s_{22}|^2 + |s_{12}|^2 + |s_{21}|^2 \\ &= \mathrm{inv} \end{aligned} \tag{3.72}$$

where inv is the notation for an invariant.

Evidently, the conditions of eqns. 3.71 and 3.72 are met for every fixed frequency ω_i of the signal bandwidth; therefore, the polarisation matrix determinant for the frequency response $\mathbf{H}(\mathrm{i}\omega) = \{H_{ij}\}$ $(i, j = 1, 2)$ and the sum of the matrix squared elements can be written as

$$\begin{aligned} \det[\mathbf{H}(j\omega)] &= H_{11}H_{22} - H_{12}H_{21} \\ &= \mathrm{inv} \end{aligned} \tag{3.73}$$

$$\begin{aligned} \mathrm{span}[\mathbf{H}(j\omega)] &= H_{11}^*H_{11} + H_{12}^*H_{12} + H_{21}^*H_{21} + H_{22}^*H_{22} \\ &= \mathrm{inv} \end{aligned} \tag{3.74}$$

Eqns. 3.73 and 3.74 correspond to the following expressions in the time domain

$$\begin{aligned} \det[\mathbf{h}(t)] &= h_{11} * h_{22} - h_{12} * h_{21} \\ &= \mathrm{inv} \end{aligned} \tag{3.75}$$

$$\begin{aligned} \mathbf{R}_h(\tau) &= \sum_{i=1}^{2} \sum_{j=1}^{2} R_{hij}(\tau) \\ &= \mathrm{inv} \end{aligned} \tag{3.76}$$

Here, eqn. 3.75 represents the convolution residue of the elements along the major and minor diagonals of the impulse response polarisation matrix $\mathbf{h}(t) = [h_{ij}]$, and eqn. 3.76 is the sum of the correlation functions $R_{hij}(\tau)$ of the same matrix elements $(i, j = 1, 2)$.

Like eqns. 3.73 and 3.74, the characteristics are invariant relative to the polarisation basis variation. Therefore, these functions have a form which does not change with the polarisation basis and may be treated as recognition criteria applicable to radar systems of any polarisation. Besides, these polarisation invariants are insensitive to the object's rotation relative to the line of sight. For these reasons, the application of the invariants may be considered as a means of diminishing the UWB standard response catalogue and of decreasing the target recognition time.

The algorithms for response polarisation recalculation can be developed within the frequency bandwidth approach, taking advantage of the analogy between the target frequency response and its polarisation scattering matrix at a fixed frequency. It is known that, as the polarisation basis changes, the polarisation scattering matrix can be changed too by using the congruent transform [82], i.e.

$$\mathbf{S}^N = \mathbf{Q} \cdot \mathbf{S}^0 \cdot \mathbf{Q}^T \tag{3.77}$$

where $\mathbf{S}^0$ and $\mathbf{S}^N$ are the scattering matrices in the initial and the recalculated polarisation bases, $\mathbf{Q} = [q_{ij}]$ is the unitary transformation matrix, $i, j = 1, 2$.

Having made the appropriate transformations, we obtain the following expression in the vector form:

$$\mathbf{S}^N = \mathbf{G} \cdot \mathbf{S}^0 \tag{3.78}$$

where

$$\mathbf{S}^0 = \begin{bmatrix} s_{11}^0 \\ s_{12}^0 \\ s_{21}^0 \\ s_{22}^0 \end{bmatrix}$$

$$\mathbf{S}^N = \begin{bmatrix} s_{11}^N \\ s_{12}^N \\ s_{21}^N \\ s_{22}^N \end{bmatrix}$$

$$\mathbf{G} = \begin{bmatrix} q_{11}q_{11} & q_{11}q_{12} & q_{11}q_{12} & q_{12}q_{12} \\ q_{11}q_{21} & q_{11}q_{22} & q_{12}q_{21} & q_{12}q_{22} \\ q_{11}q_{21} & q_{12}q_{21} & q_{11}q_{22} & q_{12}q_{22} \\ q_{21}q_{21} & q_{21}q_{22} & q_{21}q_{22} & q_{22}q_{22} \end{bmatrix}$$

The recalculation matrix $\mathbf{G} = [q_{ij}]$ is made up of elements of the transformation matrix, $\mathbf{Q}$. In the general case taking into account both the polarisation basis rotation and ellipticity variation, the **Q** matrix has the form

$$\mathbf{Q} = \mathbf{R}(\psi)\cdot\mathbf{H}(\alpha) = \begin{bmatrix} \cos\psi & -\sin\psi \\ \sin\psi & \cos\psi \end{bmatrix} \begin{bmatrix} \cos\alpha & \mathrm{i}\sin\alpha \\ \mathrm{i}\sin\alpha & \cos\alpha \end{bmatrix} \quad (3.79)$$

where $\mathbf{R}(\psi)$ is the real rotation matrix, $\mathbf{H}(\alpha)$ is the complex ellipticity matrix, ψ is the rotation angle of the polarisation basis in the plane normal to the wave propagation direction and α is the ellipticity angle of the recalculated polarisation basis (in this form the operator $\mathbf{Q}(\alpha,\psi)$ makes transformation from the linear basis system to an arbitrary orthogonal elliptic basis system) [49].

Suppose that we have obtained the target response estimate in the vector form $\mathbf{H}(\mathrm{i}\omega) = [H_{11} \quad H_{12} \quad H_{21} \quad H_{22}]^T$, and that it is necessary to find its representation in the recalculated polarisation basis. The elements of the FR polarisation matrix in this basis can be found from eqn. 3.78. They represent the sums of the FR polarisation matrix elements in the initial basis multiplied by the complex-valued coefficients, which are eventually determined by the **Q** matrix of eqn. 3.79.

Let us analyse the changes in the frequency response, $\mathbf{H}_1(\mathrm{i}\omega)$, induced by its multiplication by the complex-valued coefficient, $\dot{C}$. Obviously, the frequency response $\mathbf{H}_2(\mathrm{i}\omega)$ has the form

$$\begin{aligned} \mathbf{H}_2(\mathrm{i}\omega) &= \dot{C}\cdot\mathbf{H}_1(\mathrm{i}\omega) \\ &= [C\cdot\exp(\mathrm{i}\cdot\mathrm{sgn}\{\omega\}\cdot\varphi_0)]\cdot[\mathbf{H}_1(\omega)\cdot\exp(\mathrm{i}\Theta_1)] \\ &= \mathbf{H}_2(\omega)\cdot\exp(\mathrm{i}\Theta_2) \end{aligned} \quad (3.80)$$

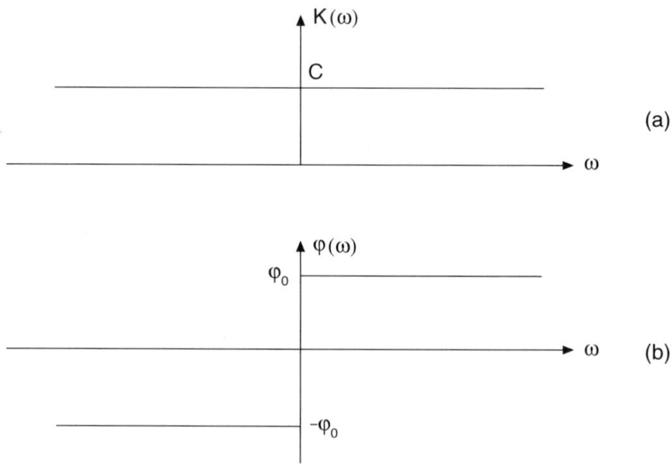

Fig. 3.5 Frequency response of polarisation transforming filter

a Amplitude-frequency response of polarisation transforming filter
b Phase frequency response

where $\mathbf{H}_2(\omega) = C \cdot \mathbf{H}_1(\omega)$, $\Theta_2 = \Theta_1 + \text{sgn}\{\omega\} \cdot \varphi_0$, C and φ_0 are the module and the argument of the number $\dot{C}$, $\mathbf{H}_2(\omega)$, $\mathbf{H}_1(\omega)$ and $\Theta_2(\omega)$, $\Theta_1(\omega)$ are the amplitude and phase responses of the target and $\text{sgn}\{\omega\} = 1$ if $\omega > 0$, $\text{sgn}\{\omega\} = -1$ if $\omega < 0$.

The process described by eqn. 3.80 corresponds to the passage of the target frequency response through a linear filter with a frequency response $K(\mathrm{i}\omega) = C \cdot \exp(\mathrm{i}\varphi_0)$. The amplitude and phase responses of this kind of filter are given in Fig. 3.5.

The filter impulse response is defined as:

$$
\begin{aligned}
h_f(t) &= \frac{C}{2\pi} \int_{-\infty}^{+\infty} \exp(\mathrm{i}\omega t + \mathrm{i} \cdot \text{sgn}\{\omega\}\varphi_0) \cdot d\omega \\
&= \mathbf{F}^{-1}\{C\}
\end{aligned}
\tag{3.81}
$$

and at $\varphi_0 = 2\pi n$ $\quad h_f = C \cdot \delta(t)$

at $\varphi_0 = \pi + 2\pi n$ $\quad h_f = -\,C \cdot \delta(t)$

at $\varphi_0 = \pi/2 + 2\pi n$ $\quad h_f = C/\pi t$

at $\varphi_0 = 3\pi/2 + 2\pi n$ $\quad h_f = -\,C/\pi t$

where $\delta(t)$ is the delta function, $n = 0, 1, 2, \ldots$.

At the other values of φ_0, the impulse responses have a form which is intermediate between the delta function and a function of the type $1/\pi t$ (the Hilbert transform) (see Fig. 3.6). Thus, the recalculation of the element $h_{ij}^0(t)$ of the target IR matrix into the new polarisation basis, $h_{ij}^N(t)$, can be represented as the sum of the IR components in the initial polarisation basis passed through linear filters with impulse responses described by eqn. 3.81.

3.5 Local UWB radar responses

It was pointed out in Section 3.1 that the local IR, $h_l(t, x, y, \Theta)$, or the local FR, $H_l(\mathrm{i}\omega, x, y, \Theta)$, characterising the properties of a surface element with the coordinates (x, y) at the target aspect angle Θ relative to the measuring antenna (Fig. 3.7) can be used as local radar responses. For simplicity, we shall examine the problem in a two-dimensional geometry and assume the target-to-antenna distance to be appreciably larger than the target size. We shall also suppose the signal wavelengths to be much less than the target linear dimensions. In this case, one can usually neglect the resonance scattering effects and represent a target as a combination of independent local scattering centres. Therefore, the total target IR can be obtained by integrating local impulse responses,

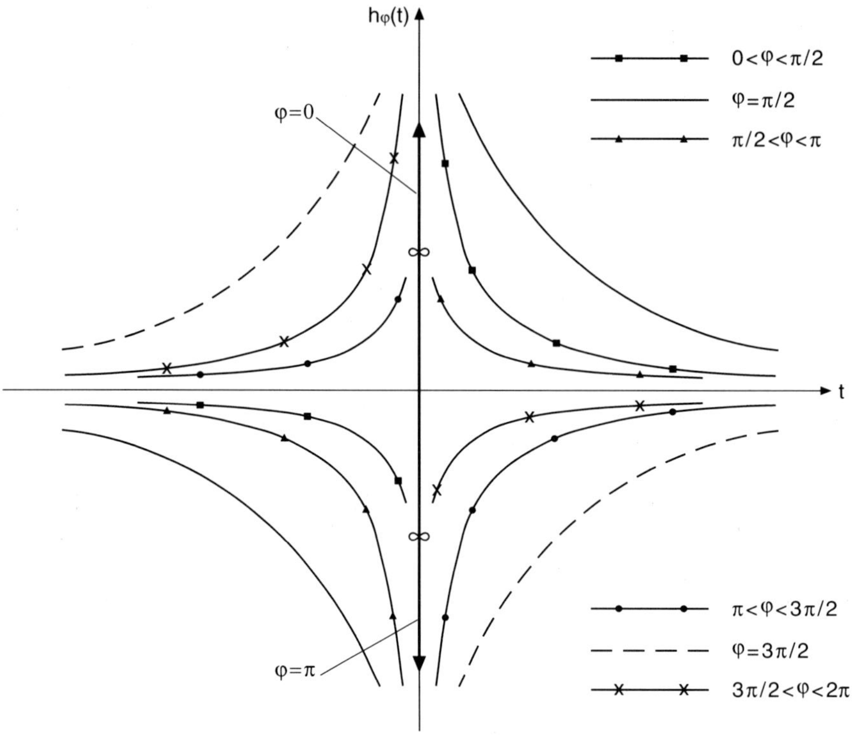

Fig. 3.6 Impulse responses of polarisation transforming filter

i.e.

$$h(t, \Theta) = k \int_{(x,y)} \int_{\mathbf{V}} h_l(t, x, y, \Theta)\, dx\, dy$$
$$= k \int_{\Delta Y} \int_{\Delta X} h_l(t, x, y, \Theta)\, dx\, dy \qquad (3.82)$$

where **V** is the space region of the target and k is a normalisation coefficient.

Eqn. 3.82 can be rewritten using the so-called rotated coordinate system $X^{\Theta}0Y^{\Theta}$, where the axis $0Y^{\Theta}$ coincides with the radiation direction

$$h(t, \Theta) = k \int_{\Delta Y^{\Theta}} \int_{\Delta X^{\Theta}} h_l(t, x^{\Theta}, y^{\Theta}, \Theta)\, dx^{\Theta}\, dy^{\Theta} \qquad (3.83)$$

where

$$x^{\Theta} = x \cos \Theta + y \sin \Theta$$
$$y^{\Theta} = y \cos \Theta - x \sin \Theta$$

The inner integral in eqn. 3.83 is a characteristic which may be termed as integral local IR, since it combines the responses of all local scattering centres with identical values of y^{Θ}, that is, with the same distance from the

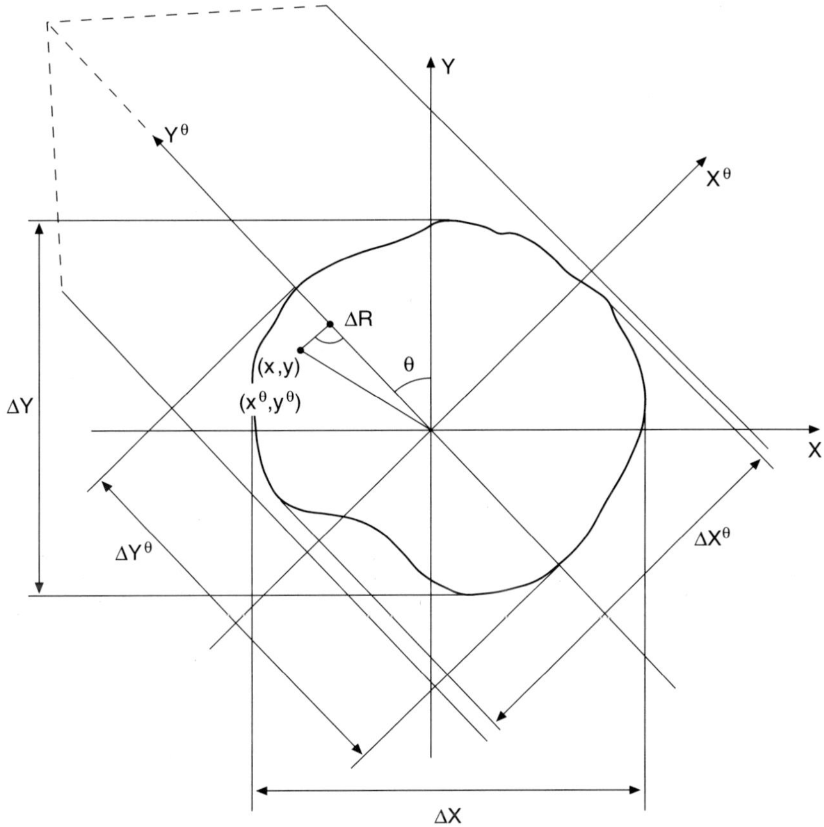

Fig. 3.7 Coordinate system by local RTR measurements

radiator. So we have

$$h_l(x, y^\Theta, \Theta) = \int_{\Delta X^\Theta} h_l(t, x^\Theta, y^\Theta, \Theta)\, dx^\Theta \tag{3.84}$$

Keeping in mind that the change in y^Θ leads to a time delay of the scattered signal by a value $2y^\Theta/c$, we can rewrite eqns. 3.84 and 3.83 as

$$h_l(x, y^\Theta, \Theta) = \int_{\Delta X^\Theta} h_{l_0}(t - 2y^\Theta/c, x^\Theta, y^\Theta, \Theta)\, dx^\Theta \tag{3.85}$$

$$\begin{aligned} h(t, \Theta) &= k \int_{\Delta Y^\Theta} \int_{\Delta X^\Theta} h_{l_0}(t - 2y^\Theta/c, x^\Theta, y^\Theta, \Theta)\, dx^\Theta\, dy^\Theta \\ &= \int_{\Delta Y^\Theta} h_l(x, y^\Theta, \Theta)\, dy^\Theta \end{aligned} \tag{3.86}$$

where $h_{l_0}(t, x^\Theta, y^\Theta, \Theta)$ is the impulse response of the local scattering centre with the coordinates (x^Θ, y^Θ) reduced to the coordinate reference.

The expressions presented allow one to solve the estimation problems for total responses from local responses. It will be shown in Chapter 5 that they can also be used to estimate local responses (on certain assumptions) from measured total responses.

A frequency analogue of a local IR is a local target FR

$$H_l(\mathrm{i}\omega, x, y, \Theta) = F\{h_l(t, x, y, \Theta)\} \tag{3.87}$$

where $F\{h_l(t, x, y, \Theta)\}$ is the Fourier transform operator.

In the general case, a local FR represents a complex-valued function. But to simplify the solution of applied problems, it may often be assumed to be a real function (without serious methodological errors) which can take positive and negative values. In this case the function may be called the local effective scattering length (ESL) of the target $l(\omega, x, y, \Theta)$, because its squared value is equal to the target RCS, $\sigma(\omega, x, y, \Theta)$

$$l^2(\omega, x, y, \Theta) = \sigma(\omega, x, y, \Theta) \tag{3.88}$$

The function $l(\omega, x, y, \Theta)$ characterises the intensity (amplitude) of a monochromatic signal of frequency ω, reflected by an element with the coordinates (x, y) at the target aspect Θ. Representation of the function $l(\omega, x, y, \Theta)$ versus space variables (x, y) at fixed ω and Θ is a way of radar target imaging. This approach permits the estimation of the reflection properties of various points and, thus, solution of the problems of analysis and reduction of target detectability.

We should keep in mind that the representation of a local FR as a real frequency function means that the reduced local IR is an even function of time.

For further simplification of the problem solutions, for example, in target imaging, additional constraints are imposed on local responses, associated with the form of their time or frequency dependence and, sometimes, with their dependence on the target aspect, at least in a certain angle sector. The following local response approximations [48] may be employed

$$\begin{cases} H_l(\mathrm{i}\omega, x, y, \Theta) = l(x, y, \Theta) \\ h_{l_0}(t, x, y, \Theta) = l(x, y, \Theta)\delta(t) \end{cases} \tag{3.89}$$

$$\begin{cases} H_l(\mathrm{i}\omega, x, y, \Theta) = l(x, y, \Theta)(\mathrm{i}\omega) \\ h_{l_0}(t, x, y, \Theta) = l(x, y, \Theta)\delta'(t) \end{cases} \tag{3.90}$$

$$\begin{cases} H_l(\mathrm{i}\omega, x, y, \Theta) = l(x, y, \Theta)\omega^2 \\ h_{l_0}(t, x, y, \Theta) = l(x, y, \Theta)\delta''(t) \end{cases} \tag{3.91}$$

The validity of these models, more exactly, their validity conditions, has not been adequately substantiated in spite of their wide application, including their use in target imaging. One exception is, perhaps, the model of eqn. 3.91. Experiments show that a model may sometimes be close to the actual mechanism of the scattered signal behaviour, but in other situations it may contradict the reality.

When imaging a radar target, especially if it is a good conductor, one often assumes that the local ESL does not depend on the location of the scattering points and that the time and frequency dependences of local responses have a form defined by eqn. 3.91. As a result, one arrives at the concept of the characteristic target function, $\gamma(x, y)$, mentioned in Section 3.2; in this case we have

$$\begin{aligned} h_{l_0}(t, x^{\Theta}, y^{\Theta}) &= l(x^{\Theta}, y^{\Theta})\delta''(t) \\ &= -\gamma(x^{\Theta}, y^{\Theta})\delta''(t) \end{aligned} \tag{3.92}$$

This approximation of local responses leads to the Kennaugh–Moffatt identity. Indeed, the substitution of eqn. 3.92 into eqn. 3.86 yields

$$h(t, \Theta) = -k\frac{d^2}{d^2y^{\Theta}}\int_{\Delta Y^{\Theta}}\left[\int\!\!\int_{\Delta X^{\Theta}} \gamma(x^{\Theta}, y^{\Theta})\, dx^{\Theta}\right]\delta(t - 2y^{\Theta}/c)\, dy^{\Theta}\Big|_{y^{\Theta}=ct/2} \tag{3.93}$$

The inner integral, evidently, represents the target profile function, $S(y^{\Theta})$. Taking into consideration the filtering property of the delta function of eqn. 3.93, we have

$$h(t, \Theta) = -k\frac{d^2}{d^2y^{\Theta}}S(y^{\Theta})\Big|_{y^{\Theta}=ct/2} \tag{3.94}$$

This expression coincides with eqn. 3.25 obtained by the physical optics method [53].

Chapter 4

Design principles and mathematical modelling of UWB radar meters

4.1 Types of meters and measurement schemes

One of the key problems in modern radar engineering and applied electrodynamics is the experimental study of the scattering characteristics of various objects, materials and targets. In recent years, the requirements for the measurement quality of radar target responses (RTR) or characteristics have greatly increased. This primarily concerns the response composition, the accuracy and speed of measurement, as well as the frequency bandwidth, resolving power, operation zone, dynamic range and the threshold sensitivity of measuring systems. As a result, the vast arsenal of measuring devices constructed in many countries during the past two or three decades has mostly become out of date.

A perspective approach to the problem of high quality measurement of RTR that has been tested experimentally in recent years is the use of automatic ultrawideband measuring and data processing (MDP) systems [5, 24, 91]. The obvious advantages of such systems are as follows:

1 UWB signals permit measurement of the total conventional RTR (RCS and polarisation scattering matrices) by single-channel devices in a frequency band of several to tens of gigahertz, instead of the awkward, multi-channel narrowband meters making measurements at a few frequencies, often different from those of actual radar signals.
2 UWB MDP systems can measure a greater variety of RTR, namely impulse responses, frequency responses, scattering resonance frequencies, etc. (see Chapter 3). They can measure not only total responses but also local responses from individual elements of the target surface.
3 UWB RTR measurements can be used in mathematical modelling to estimate the capability of available and prospective radar systems of recognising poorly detectable objects [11, 48].

4 The high space–time resolution of UWB MDP systems allows the selection of target-scattered signals from the clutter from the back wall and other parts of an anechoic chamber (AEC) or from local sources in open testing grounds. This considerably increases system sensitivity, compared with narrowband meters, because of the reduced minimum measurable RCS. High-precision UWB meters use minimal amounts of expensive radio-absorbing materials or even do without them. Moreover, there is the possibility, unachievable previously, of designing devices for the control and localisation of clutter sources and mounting these control devices during the production cycle of the objects.

Another factor stimulating the rapid development of UWB MDP systems is the need for efficient radar equipment for the detection and recognition of poorly detectable objects. The analysis of current literature shows that UWB signals are expected to offer a solution to this problem [11, 58, 59, 85]. The high effectiveness of UWB signals is due to their ability, pointed out above, to provide information on local scattering centres in a wide frequency range. Of special interest is the low frequency region, where the wavelength is comparable with the linear dimensions of an object. In this region, the effectiveness and the weight–size parameters of radio-absorbing materials have been shown experimentally to be quite low [1, 11, 95]. However, an appropriate choice of the basic parameters of probing signals, processing algorithms and other UWB equipment characteristics is impossible without a detailed investigation of the radar responses of the materials, target local centres and the target as a whole.

The improvement of RTR measurement with UWB probing signals is proceeding along the following lines:

- extending the scope of RTR to be measured;
- increasing the measurement accuracy and reducing equipment costs;
- increasing the measurement speed in a wide frequency range;
- raising the dynamic range and sensitivity of measurement.

These goals can be achieved by a variety of technical approaches to designing better devices for the generation, transmission, reception and registration of signals. Also of importance are better techniques and algorithms for signal processing and statistical estimation of various UWB RTR, because these can help to improve the measurement procedures and to solve other metrological problems.

UWB MDP systems have, currently, been made by practically all industrial countries. As an illustration, we list here some institutions and countries in which such systems have been designed and tested: the TES Laboratory at the Naval Postgraduate School (USA); Huachuca Fort (USA); Cochin University of Science and Technology (India); Eindhoven Technical University (Netherlands); ESTEC in Europe; CELAR (France); Michigan and Ohio State

Universities (USA); Messerschmidt-Bolkow-Blohm (Germany); LRIR, ALCOR, TIR, Cobra Dane, and Cobra-Gudy (USA). Some radioelectronics companies, in particular Scientific Atlanta, Emerson and Cumming, Hughes Aircraft Company, Hewlett Packard (USA) and Eclema (France), produce commercially various types of UWB MDP system and related equipment [5]. An experimental system of this kind has also been made in Russia [1].

Available and prospective MDP systems can be used to solve two interrelated problems:

(i) to study the mechanism of transient electromagnetic wave scattering by variously shaped objects made from different materials, to develop and verify mathematical scattering models;
(ii) to measure experimentally the RTR of materials, whole objects and their parts and models; to control the radar detectability of objects (also, in their production) and to localise the main scattering centres.

UWB RTR measurements can be performed in both static and dynamic conditions (Fig. 4.1). In the former case, the object or its model is placed on a rotating support, with its orientation varying with the measurement programme. Much effort is being devoted to avoiding possible clutter, the effects of the real signal environment. In the latter situation, the MDP system and the object are in natural conditions, and the measurements are made in a special-purpose experiment (on testing grounds or bases) or in real conditions involving the usual RF noise sources and distortions.

To date, remarkable progress has been made in the development of MDP systems for static measurements and, in terms of the operation zone used, such systems can be tentatively subdivided into laboratory-type and range

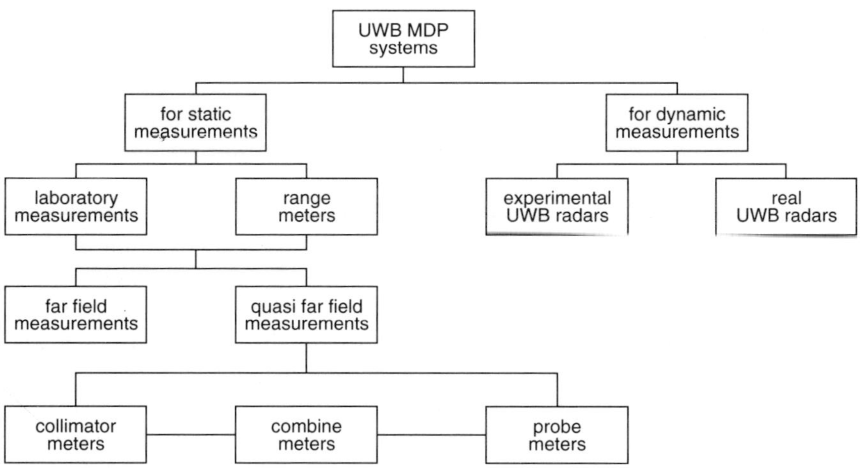

Fig. 4.1 Classification of RTR meters

systems. Laboratory MDP systems are used to study the RTR of materials, small-sized objects, units or scale models of large objects. For these, the major problems of equipment design have already been practically resolved [85]. Range MDP systems are designed for RTR measurements of natural, large-sized objects or their full-scale models. Because of the large testing area required, the problems of designing the basic functional units for such systems are more difficult to resolve [85]. This primarily concerns the formation of the desired spatial structure of the incident field – the plane amplitude and phase fronts of the probing signal. Initially, this was done by placing the object in the far field of an MDP antenna (antennas) at a distance large enough for the incident wavefront in this location to be considered as plane. This approach is quite acceptable for laboratory-type systems, both closed (with an anechoic or a screened chamber) and open. However, RTR measurements of large objects require large open testing grounds of several kilometres in length. A well-known example of such grounds is the measuring complex RATSCAT. A serious disadvantage of MDP systems is that they are very sensitive to weather conditions and are hard to disguise. To decrease the testing ground area, compact MDP systems are currently used which are provided with special radiation focusing in the near field to form a field with prescribed characteristics [26]. This is done by special hardware (lens or mirror collimators) or software (scanning probes and the focused aperture synthesis method). Combined systems are also possible, in which the transmission is performed by a collimator and the reception by a probe antenna. Combined systems may be of the laboratory type or, using anechoic or screening chambers, the open type. The respective measuring schemes are shown in Fig. 4.2.

Collimator UWB MDP systems allow one to make the RTR precision measurements necessary in radar detection tests and to obtain local RTR estimates to be used for designing devices with a lower level of detectability and RTR distortions. Probe meters can provide precision measurements (potentially, to within a methodological error) of local RTR to be used for obtaining total response estimates [48].

Both laboratory and range UWB MDP systems normally use two types of signal – nano and subnanosecond radio pulses and multifrequency signals, often with consecutive emission of the frequency components. Both types of signal are identical in their information value. Some countries, for example the USA and Russia, are currently producing well tested equipment for the generation and reception of any type of radio signal. Depending on the particular task, such an MDP system is capable of making the selection of an optimal measurement scheme. An advantage of multifrequency MDP systems is the relative simplicity of signal generation and reception at frequencies above 40–50 GHz, a frequency range where the desired pulse intensity is hard to achieve and the quality of sample receivers is not high enough. However, these systems have a disadvantage consisting in possible overload of highly sensitive receivers by powerful clutter from the chamber

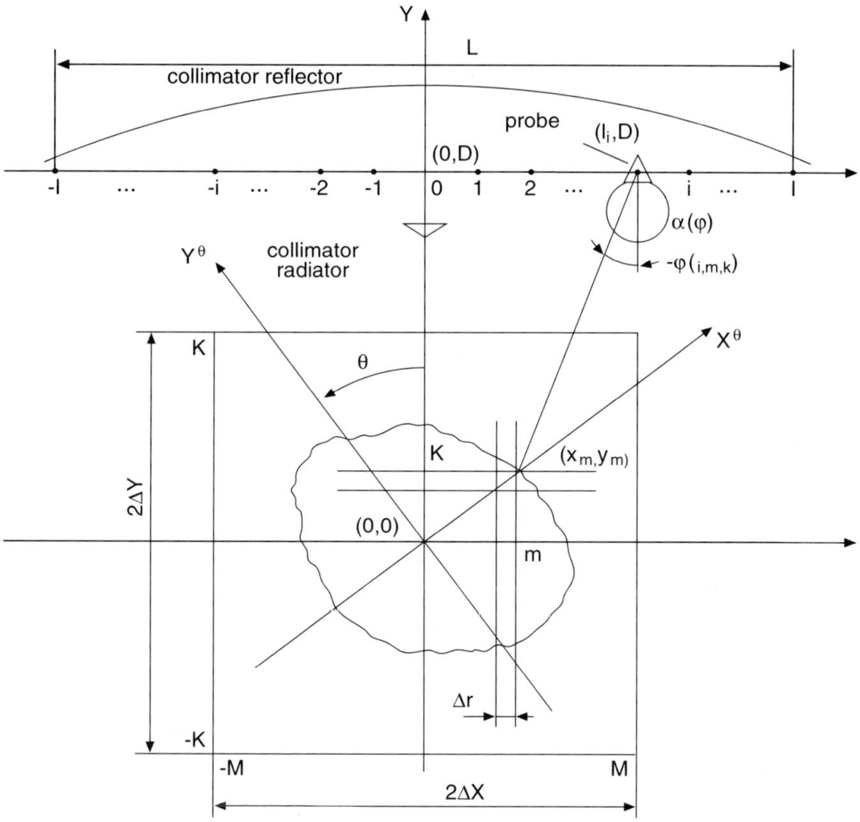

Fig. 4.2 Geometry of RTR measurements

back walls, which reduces the dynamic range and sensitivity of the meter. This difficulty is removed by software gating, where the multifrequency measurements are subjected to discrete Fourier transformation to synthesise a time signal, from which a segment matching the target echo is cut out. A similar effect can be achieved via direct signal transmission from the generator antenna to the receiving antenna.

A general diagram of a UWB MDP system is shown in Fig. 4.3. The rotating support positions the object with the prescribed accuracy and degrees of freedom. The measuring radar line consisting of a transmitting and receiving antenna system, a transmitter and a receiver, provides the shaping, transmission and reception of UWB signals. The particular design of each unit is determined by the type of measuring system and the UWB signal.

In collimator MDP systems, the antenna arrangement represents a one, two or three-mirror system with one or two emitters. The antenna system is controlled by choosing the appropriate polarisation of the probing and received signals. In probe MDP the antenna arrangement consists of a

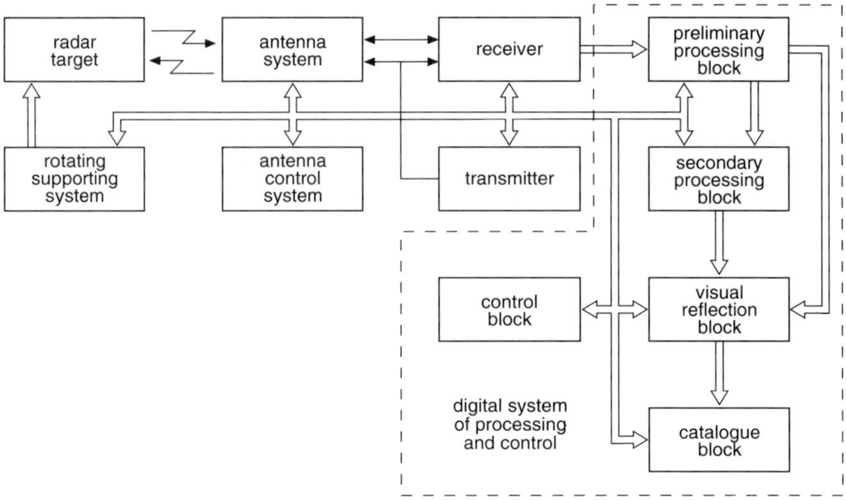

Fig. 4.3 Block diagram of UWB MDP system

loosely oriented, movable transmit–receive probe which can scan in one or two planes, as the programme requires.

The transmitting and receiving devices in multifrequency meters include a frequency synthesiser and an amplitude-phase meter with a digital control and output [1, 117]. Impulse MDP systems use semiconductor or TWT shapers with shock excitation for the generation of short duration pulses [1]. Sampling converters are employed to perform timescale conversion of the signals received.

The signal processing and control system is designed for controlling individual devices prescribed by the programme and measurement tasks, for the primary and secondary digital processing of signals, current representation and cataloguing of the measurements. These operations are implemented by one or several computers assembled into a computer network.

The details of the design and specifications of various systems have been analysed in [1, 48, 85, 116, 117].

4.2 A mathematical model of a UWB total RTR meter

Fig. 4.3 shows the general block diagram of a UWB meter used for total RTR measurements in the far field or in the quasi far field (collimator meter). For simplicity, we suppose that the transmission and reception points coincide. As an illustration, we shall discuss the case of a pulse RTR meter with a two-dimensional measurement geometry (Fig. 4.4).

For this case, Fig. 4.5 presents the equivalent block diagram of a UWB meter, which depicts the signal conversion at various cross sections of the mapping path. The elements of the diagram – linear dynamic components

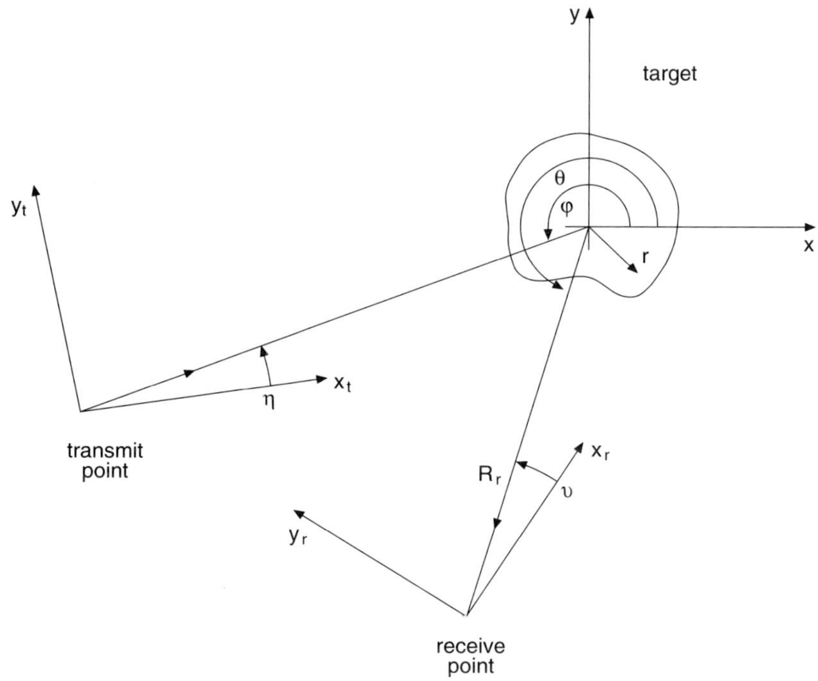

Fig. 4.4 Geometry of RTR measurements

– can be described in terms of the impulse responses (IR) and frequency responses (FR) of the meter's elements.

The IR column matrix of a probing UWB field shaper consisting of a transmitting device and an antenna is taken to be a column vector, with components which are the normalised time-dependent polarisation components of the transmitted signal

$$\mathbf{h}_t(t,\eta) = [h_{t1}(t,\eta) \quad h_{t2}(t,\eta)]^T \tag{4.1}$$

where $h_{ti} = 2\sqrt{\pi}\, R_0 e_i^{(1)}(t + R_0/c, \eta)$, $i = 1, 2$, R_0 is the distance between the

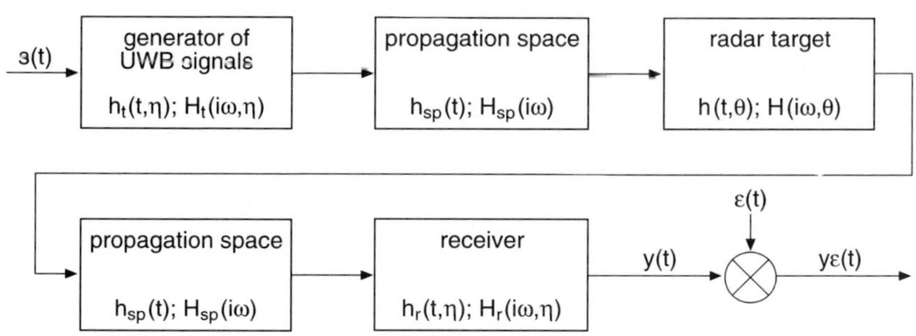

Fig. 4.5 Equivalent block diagram of UWB meter

meter and the target and η is the angle between the electrical antenna axis and the orientation towards the radar target.

We assume that the space, through which the electromagnetic waves will propagate, is close in its properties to free space. Then the IR matrices for the forward and backward space travel are

$$\mathbf{h}_{sp}(t) = \begin{bmatrix} h_{sp11}(t) & 0 \\ 0 & h_{sp22}(t) \end{bmatrix} \tag{4.2}$$

where $h_{sp11}(t) = h_{sp}(t) = \delta(t - R_0/c)/(2\sqrt{\pi}R_0)$.

It can be replaced by the scalar value $h_{sp}(t)$, and the effect of free space may be considered as being reduced to the time delay and the change in the signal intensity.

The polarisation matrix of the receiver IR represents the row vector

$$\mathbf{h}_r^T(t,\nu) = [h_{r1}(t,\nu) \quad h_{r2}(t,\nu)] \tag{4.3}$$

where the elements depend on the angle, ν, between the electrical antenna axis and the orientation to the target.

Using the IR introduced, we can write an expression for the determinate component of the receiver output signal

$$y(t) = \mathbf{h}_r^T(t,\nu) * \mathbf{h}(t,\Theta) * \mathbf{h}_t(t,\eta)/(4\pi R_0^2) \tag{4.4}$$

where $*$ denotes the convolution procedures performed very much like matrix multiplication procedures. Here, allowance is made for the fact that the time delay in the propagation space has been compensated for by the receiving device.

If the transmitting and receiving antennas are re-oriented in the orthogonal polarisation basis to perform measurements of the polarisation components of the radar target IR, the matrix for the determinate components of the received signals, $y_{ij}(t)$, corresponding to the ith polarisation of the receiving antenna and to the jth polarisation of the transmitting antenna, is

$$\begin{aligned} \mathbf{y}(t) &= [y_{ij}(t)]_{i,j=1,2} \\ &= \begin{bmatrix} h_r^T(t,\nu) \\ \hline h_r^T(t,\nu)\mathbf{J} \end{bmatrix} * \mathbf{h}(t,\Theta) * [h_t(t,\eta) \,|\, \mathbf{P}\,\mathbf{h}_t(t,\eta)]/(4\pi R_0^2) \end{aligned} \tag{4.5}$$

where

$$\mathbf{P} = \begin{bmatrix} 0 & 1 \\ 1 & 0 \end{bmatrix} \tag{4.6}$$

is the permutation matrix.

When the crossed polarisation components of the signal in the transmitting and receiving antennas can be neglected

$$\begin{aligned} y_{ij}(t) &= h_{ri}(t,\nu) * h_{tj}(t,\eta) * h_{ij}(t,\Theta)/(4\pi R_0^2) \\ &= w_{ij}(t,\eta,\nu) * h_{ij}(t,\Theta) \end{aligned} \tag{4.7}$$

where $w_{ij}(t,\eta,\nu) = h_{ri}(t,\nu) * h_{tj}(t,\eta)/(4\pi R_0^2)$ are the IR of the radar meter.

In the frequency domain, one can obtain expressions corresponding to eqns. 4.5–4.7. For example, the spectral function for the determinate component of the receiver output signal is

$$Y_{ij}(jw) = W_{ij}(\mathrm{i}\omega, \eta, \nu)H_{ij}(\mathrm{i}\omega, \Theta) \tag{4.8}$$

where $W_{ij}(\mathrm{i}\omega, \eta, \nu) = F\{w_{ij}(t, \eta, \nu)\}$ are the meter FR and $F\{\ \}$ is the direct Fourier transform operator. We shall assume that there is additive noise at the receiver output, whose spectral density of medium power $E(\omega)$ is known. Then the receiver output signal and its spectral function are realisations of random processes

$$\begin{aligned} y_{ij_\varepsilon}(t) &= w_{ij}(t, \eta, \nu) * h_{ij}(t, \Theta) + \varepsilon(t) \\ Y_{ij_\varepsilon}(\mathrm{i}\omega) &= W_{ij}(\mathrm{i}\omega, \eta, \nu)H_{ij}(\mathrm{i}\omega, \Theta) + E(\mathrm{i}\omega) \end{aligned} \tag{4.9}$$

where $\varepsilon(t)$ and $E(\mathrm{i}\omega)$ are the noise realisation and its spectral function.

Let us derive the equations for the estimation of the scattering IR and FR matrix elements in the case of polarisation distortions, which are generalisations of eqns. 4.7 and 4.8. For this, we shall make use of the general block diagram of a UWB meter shown in Fig. 4.5. To simplify the expressions to be derived, as compared to eqn. 4.5, it is convenient to use a vector representation, instead of matrices, for the target IR and FR as well as for the received signals and variable meter characteristics at various polarisations of the transmit–receive antennas. For this, we introduce the vector $\mathbf{h}_v(t)$ of the target IR containing all elements of the above IR polarisation matrix $\mathbf{h}(t)$, with the angular coordinates omitted for simplicity

$$\mathbf{h}_v(t) = [h_{11}(t)h_{12}(t)h_{21}(t)h_{22}(t)]^T \tag{4.10}$$

Let

$$\mathbf{h}_t^{(j)}(t) = [h_{t1}^{(j)}(t)h_{t2}^{(j)}(t)]^T \tag{4.11}$$

be the column vector for the probing signal shaper with controlled polarisation in its jth polarisation ($j = 1, 2$). Consider also the IR row vector of the receiving device with controlled polarisation in its ith polarisation ($i = 1, 2$)

$$\mathbf{h}_r^{T(i)}(t) = [h_{r1}^{(i)}(t)h_{r2}^{(i)}(t)] \tag{4.12}$$

Note that the shaper and receiver polarisations can be varied, say, by rotating the transmitting and receiving antennas with linear polarisation through 90° (see below).

We now introduce the vector of the receiver signal determinate components

$$\mathbf{y}_v(t) = [y_{11}(t)y_{12}(t)y_{21}(t)y_{22}(t)]^T \tag{4.13}$$

where $y_{ij}(t)$ is the signal at the receiver output in the ith polarisation of the receiving antenna and the jth polarisation of the transmitting antenna.

Under the condition of the propagation space distortions being neglected, we can obtain, for a single position of the meter, the following expression for an arbitrary element of the vector $\mathbf{y}_v(t)$

$$y_{ij}(t) = \mathbf{w}^{(ij)T}(t) * \mathbf{h}_v(e) \tag{4.14}$$

where $\mathbf{w}^{(ij)T}(t)$ is the row vector of the UWB meter IR in the ith polarisation of the receiving antenna and the jth polarisation of the transmitting antenna.

The vector

$$\mathbf{w}^{(ij)T}(t) = [w_{11}^{(ij)}(t)w_{12}^{(ij)}(t)w_{21}^{(ij)}(t)w_{22}^{(ij)}(t)]^T \tag{4.15}$$

where

$$w_{pq}^{(ij)}(t) = \frac{1}{4\pi R_0^2} h_{rp}^{(i)}(t) * h_{tq}^{(j)}(t) \tag{4.16}$$

is the IR in the ith receiving antenna polarisation and the jth transmitting antenna polarisation for the pth polarisation of the scattered electromagnetic field and the qth polarisation of the incident field $(i, j, p, q = 1, 2)$.

The total vector of the received signals is

$$\mathbf{y}_v(t) = \mathbf{w}(t) * \mathbf{h}_v(t) \tag{4.17}$$

where $\mathbf{w}(t)$ is the total IR matrix for all possible polarisations of the receiving and transmitting antennas for the incident and scattered electromagnetic fields.

In its explicit form eqn. (4.17) is as follows

$$\begin{bmatrix} y_{11}(t) \\ y_{12}(t) \\ y_{21}(t) \\ y_{22}(t) \end{bmatrix} = \begin{bmatrix} w_{11}^{(11)}(t)w_{12}^{(11)}(t)w_{21}^{(11)}(t)w_{22}^{(11)}(t) \\ w_{11}^{(12)}(t)w_{12}^{(12)}(t)w_{21}^{(12)}(t)w_{22}^{(12)}(t) \\ w_{11}^{(21)}(t)w_{12}^{(21)}(t)w_{21}^{(21)}(t)w_{22}^{(21)}(t) \\ w_{11}^{(22)}(t)w_{12}^{(22)}(t)w_{21}^{(22)}(t)w_{22}^{(22)}(t) \end{bmatrix} * \begin{bmatrix} h_{11}(t) \\ h_{12}(t) \\ h_{21}(t) \\ h_{22}(t) \end{bmatrix} \tag{4.18}$$

A polarisationally ideal antenna transmits or receives an electromagnetic wave with a polarisation coinciding with that of the respective antenna

$$h_{rp}^{(i)}(t) = \begin{cases} 0, & p \neq i \\ h_{ri}^{(i)}(t), & p = i \end{cases}$$
$$h_{tq}^{(i)}(t) = \begin{cases} 0, & q \neq j \\ h_{ti}^{(0)}(t), & q = j \end{cases} \tag{4.19}$$

In these conditions, the elements of the $\mathbf{w}(t)$ matrix are

$$w_{pq}^{(ij)}(t) = \begin{cases} 0, & p \neq i, \quad q \neq j \\ w_{ij}^{(ij)}(t), & p = i, \quad q = j \end{cases} \tag{4.20}$$

where

$$w_{ij}^{(ij)}(t) = \frac{1}{4\pi R_0^2} h_{ri}^{(i)}(t) * h_{tj}^{(j)}(t)$$

Hence, the $\mathbf{w}(t)$ matrix is diagonal ($\mathbf{w}(t) = \text{diag}\{w_{11}^{(11)}(t)w_{12}^{(12)}(t)w_{21}^{(21)}(t)w_{22}^{(22)}(t)\}$) and the set of equations (eqn. 4.18) falls into four independent equations

$$y_{ij}(t) = w_{ij}^{(ij)}(t) * h_{ij}(t) \tag{4.21}$$

which coincide with eqn. 4.7. However, in the general case one should use eqns. 4.17 or 4.18.

4.3 Mathematical models of local RTR meters

4.3.1 A probe meter

Let us consider the mathematical model of a probe meter for local response measurement. As above, we shall examine a two-dimensional problem corresponding to the measurement scheme in Fig. 4.2.

The space region in which a transmit–receive probe can be located has the size L: $X \in [-L/2, L/2]$. A reference from the $X0Y$-coordinate system is superimposed, centred on the target. The scanning plane of the probe lies at a distance D from the reference point.

For simplicity, consider scalar signals. Their polarisation structure can be accounted for as described in Section 4.2 and, keeping in mind the small electrical size of the probe and its wide directivity chart, we shall assume that the probe will have the following types of impulse and frequency responses

$$h_a(t, \varphi) = d(\varphi)h_{a0}(t) \tag{4.22}$$

$$K_a(\text{i}\omega, \varphi) = d(\varphi)K_{a0}(\text{i}\omega) \tag{4.23}$$

where φ is an angle from the normal to the scanning surface, $d(\varphi)$ is the normalised directivity chart of the probe with respect to the field strength and $h_{a0}(t)$ and $K_{a0}(\text{i}\omega)$ are the impulse and frequency responses of the probe in the axial direction.

Examine a signal received by a probe with the coordinates (X, D) from a point in space with the coordinates (x, y). In this case, the distance between the probe and the local centre with the ERL $l(x, y)$ is

$$R(X, x, y) = \sqrt{(X - x)^2 + (D - y)^2} \tag{4.24}$$

and the angle is

$$\varphi(X, x, y) = \arctan[(X - x)/(D - y)] \tag{4.25}$$

The signal $y(t, X, x, y)$, or more exactly its determinate component with no account of noise, can be represented as

$$y(t, X, x, y) = \frac{d^2(\varphi(X, x, y))l(x, y)D^2}{R^2(X, x, y)} w_0(t - 2R(X, x, y)/c) \tag{4.26}$$

Here, $w_0(t)$ is the meter impulse response normalised to the points $(0, 0)$ and $X = 0$, time delayed by $2D/c$. The experimentally measured, normalised IR must be transformed, if models different from the local scattering model (eqn. 3.85) are used. In particular, in the models described by eqns. 3.86 and 3.87, it must be singly or doubly differentiated.

The total signal at the meter output formed by the incident signal reflected from the whole space region, in which scattering centres may be located (the image shot) has the form

$$\begin{aligned} y(t, X) &= \iint_{xy} y(t, X, x, y)\, dx\, dy \\ &= \iint_{xy} k(X, x, y)l(x, y)w_0(t - 2R(X, x, y)/c)\, dx\, dy \\ &= \iint_{xy} k(X, x, y)l(x, y) \int_{-\infty}^{+\infty} w_0(\tau)\delta(\tau - (t - 2R(X, x, y)/c))\, d\tau\, dx\, dy \\ &= \int_{-\infty}^{+\infty} w_0(\tau) \left[\iint_{xy} k(X, x, y)l(x, y)\delta(\tau - (t - 2R(X, x, y)/c))\, dx\, dy \right] d\tau \\ &= \int_{-\infty}^{+\infty} w_0(\tau)h(t - \tau, X)\, d\tau \\ &= \int_{-\infty}^{+\infty} w_0(t - \tau)h(\tau, X)\, d\tau \end{aligned} \tag{4.27}$$

where $k(X, x, y) = d^2(\varphi(X, x, y)D/R^2(X, x, y))$ is a damping coefficient and

$$h(t, X) = \iint_{xy} k(X, x, y)l(x, y)\delta(2R(X, x, y) - t)\, dx\, dy \tag{4.28}$$

is the image shot IR, when the probe is located at point X.

As was pointed out, the observable meter output signal contains a noise component

$$y_\varepsilon(t, X) = y(t, X) + \varepsilon(t) \tag{4.29}$$

It is clear from the above expressions that local UWB responses, for instance local ERL, are determined by solving a set of first-kind integral equations for various positions of the measuring probe.

4.3.2 A collimator meter

There may be several approaches to radar imaging with collimator meters. One of them (see, for example, [26, 48, 85, 112]) is to process instantaneous signal values at a collimator meter output and to estimate the total impulse and ramp responses. These stages are followed by the transition from ramp responses to profile functions at various aspect angles, and these functions are related to the characteristic target function which is found from eqns 3.27 and 3.31. In this case the relationship between the observable output signal and the target impulse response at various aspects does not practically differ from that discussed in Section 4.2.

Here we shall consider a meter model, in which mean square averaging of signals is performed in order to increase the clutter resistance. We shall assume the scattering model described by eqn. 3.87 to be valid and the function $l(x, y)$ to be unipolar, which holds true for convex objects which are good conductors [55].

Suppose that a transmitter radiates a signal, $s_i(t)$, of duration t_i. The average power within a pulse is

$$\rho_t = \left(\int_0^{t_i} s_i^2(t)\, dt \right) \Big/ t_i \tag{4.30}$$

The signal power in the meter operation range at the moment the incident signal is crossing it, is

$$p_i = \rho_t \eta_a^2 \tag{4.31}$$

where η_a is a utilisation coefficient for the antenna feeder channel including the collimator.

Since the power flux density in the operation range is the ratio of p_i to the range cross section, S, the averaged power of the signal reflected by a local scattering centre with the coordinates (x, y) at the target aspect Θ will be described by the expression

$$\begin{aligned} \rho_r(x, y, \Theta) &= \rho_i \sigma(x, y)/S \\ &= \sqrt{\rho_i}\, l(x, y)/S \end{aligned} \tag{4.32}$$

Now we introduce a signal observed at the preprocessor output

$$\tilde{y}(t, \Theta) = \sqrt{\frac{1}{t_i} \int_{t-t_i}^{t} y^2(t, \Theta)\, dt} \tag{4.33}$$

It is easy to show from eqns 4.32 and 4.33 that

$$\tilde{y}(t, \Theta) = k \int_\Theta l(x^\Theta, y^\Theta)\, dx^\Theta \Bigg|_{t = 2y^\Theta / c'}, \qquad x \in V \tag{4.34}$$

where k is a proportionality factor.

In real conditions, the observable smoothed signal, $\tilde{y}(t, \Theta)$, has superimposed noise, $\varepsilon(t)$, resulting from clutter smoothing in the initially registered signal.

As in the estimation of local RTR by a probe meter, our task reduces to solving an inverse problem, namely, a set of integral equations of the type of eqn. 4.34 for various target aspects. This problem is to be solved under the measurement noise conditions.

4.4 Calculation of the energy potential of impulse meters

4.4.1 Impulse meters with a TWT transmitter

Here we shall discuss the calculation methods for the energy potential, namely for the power or amplitude of the transmitting device, when the receiver characteristics and the measurement accuracy are known. This will be done with reference to collimator meters, although the methods suggested can be adapted to other types of meter, too.

The following initial parameters will be used:

1 Minimum target RCS, σ_0, measured with a normalised error, or the effective radar length, $l_0 = \sqrt{\sigma_0}$.
2 The frequency range $[f_{min}, f_{max}]$ of the measurements to be made.
3 The maximum admissible relative mean square error of the RCS measurement in a given range, γ_0.
4 The mean square value of the clutter normalised to the receiver input, σ_ε.
5 The cross section of the meter's operation zone, S, and its transverse dimension, $L = \sqrt{S}$.
6 The longitudinal dimension of the meter's operation zone, P.

In order to perform analytical calculations, it is necessary to introduce a probing signal model. Investigations have shown that each type of transmitter (TWT or semiconductor-based) requires a specific signal model [1, 8].

For systems with shock-excited TWT, approximate calculations conveniently use an exponential model of the probing signal. The signal is assumed to have a Gaussian envelope and harmonic fill in

$$u(t) = U_0(t) \cos 2\pi f_0 t \tag{4.35}$$

where $U_0(t) = U \exp[-\eta_\alpha^2 (t/\tau_\alpha)^2]$ is the envelope and U is the signal amplitude. Here, τ_α is the pulse duration determined at the distance α from the envelope maximum, and $\eta_\alpha = 2\sqrt{(-\ln \alpha)}$ is a coefficient defined by the α value used.

From the pulse duration, τ_α, it is convenient to turn to the equivalent transmitter pulse duration independent of the choice of α

$$\tau_{eq} = \tau_\alpha / \eta_\alpha \tag{4.36}$$

The value of τ_{eq} approximately corresponds to the pulse duration on the envelope, τ_α, at the level of $\alpha = 0.78$.

Thus, the transmitter signal is described as

$$\begin{aligned} u(t) &= U_0(t) \cos 2\pi f_0 t \\ &= U\, \mathrm{e}^{-(t/\tau_{eq})^2} \cos 2\pi f_0 t \end{aligned} \tag{4.37}$$

If the transverse dimensions of the operation zone remain constant or vary slightly at all frequencies over the operation range $[f_{min}, f_{max}]$, the transmitter signal, when propagating through the collimator or reflected by the target, changes by a factor of ξ

$$\xi = \eta_A l_0 / L \tag{4.38}$$

where η_A is the amplitude efficiency of the collimator antenna. As a result, the useful component of the received signal has the form

$$y(t) = Y_0(t) \cos 2\pi f_0 t \tag{4.39}$$

where $Y_0(t) = \xi U\, \mathrm{e}^{-(t/\tau_{eq})^2}$ is the signal envelope at the receiver input.

When deriving this expression, we assumed that the time delay of the signal due to its propagation to the target and back is compensated (i.e., it is taken into account at the reception point).

In reality, the detected signal contains a noise component, $\varepsilon(t)$, defined by the clutter of the receiving device (high frequency amplifier, gating converter) recalculated for its input

$$y_\varepsilon(t) = y(t) + \varepsilon(t) \tag{4.40}$$

Since the signal processing for the estimation of UWB RTR (including the frequency dependence of RCS) is performed by digital methods, further treatment will be made for discrete pulses obtained from continuous signals by discretisation with a step T. The analogue of eqn. 4.40 can then be conveniently represented by the vector form

$$\mathbf{y}_\varepsilon = \mathbf{y} + \varepsilon \tag{4.41}$$

where

$$\begin{aligned} \mathbf{y}_\varepsilon &= [y_\varepsilon(1) \ldots y_\varepsilon(N)]^T \\ \mathbf{y} &= [y(1) \ldots y(N)]^T \\ \varepsilon &= [\varepsilon(1) \ldots \varepsilon(N)]^T \end{aligned}$$

Here, N is the array dimensionality defined by the discretisation step, T, and the longitudinal dimension of the operation zone, P (see below).

The target FR and RCS are estimated for various discrete frequencies $f_k = (k-1)/(TN)$ with a frequency step $\Delta F = 1/(TN)$ using the discrete Fourier transform (DFT) of the original (time) arrays to obtain data arrays in the frequency domain

$$\mathbf{Y}_\varepsilon = \mathbf{F}\mathbf{y}_\varepsilon$$
$$\mathbf{Y} = \mathbf{F}\mathbf{y} \tag{4.42}$$
$$\mathbf{E} = \mathbf{F}\varepsilon$$

where $\mathbf{F}$ is the direct DFT matrix (see Section 5.4). The elements $Y_\varepsilon(k)$, $Y(k)$, $E(k)$, $k = \overline{1,N}$ of the vectors $\mathbf{Y}_\varepsilon$, $\mathbf{Y}$ and $\mathbf{E}$ are generally complex.

It will be shown in Chapter 5 that the relative mean square error of the target RCS estimation at the kth frequency, f_k, is written as

$$\gamma_\sigma(k) \leqslant 2/\sqrt{q_0^2(k)} \tag{4.43}$$

where $q_0^2(k) = |S(k)|^2/\sigma_E^2(k)$ is the power signal to noise ratio. Here, $\sigma_E^2(k) = M\{E(k)E^*(k)\}$ is the dispersion of the elements $E(k)$ of the vector $\mathbf{E}$, M is the sign of the meaning.

Let us find the value of $\sigma_E^2(k)$. We assume the elements $\varepsilon(n)$ of the vector ε of the receiver noise time counts to be uncorrelated and to have the zero expectation and known dispersion σ_ε^2. Using eqn. 4.42 the dispersion matrix of the vector $\mathbf{E}$ can be found by

$$\begin{aligned} M\{\mathbf{E}\mathbf{E}^{T*}\} &= M\{\mathbf{F}\varepsilon(\mathbf{F}\varepsilon)^{T*}\} \\ &= M\{\mathbf{F}\varepsilon\varepsilon^T\mathbf{F}^{T*}\} \\ &= \mathbf{F}M\{\varepsilon\varepsilon^T\}\mathbf{F}^{T*} \\ &= F\sigma_\varepsilon^2\mathbf{I}\mathbf{F}^{T*} \\ &= \sigma_\varepsilon^2 N\mathbf{I} \end{aligned} \tag{4.44}$$

The derivation of this expression has been based on the well-known property of the DFT: $\mathbf{F}\mathbf{F}^{T*} = N\mathbf{I}$. It follows from eqn. 4.44 that, under the above conditions, the counts of the vector $\mathbf{E}$ are uncorrelated and of equal accuracy, their dispersions being defined by the time count dispersions and the DFT dimensionality

$$\sigma_E^2 = N\sigma_\varepsilon^2 \tag{4.45}$$

To find the values of $|Y(k)|^2$ from expr. 4.43, we shall make use of the known relation between the DFT values of $Y(k)$ – the discrete signal $\mathbf{y}$ and the spectral function $Y(\mathrm{i}f)$ of the analogue signal $y(t)$ at discrete frequencies f_k

$$Y(k) = \frac{1}{T}Y(\mathrm{i}f_k)\Big|_{f_k=(k-1)/(NT)}, \qquad k = \overline{1,N} \tag{4.46}$$

From eqns 4.45 and 4.46 we have

$$q_0^2(k) = \frac{Y^2(f_k)\big|_{f_k=(k-1)/(NT)}}{T^2N\sigma_\varepsilon^2}, \qquad k = \overline{1,N} \tag{4.47}$$

To provide the given RCS measurement accuracy, the condition

$$q_0^2(k) \geqslant 4/\gamma_\sigma^2 \tag{4.48}$$

must be met at all frequencies $f_k[f_{min}, f_{max}]$.

The calculation of $Y(k)$ in the accepted signal model can be simplified using the complex envelope method. For this, the spectral function, $Y(\mathrm{i}f)$, of the amplitude-modulated oscillation, $y(t)$, can be found from the spectral function, $Y_0(t)$

$$Y(\mathrm{i}f) = 1/2[Y_0(\mathrm{i}(f - f_0)) + Y_0(\mathrm{i}(f + f_0))] \tag{4.49}$$

If the median frequency f_0 of the probing signal is chosen at the centre of the frequency range $[f_{min}, f_{max}]$, i.e., $f_0 = (f_{min} + f_{max})/2$, then the minimum values of the signal to noise ratio will lie at the exponential envelope boundaries of the probing signal

$$\begin{aligned} q_{min}^2 &= \frac{Y^2(f_{min})}{T^2 N \sigma_\varepsilon^2} \\ &= \frac{Y^2(f_{max})}{T^2 N \sigma_\varepsilon^2} \\ &= \frac{Y_0^2(\Delta f/2)}{4T^2 N \sigma_\varepsilon^2} \end{aligned} \tag{4.50}$$

where $\Delta f = f_{max} - f_{min}$ is the frequency range of the RCS measurements.

Within the accepted approximation, the spectral function of the received signal envelope has the form

$$\begin{aligned} Y_0(\mathrm{i}f) &= Y_0(f) \\ &= \xi U \sqrt{\pi}\, \tau_{eq}\, \mathrm{e}^{-\pi^2 \tau_{eq}^2 f^2} \end{aligned} \tag{4.51}$$

hence

$$q_{min}^2 = \xi^2 \frac{U^2 \pi \tau_{eq}^2}{4T^2 N \sigma_\varepsilon^2}\, \mathrm{e}^{-\pi^2 \tau_{eq}^2 f^2/2} \tag{4.52}$$

We should like to note, in passing, that if the spectral bandwidth of the signal envelope, ΔF, is determined at the level α from the maximum, i.e., at the same level as for the pulse duration τ, the values of τ_α and ΔF_α can easily be shown to be related as

$$\tau_\alpha \Delta F_\alpha = \eta_\alpha^2/2\pi \tag{4.53}$$

while the equivalent duration $\tau_{eq} = \tau_\alpha/\eta_\alpha$ and the bandwidth $\Delta F_{eq} = \Delta F_\alpha/\eta_\alpha$, independent of the arbitrary choice of α, are related as

$$\tau_{eq} \Delta F_{eq} = 1/2\pi \tag{4.54}$$

When passing from cyclic to circular frequencies, this relation becomes especially simple

$$\tau_{eq}\Delta\Omega_{eq} = 1 \tag{4.55}$$

where $\Delta\Omega_{eq} = 2\pi\Delta F_{eq}$.

The analysis of eqn. 4.52 shows that there is an optimal value of the probing signal duration, $\tau_{eq\ opt}$, which provides a peak of the minimal signal to noise ratio in the given frequency range

$$\tau_{eq\ opt} = \frac{\sqrt{2}}{\pi\Delta f} \tag{4.56}$$

$$(q^2_{min})_{max} = \frac{\xi^2 U^2}{2\pi\Delta f^2 T^2 \sigma_\varepsilon^2 Ne} \tag{4.57}$$

Note that at $\tau_{eq} = \tau_{eq\ opt}$, the signal energy spectrum (the squared module of the signal spectral function) at the frequency range limits has the value e^{-1} from the maximum at the centre frequency, f_0, while the spectral function at these limits has the value $1/\sqrt{e}$ from the maximum.

Eqns. 4.57 and 4.48 permit the calculation of the transmitter pulse amplitude, at which (in case of the optimal pulse duration) the given minimum RCS will be estimated in the frequency range required, with a relative error less than the given error

$$U \geqslant \frac{2\sqrt{2\pi Ne}\,\Delta f N \sigma_\varepsilon L}{\eta_A l_0 \gamma_\sigma} \tag{4.58}$$

We shall discuss in more detail the problem of optimal probing signal duration. This is necessary because the use of the parameter τ_{eq} (or τ) is, in practice, inconvenient although in principle possible with the Hilbert transform. It is much more convenient to use an equivalent but physically more meaningful parameter – the relative level of the first time sidelobe, β. For comparatively narrowband signals ($\mu_{eq} = \Delta F_{eq}/f_0 \leqslant 0.5$–$0.9$), the relation between the parameters τ_{eq} and β can be found by taking advantage of the fact that the peak (in modulus) of the first side lobe, U_1, corresponds to the instant $t_0 \approx 1/(2f_0)$. Hence, taking into account eqn. 4.37, we have

$$\begin{aligned} U_1 &= U\,\mathrm{e}^{-t_0^2/\tau_{eq}^2} \\ &= U\,\mathrm{e}^{-1/(4f_0^2\tau_{eq}^2)} \end{aligned} \tag{4.59}$$

Its relative value $\beta = U_1/U$ at the given medium frequency, f_0, is defined by the pulse duration, τ_{eq}

$$\beta = \mathrm{e}^{-1/(4f_0^2\tau_{eq}^2)} \tag{4.60}$$

With eqn. 4.55, we can write

$$\beta = \mathrm{e}^{-\pi^2\mu_{eq}^2} \tag{4.61}$$

or

$$\mu_{eq} = \pi\sqrt{(-\ln\beta)} \tag{4.62}$$

Thus, the relative level of the first sidelobe, β, is related, in a one to one manner, to the equivalent relative signal bandwidth, μ_e.

With eqns. 4.56 and 4.60, it is easy to find the optimal value, β_{opt}, of the relative level of the first sidelobe

$$\beta_{opt} = e^{-\pi^2/8\mu^2} \tag{4.63}$$

where $\mu = \Delta f/f_0$ is the relative frequency band, for which the normalised measurement accuracy must be provided.

For wider bandwidth signals ($\mu_{eq} \geqslant 0.9$), this expression gives underestimates, because in that case the sidelobe top coordinate, t_0, differs a great deal from the value $1/(2f_0)$. Analysis of eqn. 4.37 has shown that the value of t_0 must satisfy the equation

$$\frac{1}{\tau_{eq}^2\pi f_0}t + \tan 2\pi f_0 t = 0 \tag{4.64}$$

Keeping in mind that the root of eqn. 4.64, having an infinite set of solutions, lies within the interval $(1/(4f_0), 1/(2f_0))$, one can pass over to the equivalent equation

$$t' + 1/2\mu_{eq}^2\pi + \tan t' = 0 \tag{4.65}$$

where $t' = 2\pi f_0 t - \pi$, $t' \in (\pi/2, 0)$.

Since it is difficult to find its exact analytical solution, we shall use the following procedure: the function $\tan t'$ will be approximated by a quadratic function, and then we shall seek the analytical solution to the equation obtained

$$t' + 1/2\mu_{eq}^2\pi - (a(t')^2 + bt' + c) = 0 \tag{4.66}$$

The coefficients a, b and c are calculated by a least squares method to yield: $a = 4.30$; $b = -2.70$; $c = 0.393$. After a number of transformations, we can obtain the coordinate for the first time sidelobe top

$$t_0 - (1.15 - 0.19\mu_{eq}^2)/(2f_0) \tag{4.67}$$

which is valid at $\mu_e \gtrsim 0.9$. Hence

$$\begin{aligned}\beta &= e^{-t_0^2/\tau_{eq}^2}\\ &= e^{-(1.15-0.19\mu_{eq}^2)^2\pi^2\mu_{eq}^2}\end{aligned} \tag{4.68}$$

With eqn. 4.54, we obtain the optimal value, β_{opt}, of the relative level of the first signal sidelobe for large values of the bandwidth factor

$$\beta_{opt} = e^{-1.23(1.15-0.024\mu^2)^2\mu^2} \tag{4.69}$$

where $0.9 \leqslant \mu < 1.5$.

4.4.2 Impulse meters with a semiconductor transmitter

Modern semiconductor integrated circuits provide generation of bipolar single pulses close in shape to one period of a sinewave oscillation

$$u(t) = \begin{cases} -U \sin 2\pi f_0 t, & -T_0/2 \leqslant t \leqslant T_0/2 \\ 0, & \text{at other } t \end{cases} \tag{4.70}$$

where $T_0 = 1/f_0$ and f_0 is the carrier frequency.

Unfortunately this model (eqn. 4.70) is inconvenient for analytical treatment and should be replaced by more effective models based on Gaussian curves of the type $\exp(-\alpha t^4)$ (fourth-order Gaussian curve) and $\exp(-\alpha t^2)$ (second-order Gaussian curve). For example, a close approximation to the model (eqn. 4.70) can be obtained with the second-order Gaussian curves

$$\begin{aligned} y(t) &= \xi U[\mathrm{e}^{-(t/\tau_{eq}+1.4)^2} - \mathrm{e}^{-(t/\tau_{eq}-1.4)^2}] \\ &= \frac{\eta_A l U}{L}[\mathrm{e}^{-(t/\tau_{eq}+1.4)^2} - \mathrm{e}^{-(t/\tau_{eq}-1.4)^2}] \end{aligned} \tag{4.71}$$

Here, τ_{eq} is the effective duration of a unipolar pulse (see above). This example is related to the carrier frequency of a single pulse

$$f_0 = 0.179/\tau_{eq} \tag{4.72}$$

and to its duration

$$\tau_p = T_0 = 1/f_0 = 5.59\tau_{eq} \tag{4.73}$$

The energy spectrum of this signal is

$$\begin{aligned} Y^2(f) &= \xi^2 U^2 \pi \tau_{eq}^2 \exp(-2\pi^2 \tau_{eq}^2 f^2) \sin^2(8.8\tau_{eq} f) \\ &= \xi^2 U^2 \pi \tau_{eq}^2 \exp(-2\pi^2 \tau_{eq}^2 f^2)[1/2(1 - \cos(17.6\tau_{eq} f))] \end{aligned} \tag{4.74}$$

Let us use a more convenient approximation of the factor in square brackets to get an accuracy of the order of a few percent

$$1/2(1 - \cos(1.76\tau_{eq} f)) \approx \exp[-93.63(\tau_{eq} f - 0.179)^2] \tag{4.75}$$

Therefore

$$Y^2(f) = \xi^2 U^2 \pi \tau_{eq}^2 \exp(-2\pi^2 \tau_{eq}^2 f^2) \exp[-93.63(\tau_{eq} f - 0.179)^2] \tag{4.76}$$

The form of the function describing the energy spectrum (its width and displacement) is defined by one parameter, τ_{eq}, which, as follows from eqns. 4.72 and 4.76, determines the pulse centre frequency and duration.

It is advisable to choose τ_{eq} (i.e., f_0 and τ_p) such that the energy spectrum of the received signal would have the same values at the bounds of the frequency range required

$$\begin{aligned} Y^2(f_{min}) &= Y^2(f_{max}) \\ &= Y_0 \end{aligned} \tag{4.77}$$

This condition yields the equation

$$\exp(-2\pi^2\tau_{eq}^2 f_{min}^2)\exp[-93.63(\tau_{eq}f_{min} - 0.179)^2]$$
$$= \exp(-2\pi^2\tau_{eq}^2 f_{max}^2)\exp[-93.63(\tau_{eq}f_{max} - 0.179)^2] \quad (4.78)$$

Having taken the logarithm, we obtain the equation

$$[(2\pi^2 + 93.6)(f_{max}^2 - f_{min}^2)]\tau_{eq}^2 - 33.5(f_{max} - f_{min})\tau_{eq} = 0 \quad (4.79)$$

which has two roots

(i) $\tau_{eq1} = 0$ (trivial solution)

(ii) $\tau_{eq2} = 0.3/(f_{max} - f_{min})$ (4.80)

$= 0.15/f_{av}$ (optimal value of τ_{eq})

where $f_{av} = 1/2(f_{max} - f_{min})$ is the average frequency in the range $[f_{min}, f_{max}]$.

At the optimal value of $\tau_{eq} = \tau_{eq0}$, the energy spectrum at the frequency bounds is

$$Y_0^2 = 0.07\xi^2 U^2/f_{av}^2 \exp[-0.44(f_{min}/f_{av})^2]$$
$$* \exp[-2.1(f_{min}/f_{av} - 1.2)^2] \quad (4.81)$$

or

$$Y_0^2 = 0.07\xi^2 U^2/f_{av}^2 \exp[-1.76(1/(k_f + 1))]$$
$$* \exp[-8.4(1/(k_f + 1) - 0.6)^2] \quad (4.82)$$

where $k_f = f_{max}/f_{min}$ is the frequency overlap factor.

The minimum value of the signal to noise ratio at the frequency bounds must satisfy the condition

$$q_{min}^2 = \frac{Y_0^2}{T^2 N\sigma_\varepsilon^2} \geqslant 4/\gamma_\sigma^2 \quad (4.83)$$

It is easy to derive from this expression the condition for the required pulse amplitude

$$U \geqslant \frac{7.6T\sigma_\varepsilon f_{av}\sqrt{N}}{\gamma_\sigma\xi\exp[-0.22(f_{min}/f_{av})^2]\exp[-1.05(f_{min}/f_{av} - 1.2)^2]} \quad (4.84)$$

or

$$U \geqslant \frac{7.6T\sigma_\varepsilon f_{av}\sqrt{N}}{\gamma_\sigma\xi\exp\left[-0.88\dfrac{1}{(k_f+1)^2}\right]\exp\left[-4.2\left(\dfrac{1}{(k_f+1)} - 0.6\right)^2\right]} \quad (4.85)$$

4.5 Measurement equations and their incorrectness

Equations for UWB radar response measurement relate recorded signals to test responses or parameters of a model radar target. The form of the equations and solution methods varies with the type of signal, the way of response representation and the measuring equipment used (see Sections 4.3, 4.4). It has been shown [1, 57, 85] that measurement equations can be reduced to the first-kind linear operator equation

$$y = A\{x\} \tag{4.86}$$

where $y \in Y$ and $x \in X$ are the preset and desired elements of linearly normalised spaces, Y and X, and $A\{\ \}$ is a continuous restricted operator mapping x into y.

Since the initial data in practice contain an error, that is, one only knows the elements

$$\begin{aligned} y_\varepsilon &= y + \varepsilon \\ A_\Delta\{\ \} &= A\{\ \} + \Delta A\{\ \} \end{aligned} \tag{4.87}$$

this equation takes the form

$$y_\varepsilon = A_\Delta\{x\} \tag{4.88}$$

It will be shown below that for the estimation of total responses this operator equation can be represented as a vector-matrix linear equation

$$\mathbf{y}_\varepsilon = \mathbf{A}_\Delta \mathbf{x} \tag{4.89}$$

which is known as the linear regression equation in the statistical description of measurement errors. It is commonly written as

$$\mathbf{y}_\varepsilon = \mathbf{A}_\Delta \mathbf{x} + \varepsilon \tag{4.90}$$

It is important to note that the initial data error, ε, and the transformation operator error, $\Delta A\{\ \}$, cannot be found directly, so they are often said to be unobservable.

We shall give a few examples with reference to total RTR meters. For details the reader is referred to [1, 85].

Let us start with the estimation of impulse and frequency responses of radar targets. It follows from eqn. 4.7 that in the absence of polarisation distortions, which will be analysed in Chapter 5, the measurement equation for the impulse response is (for simplicity, the polarisation subscripts and angular coordinates will be omitted)

$$y_\varepsilon(t) = w(t) * h(t) + \varepsilon(t) \tag{4.91}$$

The real impulse response, $w(t)$, is known to have a certain error

$$w(t) = w(t) + \Delta w(t) \tag{4.92}$$

This leads to the operator perturbation in eqn. 4.91. Taking account of this fact, eqn. 4.91 takes, in the discrete case, the form of a linear matrix equation

$$\mathbf{y}_\varepsilon = \mathbf{A}_\Delta \mathbf{h} + \varepsilon \tag{4.93}$$

where

$$\mathbf{y}_\varepsilon = [y(0) \ldots y((k-1)T)]^T$$
$$\mathbf{h} = [h(0) \ldots h((k-1)T)]^T$$

and T is a discretisation step.

The matrix $\mathbf{A}_\Delta$ (analogue of the operator $A_\Delta\{\ \}$) is formed by counts of the radar system impulse response, $w_\Delta(t)$

$$\mathbf{A}_\Delta = \begin{bmatrix} Tw_\Delta(0) & 0 & 0 & \ldots & 0 \\ Tw_\Delta(T) & Tw_\Delta(0) & 0 & \ldots & 0 \\ \vdots & \vdots & \vdots & & \vdots \\ Tw_\Delta((k-1)T) & Tw_\Delta((k-2)T) & Tw_\Delta((k-3)T) & \ldots & 0 \\ 0 & Tw_\Delta((k-1)T) & Tw_\Delta((k-2)T) & \ldots & 0 \\ \vdots & \vdots & \vdots & & \vdots \\ 0 & 0 & 0 & \ldots & Tw_\Delta(0) \\ 0 & 0 & 0 & \ldots & Tw_\Delta((k-1)T) \end{bmatrix}$$

The meaning of eqns. 4.93 and 4.90 is the same.

The frequency response can be found from the estimated vector $\mathbf{h}$ using the discrete Fourier transform.

Now we shall turn to the estimation of coefficients for the scattering differential equation and of resonance frequencies. Suppose that the differential equation for a radar target has the form

$$y^{(n)} + \sum_{i=0}^{n-1} a_i y^{(i)}(t) = \sum_{i=0}^{m-1} b'_i x^{(i)}(t) \tag{4.94}$$

where $y(t)$ and $x(t)$ are a scattered and a probing signal, and a_i and b_i are the coefficients to be calculated.

From the differential equation (eqn. 4.94) we can pass to an equivalent difference equation relating the discrete counts of a scattered and an incident signal

$$y(k) + \sum_{i=1}^{N-1} a_i y(k-i) = \sum_{i=0}^{P-1} b_i x(k-i), \qquad k = \overline{1, N} \leqslant n+1 \tag{4.95}$$

This transition can be made in various ways, in particular, by the finite difference approximation of derivatives.

Suppose now that we have registered M counts of the input and output signals ($M \geqslant 2N$)

$$\{y(1), \ldots, y(M)\}, \qquad \{x(1), \ldots, x(M)\}$$

For the time being, we shall assume the measurements to contain no errors, and we put $P = N$. From eqn. 4.95, each kth count of the registered scattered signal, starting with the Nth count, must satisfy the equation

$$\begin{aligned} y(k) = &- a_1 y(k-1) - \ldots - a_N y(k-N+1) \\ &+ b_0 x(k) + \ldots + b_N x(k-N+1), \qquad k = \overline{N, M} \end{aligned} \tag{4.96}$$

The coefficients of the difference equation must satisfy the set of equations (eqn. 4.96) which can be written in the matrix form

$$\mathbf{y} = \mathbf{Ac} \tag{4.97}$$

where

$$\mathbf{y} = [y(N) \quad \ldots \quad y(M)]^T$$

$$\mathbf{c} = [a_1 \quad \ldots \quad a_{N-1} \quad b_0 \quad \ldots \quad b_{N-1}]^T$$

$$\mathbf{A} = \begin{bmatrix} -y(N-1) & -y(N-2) & \ldots & y(1) & x(N) & \ldots & x(1) \\ \vdots & \vdots & & \vdots & \vdots & & \vdots \\ -y(m-1) & -y(m-2) & \ldots & y(m-N+1) & x(M) & \ldots & x(M-N+1) \end{bmatrix}$$

is the $[(M-N+1) \times (2N-1)]$ matrix.

Since the input and output signals of a radar system actually contain the errors $y_\varepsilon(k) = y(k) + \varepsilon(k)$ and $x_\xi(k) = x(k) + \xi(k)$, $k = \overline{1, M}$, this leads to the perturbation of the vector $\mathbf{y}$ and the matrix $\mathbf{A}$

$$\begin{aligned} \mathbf{y}_\varepsilon &= \mathbf{y} + \varepsilon \\ \mathbf{A}_\Delta &= \mathbf{A} + \Delta\mathbf{A} \end{aligned} \tag{4.98}$$

where

$$\varepsilon = [\varepsilon(N) \quad \ldots \quad \varepsilon(M)]^T$$

$$\Delta\mathbf{A} = \begin{bmatrix} -\varepsilon(N-1) & \ldots & -\varepsilon(1) & \xi(N) & \ldots & \xi(0) \\ \vdots & & \vdots & \vdots & & \vdots \\ -\varepsilon(m-1) & \ldots & -\varepsilon(m-N+1) & \xi(M) & \ldots & \xi(m-N+1) \end{bmatrix}$$

Therefore, the problem of finding the difference equation coefficients reduces to the solving of a linear vector-matrix equation of the type of eqn. 4.90.

Analysing the free component of a scattered signal (see Chapter 3), we can formulate the special problem of finding the vector of the left-hand side coefficients in eqn. 4.95. Then eqn. 4.97 can be rewritten as

$$\mathbf{y} = \mathbf{A}'\mathbf{c}' \tag{4.99}$$

where

$$\mathbf{A}' = \begin{bmatrix} -y(N-1) & \dots & -y(1) \\ \vdots & & \vdots \\ -y(M-1) & \dots & -y(m-N+1) \end{bmatrix}$$

is the $[(M-N+1)\times(N-1)]$ matrix; $\mathbf{c}' = [a_1, \dots, a_{N-1}]$.

Measurement noise produces, as in the above case, perturbations of the vector $\mathbf{y}$ and the matrix $\mathbf{A}'$. As a result, the problem of partial identification reduces to a set of linear equations of the type of eqn. 4.89 or to a linear regression equation, like eqn. 4.90, for a static description of measurement noise.

Note that with the difference equation coefficients (eqn. 4.95), we can find the roots, z_k, of the equations

$$z^N + \sum a_k z^{-k} = 0 \tag{4.100}$$

$$\sum b_k z^{-k} = 0 \tag{4.101}$$

which represent the poles and zeros of the discrete transfer function of the target. Then, using the known ratio

$$s_k = \mathrm{Ln}\, z_k / T \tag{4.102}$$

relating the points of the z and s planes, we find the poles and zeros of this transfer function and from them the numerator and denominator polynomials of the transfer function, whose coefficients are the coefficients of eqn. 4.94.

Since neither the input, incident signal nor the scattered signal can be observed directly, the above equations describe the series connection of the pulse shaper, target and receiver. In order to calculate the differential equation coefficients for the target proper, it is expedient to make use of the following procedure. The signal shaper and receiver poles, predetermined by a calibration scatterer with known zeros and poles, are cancelled from the found zeros and poles of the continuous or discrete transfer function. The remaining ones are used to find the desired coefficients.

After the natural target frequencies have been determined, the expansion coefficients (eqns. 3.47 or 3.49) can be found. For this, the calculated values of the resonance frequencies, s_m or z_m $(m = 1, M)$, are substituted into the discrete analogue of eqn. 3.49

$$\begin{aligned} y(k) &= \sum B_m \exp(s_m Th) \\ &= \sum B_m z_m^k, \qquad k = \overline{1, M} \end{aligned} \tag{4.103}$$

to get a set of equations

$$\begin{bmatrix} 1 & 1 & \dots & 1 \\ z_1 & z_2 & \dots & z_N \\ \vdots & \vdots & & \vdots \\ z_1^{m-1} & z_2^{m-1} & \dots & z_N^{m-1} \end{bmatrix} \begin{bmatrix} B_1 \\ B_2 \\ \vdots \\ B_N \end{bmatrix} = \begin{bmatrix} y(1) \\ y(z) \\ \vdots \\ y(M) \end{bmatrix} \tag{4.104}$$

or

$$\mathbf{Za} = \mathbf{h} \tag{4.105}$$

The above procedures for defining resonance frequencies and residues of the rational fractional TF of the target by solving two sets of linear algebraic equations (eqns. 4.97 and 4.105) together with the evaluation of the characteristic equation roots (eqn. 4.100) are the essence of the well known method by Prony.

In fact, resonance frequencies and other parameters within the resonance scattering model can be evaluated not only by the indirect method of finding the differential (difference) equation coefficients but also by other methods exhaustively discussed in [30, 31, 57, 85, 87, 88].

Let us examine the equation for evaluation of glittering point parameters. Suppose that the forced component of the target impulse response is described by the expression

$$h(t) = \sum_{l=1}^{L} a_l \delta(t - T_l) \tag{4.106}$$

which corresponds to the frequency response

$$H(\mathrm{i}\omega) = \sum_{l=1}^{L} a_l \exp(-\mathrm{i}\omega T_l) \tag{4.107}$$

Consider the frequency response with a ($\Omega = \omega_{k+1} - \omega_k$) discretisation step

$$\begin{aligned} H(k) &= H(\mathrm{i}k\Omega) \\ &= \sum_{l=1}^{L} a_l \exp(-\mathrm{i}k\Omega T_l), \qquad k = \overline{0, M-1} \end{aligned} \tag{4.108}$$

By making a Z transformation of the function $H(k)$, we can find its Z representation

$$\begin{aligned} H(z) &= \sum a_l \frac{z}{z - \lambda_l} \\ &= \frac{d_L z^L + d_{L-1} z^{L-1} + \ldots + d_0}{z^L + c_{L-1} z^{L-1} + \ldots + c_0} \end{aligned} \tag{4.109}$$

where $\lambda_l = \exp(-\mathrm{i}\Omega T_l)$ and d_j and c_j are coefficients related to the values of λ_l and a_l ($l = \overline{1, L}$).

The latter expression shows that the discretised frequency response adequately representable by a set of glittering points satisfies the Lth order homogeneous difference equation for the complex-valued variable

$$H(k+L) = -c_{L-1}H(k+L-1) - c_{L-2}H(k+L-2) - \ldots - c_0 H(k), \qquad k \gg 0 \tag{4.110}$$

If we have M counts of the target frequency response ($m \gg L$), we can use this equation to construct a set of linear algebraic equations for unknown coefficients c_i $(i = \overline{0, L-1})$ of the difference equation

$$\mathbf{h} = \mathbf{A}\mathbf{c} \tag{4.111}$$

where

$$\mathbf{h} = [H(L) \quad \ldots \quad H(L+N-1)]^T$$

$$\mathbf{c} = [c_0 \quad \ldots \quad c_{L-1}]^T$$

$$\mathbf{A} = \begin{bmatrix} -H(0) & -H(1) & \ldots & -H(L-1) \\ \vdots & \vdots & & \vdots \\ -H(M-1) & -H(M) & \ldots & -H(L+M-2) \end{bmatrix}$$

Since the frequency response contains an error $(H_\varepsilon(k) = H(k) + \varepsilon(k))$, both the vector, $\mathbf{h}$, and the matrix, $\mathbf{A}$, are perturbed

$$\begin{aligned} \mathbf{h}_\varepsilon &= \mathbf{h} + \varepsilon \\ \mathbf{A}_\Delta &= \mathbf{A} + \Delta\mathbf{A} \end{aligned} \tag{4.112}$$

where

$$\varepsilon = [\varepsilon(L) \quad \ldots \quad \varepsilon(L+M-1)]^T$$

$$\Delta\mathbf{A} = \begin{bmatrix} -\varepsilon(0) & \ldots & -\varepsilon(L-1) \\ \vdots & & \vdots \\ -\varepsilon(m-1) & \ldots & -\varepsilon(L+M-2) \end{bmatrix}$$

With the known coefficients c_i $(i = \overline{0, L-1})$, we can find the roots, α_l, of the polynome

$$z_L + c_{L-1} z^{L-1} + \ldots + c_0 = 0 \tag{4.113}$$

which help to define the parameters T_l

$$T_l = -\frac{1}{\Omega} \arg\{\alpha_l\} \equiv -\frac{1}{\Omega} \arctan(\alpha_{Il}/\alpha_{Rl}) \tag{4.114}$$

where α_{Il} and α_{Rl} are the imaginary and real parts of the root α_l (note that $|\alpha_l| = 1$).

The next step is to find the intensity of glittering points α_l $(l = \overline{1, L})$ with known T_l. For this we solve a set of linear algebraic equations

$$\mathbf{h} = \mathbf{Z}\mathbf{a} \tag{4.115}$$

where

$$\mathbf{a} = [a_1 \quad \ldots \quad a_L]^T$$

$$\mathbf{Z} = \begin{bmatrix} z_1^0 & z_1^0 & \ldots & z_L^0 \\ z_1^1 & z_1^1 & \ldots & z_L^1 \\ \vdots & \vdots & & \vdots \\ z_1^{N-1} & z_1^{N-1} & \ldots & z_L^{N-1} \end{bmatrix}$$

Here, $z_l = \exp(\mathrm{i}\Omega T_l)$ $(l = \overline{1, L})$. One should keep in mind that the left-hand side of eqn. 4.115 contains observation errors.

The two sets of equations (eqns. 4.111 and 4.115) contain complex matrices and vectors, while eqn. 4.113 has complex-valued coefficients, which may create certain computational difficulties. It is possible to pass to real expressions by using the real part of the frequency response [85]

$$H_R(\omega) = R_l\{H(\mathrm{i}\omega)\}$$

$$= \sum_{l=1}^{L} a_l \cos \omega T_l$$

A specific feature of inverse problems for the estimation of UWB target responses from experimental signals is that they are ill posed mathematically. A mathematical problem is taken to be well posed if the following conditions are met [102, 103, 104]:

(i) the problem is solvable;
(ii) the solution is unique;
(iii) the solution is resistant to initial data errors, that is, the errors in prescribing the operator and in making measurements do not lead to serious solution errors.

If any of these conditions cannot be met, the problem is believed to be ill posed.

That eqn. 4.86 is ill posed is obvious because the operator $A^{-1}\{\ \}$ is inverse to the direct operator $A\{\ \}$. Sometimes, the generalised inverse operator $A^{-1}\{\ \}$ does exist, but it is not continuous, which violates condition (iii). Besides, the operator $A^{-1}\{\ \}$ often does not exist, violating condition (i), or it cannot provide a unique solution, violating condition (ii).

Problems associated with the first-kind integral equation being ill posed are discussed, for example, in the book [102]. We shall analyse in some detail the problems formulated by this type of equation (eqns. 4.89 and 4.90).

The existence of errors, $\Delta\mathbf{A}$ and ε, results in the absence of exact solutions to these equations. In other words, there is no vector **xX**, which could be substituted into, say, eqn. 4.89 to turn it to an identity. This means the violation of condition (i). The solution residue vector is practically always nonzero

$$\mathbf{y}_\varepsilon - \mathbf{A}_\Delta \mathbf{x} \neq 0 \tag{4.116}$$

For this reason, the necessity for searching for an approximate solution, $\mathbf{x}^*$, becomes inevitable.

In some cases the $\mathbf{A}_\Delta$ matrix may be of partial rank, leading to nonunique solutions of eqns. 4.89 and 4.90. This violates condition (ii). Quite often, the $\mathbf{A}_\Delta$ matrix turns out to be poorly conditioned. Then small errors in the initial data lead to substantial errors in the approximate solution, $\mathbf{x}^*$, obtained by the inverse operator $\mathbf{A}_\Delta^-$: $\mathbf{x}^* = \mathbf{A}_\Delta^- \mathbf{y}_\varepsilon$. This results in the violation of condition (iii). A measure of error in the determinate case may be, for example, the vector norm of the error $||\mathbf{x}^* - \mathbf{x}_T||$, where $\mathbf{x}^T$ is the true (exact) value of vector $\mathbf{x}$. In a statistical formulation these may be parameters of the dispersion matrix of the error vector.

To illustrate the necessity for finding nontrivial solution methods for equations of the type indicated, we shall consider the classical solution with a general inverse matrix. Assume for definiteness that the dimension of vector $\mathbf{y}$ is larger than that of $\mathbf{x}$ ($n > m$).

In the deterministic case we can pass from eqn. 4.89 to the following equation (so called set of normal equations)

$$\mathbf{A}_\Delta^T \mathbf{A}_\Delta \mathbf{x} = \mathbf{A}_\Delta^T \mathbf{y}_\varepsilon \tag{4.117}$$

In contrast to eqn. 4.89, this equation does have a solution, that is, there is a vector $\mathbf{x}^*$ such that

$$\mathbf{A}_\Delta^T \mathbf{A}_\Delta \mathbf{x}^* = \mathbf{A}_\Delta^T \mathbf{y}_\varepsilon \tag{4.118}$$

Since the $\mathbf{A}^T\mathbf{A}$ matrix is a square matrix and is assumed to be singular, the solution of eqn. 4.118 is given in the form

$$\mathbf{x}^* = (\mathbf{A}_\Delta^T \mathbf{A}_\Delta)^{-1} \mathbf{A}_\Delta^T \mathbf{y}_\varepsilon \tag{4.119}$$

It is known that eqn. 4.119 provides the square minimum of the residue norm [102]

$$\begin{aligned} Q &= ||\mathbf{y}_\varepsilon - \mathbf{A}_\Delta \mathbf{x}||^2 \\ &= (\mathbf{y}_\varepsilon - \mathbf{A}_\Delta \mathbf{x})^T (\mathbf{y}_\varepsilon - \mathbf{A}_\Delta \mathbf{x}) \end{aligned} \tag{4.120}$$

So this solution is often referred to as the least squares solution to eqn. 4.89 and has the form

$$\hat{\mathbf{x}} = (\mathbf{A}_\Delta^T \mathbf{A}_\Delta)^{-1} \mathbf{A}_\Delta^T \mathbf{y}_\varepsilon \tag{4.121}$$

The matrix $\mathbf{A}_\Delta^- = (\mathbf{A}_\Delta^T \mathbf{A}_\Delta)^{-1} \mathbf{A}_\Delta^T \mathbf{y}_\varepsilon$ is a general inverse with respect to the $\mathbf{A}$ matrix. Let us evaluate the solution accuracy. For this, we shall accept some criteria for the initial data accuracy, using the maximum norms for the error vector $||\varepsilon||_{max}$ and the perturbation matrix $||\Delta\mathbf{A}||_{max}$. It can be shown that the error norm of the solution to eqn. 4.121 will satisfy the following inequality [62]

$$\frac{||\mathbf{x}_T - \hat{\mathbf{x}}||}{||\mathbf{x}_T||} \leqslant \frac{\text{cond}(\mathbf{A}_\Delta^T \mathbf{A}_\Delta)}{1 - \text{cond}(\mathbf{A}_\Delta^T \mathbf{A}_\Delta)||\Delta\mathbf{A}||_{max}/||\mathbf{A}||_{max}} \left(\frac{||\Delta\mathbf{A}||_{max}}{||\mathbf{A}||} + \frac{||\varepsilon||_{max}}{||\mathbf{y}||} \right) \tag{4.122}$$

where $\text{cond}(\mathbf{A}_\Delta^T \mathbf{A}_\Delta)$ is the condition number of the matrix $\mathbf{A}_\Delta^T \mathbf{A}_\Delta$, which is the ratio of the maximal to the minimal eigenvalues

$$\begin{aligned} \text{cond}(\mathbf{A}_\Delta^T \mathbf{A}_\Delta) &= ||\mathbf{A}_\Delta^T \mathbf{A}_\Delta|| / ||(\mathbf{A}_\Delta^T \mathbf{A}_\Delta)^{-1}|| \\ &= \mu_{max}/\mu_{min} \geqslant 1 \end{aligned} \tag{4.123}$$

In deriving expr. 4.122, we put

$$\mathrm{cond}(\mathbf{A}_\Delta^T\mathbf{A}_\Delta)||\Delta\mathbf{A}||_{max}/||\mathbf{A}||_{max} < 1 \tag{4.124}$$

Hence, the maximum possible relative error of the solution to eqn. 4.119, $\gamma_{x\ max} = (||\mathbf{x}_T - \hat{\mathbf{x}}||/||\mathbf{x}_T||)$ can be found in the following way

$$\gamma_{x\ max} \approx \frac{\mathrm{cond}(\mathbf{A}_\Delta^T\mathbf{A}_\Delta)(\gamma_{A\ max} + \gamma_{y\ max})}{1 - \mathrm{cond}(\mathbf{A}_\Delta^T\mathbf{A}_\Delta)\gamma_{A\ max})} \tag{4.125}$$

where $\gamma_{A\ max}$ and $\gamma_{y\ max}$ are the maximal relative errors of the given initial data.

Clearly, if the matrix $(\mathbf{A}_\Delta^T\mathbf{A}_\Delta)$ is ill conditioned, the relative error of the solution derived by the generalised inverse operator may be excessively large, violating the stability criterion.

We shall now demonstrate the incorrectness of the linear regression equation (eqn. 4.90). This issue will be discussed with reference to the equation with a known matrix. The case with a perturbed matrix can be demonstrated in a similar way. It will be assumed that the mathematical error expectation is equal to zero, and that its elements are uncorrelated and have equal dispersions

$$\mathbf{M}\{\varepsilon\} = \mathbf{0}$$

$$\mathbf{D}\{\varepsilon\} = \mathbf{M}\{\varepsilon\varepsilon^T\} = \sigma_\varepsilon^2\mathbf{I} \tag{4.126}$$

where $\mathbf{I}$ is a unit matrix.

The classical least squares approximation to eqn. 4.90 also uses a generalised operator [22, 64]

$$\hat{\mathbf{x}} = (\mathbf{A}_\Delta^T\mathbf{A}_\Delta)^{-1}\mathbf{A}_\Delta^T\mathbf{y}_\varepsilon \tag{4.127}$$

In the framework of the Gauss–Markov theorem, this estimation of the vectors of linear regression parameters is unbiased and efficient for the class of unbiased estimates linear in $\mathbf{y}$ [22]. When the vector ε has a Gaussian distribution, the estimation (eqn. 4.127) is efficient in the class of all unbiased estimates, both linear and nonlinear, and coincides with the maximum likelihood estimate [22].

Let us analyse the stability of the generalised inverse solution. The estimate error, $\hat{\mathbf{x}}$, will be characterised by the average sum of the squared estimate errors for the vector elements

$$L(\hat{\mathbf{x}}) = \mathbf{M}\{(\hat{\mathbf{x}} - \mathbf{x}_T)^T(\hat{\mathbf{x}} - \mathbf{x}_T)\} \tag{4.128}$$

This quantity can also be defined as the sum of diagonal elements of the correlation rating matrix: $L(\hat{\mathbf{x}}) = \mathrm{tr}\,\mathbf{D}\{\hat{\mathbf{x}}\}$, where

$$\mathbf{D}\{\hat{\mathbf{x}}\} = \mathbf{M}\{(\hat{\mathbf{x}} - \mathbf{x}_T)^T(\hat{\mathbf{x}} - \mathbf{x}_T)\}.$$

Because $(\hat{\mathbf{x}} - \mathbf{x}_T) = (\mathbf{A}_\Delta^T\mathbf{A}_\Delta)^{-1}\mathbf{A}_\Delta^T\mathbf{y}_\varepsilon$

$$\mathbf{D}\{\hat{\mathbf{x}}\} = \sigma_\varepsilon^2(\mathbf{A}_\Delta^T\mathbf{A}_\Delta)^{-1} \tag{4.129}$$

Since

$$\mathrm{tr}(\mathbf{A}_\Delta^T\mathbf{A}_\Delta)^{-1} = \sum_{i=0}^{m} \frac{1}{\lambda_i}$$

where λ_i are the eigenvalues of the matrix $(\mathbf{A}_\Delta^T\mathbf{A}_\Delta)$

$$L(\hat{\mathbf{x}}) = \sigma_\varepsilon^2 \sum_{i=0}^{m} \frac{1}{\lambda_i} \tag{4.130}$$

When the matrix $(\mathbf{A}_\Delta^T\mathbf{A}_\Delta)$ is ill conditioned (which happens quite often in practice), the minimum eigenvalue tends to zero. Therefore, the sum of $\hat{\mathbf{x}}$ estimation dispersions tends to infinity, indicating the unstable character of the classical generalised solution (eqn. 4.127). Note that the correctness criterion (iii) may be violated even if the matrix $(\mathbf{A}^T\mathbf{A})$ is well conditioned, when the approximate operator $\mathbf{A}_\Delta$ is significantly different from the exact one.

4.6 Solution methods for measurement equations

We have shown that the initial system of linear algebraic equations (SLAE), like eqn. 4.89, is ill conditioned. The conditioning of normal equations, like eqn. 4.117, has been shown to be still poorer [85]. This makes most of the popular SLAE methods inapplicable to measurement equations for UWB radar target responses.

There are only two exceptions – the singularity expansion method (SEM) and the regularisation method to be discussed below. These can serve as a good basis for constructing highly efficient signal processing algorithms well adapted to ill-conditioned matrices and large initial data errors. We shall first discuss the SEM [64].

This method is based on a factorisation of the system of linear algebraic equations matrix which permits the transition to an equivalent set of equations with a diagonal matrix. This should be a set of n equations with one unknown in each equation. Let us introduce the basic concepts and definitions [85].

Suppose the $\mathbf{A}$ is an arbitrary $n \times m$ matrix. We have to examine the square matrices $\mathbf{A}^T\mathbf{A}$ and $\mathbf{A}\mathbf{A}^T$ of orders m and n, respectively. It is possible to show that the zero eigenvalues of the matrices, λ_i, coincide ($\lambda_i > 0$, $i = 1, r$, where r is the rank of the matrices $\mathbf{A}$, $\mathbf{A}^T\mathbf{A}$, $\mathbf{A}\mathbf{A}^T$). The arithmetic values of the square roots of the common eigenvalues of the $\mathbf{A}^T\mathbf{A}$ and $\mathbf{A}\mathbf{A}^T$ matrices, $\mu_i = \sqrt{\lambda_i}$, are referred to as singular (principal) numbers of the $\mathbf{A}$ matrix. Suppose, also, that the first rs of the singular number have been indexed and arranged in a nonascending order: $\mu_1 \geqslant \mu_2 \geqslant \ldots \geqslant \mu_r > 0$. The other singular numbers are taken to be equal to zero: $\mu_{r+1} = \ldots = 0$.

Singular numbers are important characteristics of a matrix and possess the following properties:

(i) singular numbers do not change when a matrix is multiplied by an orthogonal matrix on the right or on the left;
(ii) a square matrix is singular when, and only when, its singular numbers are nonzero;
(iii) the determinant modulus of the **A** matrix is equal to the product of all singular numbers.

The following theorem forms the basis for the system of linear algebraic equations solution method for the case in question [64]. Any $n \times m$ **A** matrix can be decomposed as

$$\mathbf{A} = \mathbf{U}\mathbf{D}\mathbf{V}^T \tag{4.131}$$

where **U** and **V** are orthogonal matrices of order n and m, respectively, and **D** is a rectangular, diagonal $n \times m$ matrix with nonascending elements. The decomposition (eqn. 4.131) is called singular decomposition of the **A** matrix. For a real **A** matrix, singular decomposition is real, that is, the **U**, **V** and **D** matrices have real elements.

Matrices entering the singular decomposition (eqn. 4.131) have the form

$$\mathbf{D} = \mathrm{diag}(\mu_1, \ldots, \mu_r, 0, \ldots, 0) \tag{4.132}$$

$$\begin{aligned} \mathbf{U} &= [\mathbf{u}_i], \qquad i = 1, n \\ \mathbf{V} &= [\mathbf{v}_i], \qquad i = 1, m \end{aligned} \tag{4.133}$$

Here, $\mathbf{u}_i$ and $\mathbf{v}_i$ are the eigenvectors of the matrices $\mathbf{A}^T\mathbf{A}$ and $\mathbf{A}\mathbf{A}^T$, respectively.

The singular decomposition of the matrix of eqn. 4.89 allows the transition to the equivalent set, if the **A** matrix is square and nonsingular (the indices with vectors and matrices will be, for the time being, omitted)

$$\mathbf{D}\mathbf{V}^T\mathbf{x} = \mathbf{U}^T\mathbf{y} \equiv \mathbf{y} \tag{4.134}$$

Hence, the solution of the initial SLAE can be reduced to a consecutive solution of two sets of equations

$$\mathbf{D}\mathbf{z} = \mathbf{y}' \tag{4.135}$$

$$\mathbf{V}^T\mathbf{x} = \mathbf{z} \tag{4.136}$$

The first of these equations has a diagonal matrix, and its solution is obvious. In solving other sets of equations, one should keep in mind that $(\mathbf{V}^T)^{-1} = \mathbf{V}$. The solution of eqn. 4.89 can eventually be represented as

$$\hat{\mathbf{x}} = \mathbf{V}\mathbf{D}^{-1}\mathbf{U}^T\mathbf{y} = \sum_{i=1}^{r} \mu_i \mathbf{V}_i \mathbf{U}_i^T \mathbf{y} \tag{4.137}$$

or in the recursive form

$$\hat{\mathbf{x}}^{(i)} = \hat{\mathbf{x}}^{(i-1)} + \mu_i \mathbf{V}_i \mathbf{U}_i^T \mathbf{y}, \qquad i = \overline{1, r}$$
$$\mathbf{x}^{(0)} = 0 \tag{4.138}$$

The singular decomposition of the SLAE matrix can be performed by the well-known procedures for determining the eigenvalues and eigenvectors of the matrices $\mathbf{A}^T\mathbf{A}$ and $\mathbf{A}\mathbf{A}^T$ [64].

The classical version of the singular expansion method has no advantages of computational stability over QR decomposition. However, it is a good basis for effective inversion.

We should like to give a few definitions. The normal pseudosolution of the SLAE ($\mathbf{Ax} = \mathbf{y}$) is the vector $\mathbf{x}_N$ providing the minimum residue norm $\|\mathbf{y} - \mathbf{Ax}\|$ and having the minimum norm $\|\mathbf{x}\|$ [64]. Clearly, if the matrix $\mathbf{A}^T\mathbf{A}$ is nonsingular, the normal pseudosolution coincides with the generalised least squares solution of the SLAE. A singular $\mathbf{A}^T\mathbf{A}$ matrix may have an infinite number of generalised solutions, and the normal pseudosolution coincides with the one having the lowest norm.

The normal pseudosolution can be obtained formally using a pseudohardware matrix $\mathbf{A}^+$

$$\mathbf{x}_N = \mathbf{A}^+\mathbf{y} \tag{4.139}$$

An $m \times n$ $\mathbf{A}^+$ matrix is called a pseudoinverse (Moor–Penrose, generalised inverse matrix) for the $n \times m$ $\mathbf{A}$ matrix, if it meets the following conditions

$$\mathbf{A}^+\mathbf{AA} = \mathbf{A}$$
$$\mathbf{A}^+\mathbf{AA}^+ = \mathbf{A}$$
$$\mathbf{AA}^+ = (\mathbf{AA}^+)^T$$
$$\mathbf{A}^+\mathbf{A} = (\mathbf{A}^+\mathbf{A})^T \tag{4.140}$$

It can be shown that such a matrix always exists and is unique [64]. A pseudosimilar matrix possesses the following property extremal to any $m \times n$ $\mathbf{X}$ matrix, i.e.

$$\|\mathbf{AA}^+ \quad \mathbf{I}\| \leqslant \|\mathbf{AX} - \mathbf{I}\| \tag{4.141}$$

If this expression is an equality for any of the $\mathbf{X}$ matrices different from $\mathbf{A}^+$, then

$$\|\mathbf{A}^+\| < \|\mathbf{X}\| \tag{4.142}$$

If the $\mathbf{A}$ matrix is square and nonsingular, $\mathbf{A}^+ = \mathbf{A}^{-1}$. If it is rectangular and has a complete rank, $r = m$, then

$$\mathbf{A}^+ = (\mathbf{A}^+\mathbf{A})^{-1}\mathbf{A}^T \tag{4.143}$$

and the normal pseudosolution coincides with the least squares solution (eqn. 4.117). If the **A** matrix has an incomplete rank, there is no simple representation of a pseudoinverse matrix.

At present, matrix pseudoinversion for sets of equations, especially when they are ill conditioned, is performed by the singular decomposition method [62]. In this method, one does not have to pass to a normal set of equations, thus increasing the ill conditioning, but can obtain algorithms generalised for the case of a singular SLAE.

Consider a pseudoinverse matrix for the rectangular $n \times m$ matrix

$$\mathbf{D} = \begin{bmatrix} d_1 & 0 & 0 & \dots & 0 \\ 0 & d_2 & 0 & \dots & 0 \\ \vdots & \vdots & \vdots & & \vdots \\ 0 & 0 & 0 & \dots & d_m \\ \vdots & \vdots & \vdots & & \vdots \\ 0 & 0 & 0 & \dots & 0 \end{bmatrix} \tag{4.144}$$

If it has a rank $r \leqslant m$, then r of its diagonal elements are nonzero. It is easy to see that its pseudoinverse matrix is the $\mathbf{D}^+$ $m \times n$ matrix

$$\mathbf{D}^+ = \begin{bmatrix} d_1^+ & 0 & \dots & 0 & \dots & 0 \\ 0 & d_2^+ & \dots & 0 & \dots & 0 \\ \vdots & \vdots & & \vdots & & \vdots \\ 0 & 0 & \dots & d_m^+ & \dots & 0 \end{bmatrix} \tag{4.145}$$

where

$$d_i^+ = \begin{cases} 1/d_i, & d_i \neq 0 \\ 0, & d_i = 0 \end{cases}$$

Note that the $\mathbf{DD}^+$ matrix represents a matrix of order m, which has on its principal diagonal r units, the other elements being zeros.

Consider now a pseudoinverse matrix for an arbitrary **A** matrix. Since it may always have, including the singular case, a singular decomposition

$$\mathbf{A} = \mathbf{UDV}^T \tag{4.146}$$

its pseudoinverse matrix has the form

$$\mathbf{A}^+ = \mathbf{VD}^+\mathbf{V}^T \tag{4.147}$$

which has been defined for both singular and nonsingular SLAE.

The normal pseudosolution can be obtained either in the conventional form

$$\hat{\mathbf{x}} = \mathbf{VD}^+\mathbf{U}^T\mathbf{y} = \sum_{i=1}^{r} \mu_i^{-1}\mathbf{V}_i\mathbf{U}_i^T\mathbf{y} \tag{4.148}$$

or as a recursive relation

$$\begin{aligned} \hat{\mathbf{x}}^{(i)} &= \hat{\mathbf{x}}^{(i-1)} + \mu_i^{-1}\mathbf{V}_i\mathbf{U}_i^T\mathbf{y}, \qquad i = \overline{1, r} \\ \hat{\mathbf{x}}^{(0)} &= 0 \end{aligned} \tag{4.149}$$

where r is the $\mathbf{A}$ matrix rank and μ_i are its singular numbers.

Direct application of pseudoinversion gives satisfactory results only if the matrix is well conditioned and the initial data errors are small. Otherwise, this method is inefficient primarily because the error vector norm $||\mathbf{x}_T - \mathbf{A}^+\mathbf{y}||$ is usually unacceptably large.

For a perturbed SLAE, one introduces an effective pseudoinverse $\mathbf{A}_\tau^+$ matrix, which has a pseudoinverse matrix structure, but the elements of the $\mathbf{D}_\tau^+$ matrix are defined differently

$$d_{\tau_1}^+ = \begin{cases} 1/d_i, & d_i \geqslant \tau \\ 0, & d_i < \tau \end{cases} \tag{4.150}$$

It has been taken into account that $d_i \geqslant 0$, since $d_i = \mu_i$, $i = \overline{1, m}$, where μ_i are the singular numbers of the $\mathbf{A}$ matrix.

The $\mathbf{A}_\tau^+$ matrix only meets the last three of the Moor–Penrose conditions, while the first one is replaced by the equality

$$||\mathbf{A}\mathbf{A}_\tau^+\mathbf{A} - \mathbf{A}|| \leqslant \tau \tag{4.151}$$

The quantity τ must be matched with the error level of the given initial data; with increasing perturbations it must be increased, but at higher data accuracy $\tau \to 0$.

An effective pseudosolution is

$$\hat{\mathbf{x}}_\tau = \mathbf{A}\tau\mathbf{V}\mathbf{D}_\tau^+\mathbf{U}^T\mathbf{y} + \mathbf{y} = \sum_{i=1}^{r'} \mu_i^{-1}\mathbf{V}_i\mathbf{U}_i^T\mathbf{y} \tag{4.152}$$

or

$$\begin{aligned} \hat{\mathbf{x}}_\tau^{(i)} &= \hat{\mathbf{x}}_\tau^{(i-1)} + \mu_i^{-1}\mathbf{V}_i\mathbf{U}_i^T\mathbf{y}, \qquad i = \overline{1, r'} \\ \hat{\mathbf{x}}_\tau^{(0)} &= 0 \end{aligned} \tag{4.153}$$

where r' is a number smaller than the matrix rank, r, and is matched, as τ, with the initial data error level.

Obviously, at a low level of perturbations ($\tau \to 0$, $r' \to r$), $\mathbf{x} \approx \mathbf{x}_N \approx \mathbf{x}_T$. As the error level grows, $\mathbf{x} \neq \mathbf{x}_N \neq \mathbf{x}_T$. However, if the parameter τ is given correctly, the effective pseudosolution accuracy may appreciably exceed the normal pseudosolution accuracy: $||\mathbf{x}_T - \mathbf{A}_\tau^+\mathbf{y}|| \leqslant ||\mathbf{x}_T - \mathbf{A}^+\mathbf{y}||$. The parameters r' can be chosen by the residue method. Its disadvantage, however, in the case of the target IR estimation problem is that it completely eliminates individual frequency components. It also leads to the meter FR bends (including the processing), gives rise to numerous time sidelobes of the meter IR and, as a result, to its low RCS.

On the contrary, the regularisation method is free from most of these demerits. The foundation for this method was laid by A.N. Tikhonov in the mid-1950s. Today, it is quite popular for solving incorrect ill-posed problems in various fields of science and technology [102]. This method combines a number of specific procedures for incorrectness control, for example, the quasi solution procedure, pseudoinversion, ridge estimation, stochastically reduced estimation, etc. We shall illustrate the principal idea of this method and the regularised solution scheme with reference to the operator equation (eqn. 4.88) for the case when the operator $A\{\ \}$ is known exactly.

In the regularisation method, the approximate solution of an equation with an approximately known left-hand side y is taken in the element $\hat{x}_\alpha \in X$, on which the functional

$$\Phi_\alpha(\hat{x}, y_\varepsilon) = Q(A\{\hat{x}\}, y_\varepsilon) + \alpha F(\hat{x}) \tag{4.154}$$

reaches its lower bound. The parameter α is known as the regularisation parameter and must be matched, in a definite way, with the given initial data error.

The functional (eqn. 4.154) is termed as smoothing. The term

$$Q \equiv Q(A\{x\}, y) = ||\mathbf{y}_\varepsilon - \mathbf{A}\hat{\mathbf{x}}||$$

entering into it (for eqn. 4.89) represents the squared residue norm of the equation on the element $\mathbf{x}$.

The functional $F(\mathbf{x})$ is usually called a stabiliser. Note that when Q is used independently, the search for its minimum leads to conventional unregularised solutions by means of inverse (generalised inverse) operators. For instance, for a linear equation (eqn. 4.89) the value minimisation results in the well-known least squares solution (eqn. 4.121).

The approximate solution, $\hat{\mathbf{x}}_\alpha$, obtained by minimising the smoothing functional (eqn. 4.154) possesses an important property: the condition $\hat{\mathbf{x}}_\alpha \to \mathbf{x}_T$ at $||\varepsilon||_{max} \to 0$ or $\sigma_\varepsilon^2 \to 0$ is valid for it, if the regularisation parameter is matched with the initial data error in the deterministic approach $(\alpha = \alpha(||\varepsilon||_{max}))$ or in the statistical problem formulation $(\alpha = \alpha(\sigma_\varepsilon^2))$.

The approximate solution, $\hat{\mathbf{x}}_\alpha$, is known as a regularised solution. Generally, it is not unique, because different stabilisers, $F(\mathbf{x})$, produce different solutions, $\hat{\mathbf{x}}_\alpha$. The choice of the functional itself, $F(\mathbf{x})$, is often prompted by the nature of the problem to be solved and is, naturally, based on *a priori* information.

We shall examine one variant of presetting a stabiliser in radio engineering problems [1], confining ourselves to inverse problems described by eqns. 4.89 and 4.90, with the exactly known operator $\mathbf{A}$ and having in mind, also, more complex tasks.

Suppose that a certain fixed vector, $\mathbf{x}_0$, is close to the exact, or one of the exact, solutions of these equations. Then it is reasonable to preset the stabiliser in eqn. 4.154 as the squared difference vector norm

$$F(\hat{\mathbf{x}}) = ||\mathbf{x}_0 - \hat{\mathbf{x}}|| = (\mathbf{x}_0 - \hat{\mathbf{x}})^T(\mathbf{x}_0 - \hat{\mathbf{x}}) \tag{4.155}$$

and seek a regularised solution by minimising the smoothing functional

$$\Phi_\alpha(\hat{\mathbf{x}}, \mathbf{y}_\varepsilon) = (\mathbf{y}_\varepsilon - \mathbf{A}\hat{\mathbf{x}})^T(\mathbf{y}_\varepsilon - \mathbf{A}\hat{\mathbf{x}}) + \alpha(\mathbf{x}_0 - \hat{\mathbf{x}})^T(\mathbf{x}_0 - \hat{\mathbf{x}}) \tag{4.156}$$

From the condition of minimum $\Phi_\alpha(\mathbf{x}, \mathbf{y})$, we derive an approximate estimate

$$\hat{\mathbf{x}}_\alpha = (\mathbf{A}^T\mathbf{A} + \alpha\mathbf{I})^{-1}(\mathbf{A}^T\mathbf{y}_\varepsilon + \alpha\mathbf{x}_0) \tag{4.157}$$

which is the regularised solution to eqn. 4.89 normal relative to the vector $\mathbf{x}_0$. Unlike the least squares estimate, not only does this estimate minimise the squared residue norm, but it is the closest of all approximate solutions, $\hat{\mathbf{x}}_\alpha \in \mathbf{X}$, to the *a priori* known vector, $\mathbf{x}_0$. If one fails to preset the vector $\mathbf{x}_0$, it is natural to put it equal to zero. As a result, we arrive at an estimate with a minimum norm (normal estimate)

$$\hat{\mathbf{x}}_\alpha = (\mathbf{A}^T\mathbf{A} + \alpha\mathbf{I})^{-1}\mathbf{A}^T\mathbf{y}_\varepsilon \tag{4.158}$$

which is termed a ridge estimate in the linear regression theory.

By giving the stabiliser in a more complicated form, for example as

$$F(\hat{\mathbf{x}}) = (\mathbf{x}_0 - \hat{\mathbf{x}})^T\mathbf{C}(\mathbf{x}_0 - \hat{\mathbf{x}})$$

where $\mathbf{C}$ is a symmetric, positively definite matrix, we arrive at the estimate

$$\hat{\mathbf{x}}_\alpha = (\mathbf{A}^T\mathbf{A} + \alpha\mathbf{C})^{-1}(\mathbf{A}^T\mathbf{y}_\varepsilon + \alpha\mathbf{C}\mathbf{x}_0)$$

where at $\mathbf{x}_0 = 0$ is

$$\hat{\mathbf{x}}_\alpha = (\mathbf{A}^T\mathbf{A} + \alpha\mathbf{C})^{-1}\mathbf{A}^T\mathbf{y}_\varepsilon \tag{4.159}$$

This is a generalised ridge-type estimate.

Another common type of regularised estimate is the James–Stein (reduced or squeezing) estimate [22]. With the stabiliser

$$F(\hat{\mathbf{x}}) = ||\mathbf{y}_\varepsilon - \mathbf{A}\hat{\mathbf{x}}||^2 = (\mathbf{y}_\varepsilon - \mathbf{A}\hat{\mathbf{x}})^T(\mathbf{y}_\varepsilon - \mathbf{A}\hat{\mathbf{x}}) \tag{4.160}$$

we obtain

$$\begin{aligned}\hat{\mathbf{x}}_\alpha &= \frac{1}{1+\alpha}(\mathbf{A}^T\mathbf{A})^{-1}\mathbf{A}^T\mathbf{y}_\varepsilon + \frac{\alpha}{1+\alpha}\mathbf{x}_0 \\ &= \frac{1}{1+\alpha}\hat{\mathbf{x}} + \frac{\alpha}{1+\alpha}\mathbf{x}_0\end{aligned} \tag{4.161}$$

Putting $\mathbf{x}_0 = 0$, we have

$$\begin{aligned}\hat{\mathbf{x}}_\alpha &= \frac{1}{1+\alpha}(\mathbf{A}^T\mathbf{A})^{-1}\mathbf{A}^T\mathbf{y}_\varepsilon \\ &= r\hat{\mathbf{x}}\end{aligned} \tag{4.162}$$

Having introduced $\beta = 1 + \alpha$, we can write this expression as

$$\hat{\mathbf{x}}_\beta = \frac{1}{\beta}\hat{\mathbf{x}} \tag{4.163}$$

where β is a reduction coefficient (sometimes, this coefficient is expressed as r).

Ridge and reduced estimates are interrelated. For example, one can easily pass to the estimate of eqn. 4.162 by presetting the matrix $\mathbf{C} = \mathbf{A}^T\mathbf{A}$ in eqn. 4.159.

A ridge estimate (eqn. 4.158) minimises the squared residue norm, $||\mathbf{y}_\varepsilon - \mathbf{A}\hat{\mathbf{x}}_\alpha||^2$, at a fixed squared solution norm, $||\hat{\mathbf{x}}_\alpha||$. The opposite is also correct: a ridge estimate has a minimum squared norm in the class of estimates with a preset residue norm. Our analysis of ridge estimate properties will be made for a statistical problem, for example, the one described in [85].

A ridge estimate possesses the following properties:

1 A ridge estimate, $\hat{\mathbf{x}}_\alpha$, is a biased estimate, because at $\alpha \neq 0$

$$\begin{aligned} M\{\hat{\mathbf{x}}_\alpha\} &= \mathbf{M}\{(\mathbf{A}^T\mathbf{A} + \alpha\mathbf{I})^{-1}\mathbf{A}^T(\mathbf{A}\mathbf{x}_T + \xi)\} \\ &= (\mathbf{A}^T\mathbf{A} + \alpha\mathbf{I})^{-1}\mathbf{A}^T\mathbf{A}\mathbf{x}_T \\ &= (\mathbf{I} + \alpha(\mathbf{A}^T\mathbf{A})^{-1})^{-1}\mathbf{x}_T \neq \mathbf{x}_T \end{aligned} \tag{4.164}$$

2 Its bias is

$$\begin{aligned} \mathbf{M}\{\mathbf{x}_T - \hat{\mathbf{x}}_\alpha\} &= \mathbf{M}\{\Delta\hat{\mathbf{x}}_\alpha\} \\ &= \{\mathbf{I} - (\mathbf{I} - \alpha(\mathbf{A}^T\mathbf{A})^{-1})^{-1}\}\mathbf{x}_T \end{aligned} \tag{4.165}$$

3 A ridge estimate results from a linear transformation of a least squares estimate. Indeed, since

$$\mathbf{A}^T\mathbf{y}_\varepsilon = \mathbf{A}^T\mathbf{A}\hat{\mathbf{x}}$$

then

$$\hat{\mathbf{x}}_\alpha = (\mathbf{A}^T\mathbf{A} + \alpha\mathbf{I})^{-1}\mathbf{A}^T\mathbf{A}\hat{\mathbf{x}} = \mathbf{R}\hat{\mathbf{x}} \tag{4.166}$$

where $\mathbf{R} = \mathbf{I} + \alpha(\mathbf{A}^T\mathbf{A})^{-1}$, and $\mathbf{R} \neq \mathbf{I}$ at $\alpha > 0$. Hence, for a nonsingular matrix $(\mathbf{A}^T\mathbf{A})$

$$\lim_{\alpha \to 0} \hat{\mathbf{x}}_\alpha = \hat{\mathbf{x}}$$

4 For any given $\mathbf{A}$ matrix of the regression equation (eqn. 4.89), there is a regularisation parameter, α, at which the mean square sum of the estimate errors (mean squared error norm) will be less than the total dispersion of the least squares estimate

$$L(\hat{\mathbf{x}}_\alpha) \leqslant L(\hat{\mathbf{x}}) \tag{4.167}$$

where

$$L(\hat{\mathbf{x}}) = M\{||\hat{\mathbf{x}} - \mathbf{x}_T||^2\} \equiv \operatorname{tr}\mathbf{D}\{\hat{\mathbf{x}}\} \equiv \sigma_\varepsilon^2 \operatorname{tr}[(\mathbf{A}^T\mathbf{A})^{-1}] \gg 0$$

$$L(\hat{\mathbf{x}}_\alpha) = M\{||\hat{\mathbf{x}}_\alpha - \mathbf{x}_T||^2\} \equiv \operatorname{tr}\mathbf{D}\{\hat{\mathbf{x}}_\alpha\} + (\mathbf{R}\mathbf{x}_T - \mathbf{x}_T)^T(\mathbf{R}\mathbf{x}_T - \mathbf{x}_T)$$

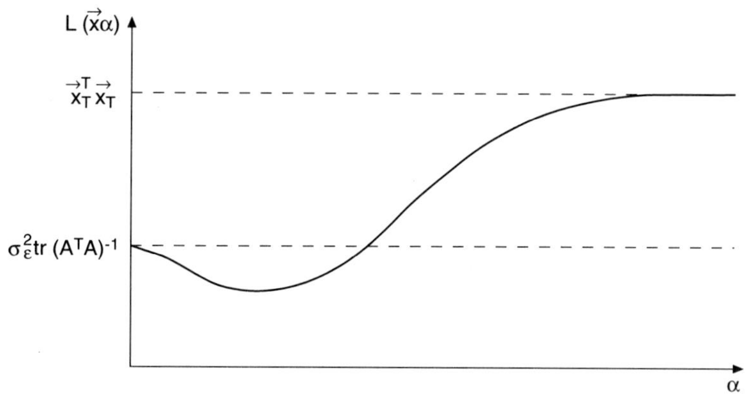

Fig. 4.6 Total mean square error of estimation versus the regularisation parameter

In contrast to the least squares unbiased estimate, the mean square error norm of a ridge estimate contains two components

$$L(\hat{\mathbf{x}}_\alpha) = \operatorname{tr}\mathbf{D}\{\hat{\mathbf{x}}_\alpha\} + ||\mathbf{R}\mathbf{x}_T - \mathbf{x}_T||^2 \tag{4.168}$$

The first component represents the total ridge dispersion strictly descending with increasing α; the second component is the perfect squared bias, which grows strictly with increasing α. Hence, at a certain value of α, $\alpha = \alpha_0$, the mean square sum of the estimate error will be minimal (Fig. 4.6). In other words, it follows from eqn. 4.168 that, on assumption of a small estimate bias, one can reduce the regularised estimate dispersion so much that the perfect mean squared error will also be smaller.

The most difficult thing in dealing with ridge estimates, as with regularised estimates in general, is the choice of the regularisation parameter. In this connection, the residue method may be very helpful [85, 104]. The regularisation parameter that meets the condition

$$\rho(\alpha) \equiv ||\mathbf{y}_\varepsilon - \mathbf{A}\hat{\mathbf{x}}_\alpha|| = ||\varepsilon|| \tag{4.169}$$

can provide estimates close, in order of magnitude, to optimal values. Since the residue norm, $\rho(\alpha)$, represents a single-valued, bounded and strictly ascending function of the parameter, the problem solution does exist and is unique. Any of the available methods for solving linear equations is applicable for finding this solution. The most correct way is to use a statistical approach, in which α is gradually reduced starting each time with a check on the assumption of equality of the vectors $\mathbf{y}_\varepsilon$ and $\mathbf{A}\hat{\mathbf{x}}_\alpha$. If the assumption is accepted at a certain value of α, the estimate will be close to optimal.

Other procedures for determining regularisation parameters have been described elsewhere [80, 85] and will be discussed in Chapter 5 of this book.

5 The ridge dispersion has the form

$$\begin{aligned}\mathbf{D}\{\hat{\mathbf{x}}_\alpha\} &= \mathbf{D}\{(\mathbf{A}^T\mathbf{A}+\alpha\mathbf{I})^{-1}\mathbf{A}^T\mathbf{y}_\varepsilon\} \\ &= [(\mathbf{A}^T\mathbf{A}+\alpha\mathbf{I})^{-1}\mathbf{A}^T]D\{\mathbf{y}_\varepsilon\}[(\mathbf{A}^T\mathbf{A}+\alpha\mathbf{I})^{-1}\mathbf{A}^T]^T \\ &= (\mathbf{A}^T\mathbf{A}+\alpha\mathbf{I})^{-1}\mathbf{A}^T\sigma_\varepsilon^2\mathbf{A}(\mathbf{A}^T\mathbf{A}+\alpha\mathbf{I})^{-1} \\ &= \sigma_\varepsilon^2(\mathbf{A}^T\mathbf{A}+\alpha\mathbf{I})^{-1}\mathbf{A}^T\mathbf{A}(\mathbf{A}^T\mathbf{A}+\alpha\mathbf{I})^{-1} \\ &= \sigma_\varepsilon^2\mathbf{R}(\mathbf{A}^T\mathbf{A}+\alpha\mathbf{I})^{-1} \end{aligned} \tag{4.170}$$

It can be seen that the regularised estimate dispersion can be made as small as necessary by increasing the regularisation. However, the bias norm will also increase. At $\alpha = \alpha_0$, a parity is achieved between the bias and dispersion values, providing the minimum value of $L(\hat{\mathbf{x}}_\alpha)$.

More general ridge estimates and their properties are discussed in [58, 80, 85].

The types of estimate we have analysed are widely used for finding UWB radar target responses. Taking account of the physical meaning of parametric target models, the singular decomposition method serves as a basis for most estimation algorithms. These are used to obtain the parameters of model response scattering and glittering points, differential equation coefficients, etc. The regularisation method is employed for the estimation of total and local target responses. Appropriate algorithms will be discussed in Chapter 5.

Chapter 5

UWB signal processing

5.1 Aims and stages of signal processing

UWB signal processing by radar meters can be subdivided into the preliminary, primary and secondary stages. Preliminary processing is aimed at signal extraction by rejecting various perturbations arising during the measurement. At this stage, however, only initial information necessary for solving inverse problems is collected (for details, see Section 5.2). This stage is followed by the primary processing of individual signals measured at a fixed position of the object with respect to the measuring antenna (probe). Its aim is to correct the dynamic characteristics of the meter's transmit–receive channel. Secondary processing represents a simultaneous processing of the whole array of signal measurements made at various positions of the object and the antenna (probe).

The volume of problems to be solved at the primary and secondary processing stages varies with the type of UWB response being measured (total or local) and with the measuring system (collimator or probe type). If the total response is measured by a collimator meter, primary processing solves all basic processing problems, so that the secondary stage involves only subsidiary operations associated with response data collection and filing (data base). Analysis of local responses in the secondary processing involves the solution of far more complex problems (see below). In primary signal processing, UWB probe meters provide only initial data to be used at the next stage for evaluation of the objects' local or total responses.

The correction procedure in primary processing is performed in order to compensate for the distortions from the UWB measuring channel, which has characteristics that are essentially different from those of an ideal, unperturbed channel. As a result, the received signal is compressed and the level of its time sidelobes decreased, thus increasing the resolution and dynamic range of the measuring system. Also, the correction procedure involves response scaling by comparison with a pre-calibrated standard.

Primary processing by UWB collimator meters will be discussed in Section 5.2.

Primary signal processing in the probe mode has some specificity, because the dynamic characteristics of the measuring channel depend not only on the position of each point in the meter's operation range but also on each position of the measuring probe. Therefore, the problem of correcting dynamic characteristics of the measuring channel can be effectively solved by the secondary processing only (see below). But taking into account the large body of computations to be made, it seems reasonable to carry out a partial correction. This may be feasible if one neglects the waveform variation of the signals from the object's measurements, that is, if the following condition is fulfilled: $w(R, \Theta, t) = d(\Theta)k(R)w_0(t)$, where $d(\Theta)$ is the normalised probe directivity, $k(R)$ is a coefficient varying with the probe to object distance, and $w_0(t)$ is the normalised IR of the meter describing its dynamic characteristics relative to a point with the coordinates $R = 0$ and $\Theta = 0$. In this case, the correlation algorithms of the channel dynamic characteristics for collimator and probe meters coincide.

Secondary processing issues for the evaluation of local responses by various meters are discussed in Sections 5.7 and 5.8.

5.2 Preliminary processing

The principal tasks of preliminary signal processing are:

- accumulation of received signals in order to decrease the additive noise level;
- rejection of malfunctions and anomalous measurements;
- recording of clutter from the rotating support and other local scatterers with their subsequent subtraction from the radar signals returning from a calibration or test target;
- integration of partial signals from individual local centres or areas of large-sized targets, which do not allow direct observation of the total signal on the oscilloscope sweep;
- correction of the measuring antenna wave fronts.

In addition, preliminary processing may involve restoration of any missing data, information compression, investigation of the distribution laws, estimation of random errors due to the measurement noises, as well as quantisation and rounding-off in digital recording. Successful solution of these problems may essentially refine the results of the subsequent processing stages. For example, the simple operation of averaging N signal samples in the case of uncorrelated sampling converter noises decreases the mean square noise $\sqrt{N}$ times. This permits, other conditions being equal, a considerable reduction in the transmitter power required. Coherent averaging can be performed in two modes – point by point and transverse. In the first case, each time count of the signal is read out and averaged N times before going over to the next count. In the other mode, the whole count array is read out and averaged.

The subtraction from the return signals of the background clutter and other local noise significantly reduces the residual noise, which is largely due to the reradiation scatterers which interfere by lying between the target and the radar and which, therefore, correspond to secondary diffraction signals. It is important to note that the signal difference dispersion increases by a factor of two. The procedures normally employed for anomalous count rejection and partial signal integration do not seem to have much specificity, so they will not be discussed in this book.

A source of regular errors in return signal measurement is the nonuniformity (nonplanarity) of signal waveforms received from a target, especially if the target has a large size. A common way of controlling this type of error is by increasing the distance from the measuring antenna to the target of interest. One can also design antenna systems with wavefronts as plane as is possible in the given aspect sector. In the first approach, the measuring system has to be made larger and its power potential must be raised. But then the radar cross section with its interfering reflections will also become larger. The other approach offers more promise, but it is more complicated. There are certain limitations on the capabilities of classical antenna systems with the required plane directivity patterns having sharp cut-offs in amplitude beyond the operation sector (apertures, horns and other types). But it is possible to design signal processing antennas in line with this major trend in modern antenna engineering. This kind of antenna can be made either as a specific beam antenna array or be based on aperture synthesis involving digital processing of signals received by a conventional UWB antenna.

5.3 Calibration (identification) of meters

5.3.1 Identification of meters by calibration scatterers

The impulse response of a radar target can be found from eqn. 4.93 by the regularisation method, in particular, by ridge estimation as described in Section 4.6. This, however, requires a preliminary evaluation of the dynamic characteristics (impulse and frequency responses) of the UWB meter used. This procedure represents a generalisation of the conventional calibration of narrowband radar target response (RTR) meters, in which one finds, from simple RCS measurements, the proportionality factor for the mean power of a received signal and the quantity being measured. The problem to be solved for the calibration of UWB meters is more sophisticated, because it reduces to an ill-posed inverse problem similar to that for target IR estimation.

Calibration of individual UWB meters should be carried out experimentally, since theoretical treatments cannot yield the necessary accuracy of parameter estimation for a complex object, including elements and units of wave (antennas) and oscillation (amplifiers, feeders, etc.) types.

In laboratory and testing-ground conditions, it is better to identify the whole meter by direct calibration. For this purpose, one uses a signal scattered by a calibration object, which serves as a target, with the UWB RTR known *a priori* with a high accuracy. The calibration target should be a scatterer, whose FR in the meter's frequency band does not have essential gaps. For investigation of targets with a large dynamic range of parameters, the calibration and verification of meters should be made with a set of calibration targets. To analyse the polarisation distortions (errors) and to compensate for them (see Section 5.5), one should use depolarising calibration scatterers. Metallic spheres, cylinders, corner reflectors and plates are commonly used for the calibration of meters. A convenient calibration object is a sphere made from a highly conductive material, with the frequency responses calculated and reported in the literature [4]. If its largest circumference considerably exceeds the meter's maximum operational wavelengths, the responses can be evaluated with a maximum accuracy and a minimum computation cost, because its FR is

$$H_c(\mathrm{i}\omega) = -\sqrt{\pi R^2} \tag{5.1}$$

If the sphere radius is $R \geqslant 2\lambda_{max}$, the maximum FR deviation from the asymptotic value is below 5%, and the mean square error is less than 3%.

The calibration must meet certain requirements. First, the calibration object must be placed at the same point in space as the object under study. Then the radar dynamic characteristics in the measurement and calibration regimes must coincide. Secondly, if space separation of the calibration object and the test object is necessary, it should be made by changing the distance but not the angular coordinates. The radar system FR, $W_c(\mathrm{i}\omega)$, measured in the calibration regime for the range R_c, can be recalculated to FR corresponding to the range, R_0, of the object under study, $W(\mathrm{i}\omega)$

$$W(\mathrm{i}\omega) = \frac{R_c}{R_0} W_c(\mathrm{i}\omega) \tag{5.2}$$

Thus, for the evaluation of the radar dynamic range, the meter must include a unit magnitude of the parameter being measured preset by the calibration scatterer (see Fig. 5.1). The measurement involves metrological operations associated with the recording and processing of calibration signals. The time between calibrations is defined by the degree of stability of the measuring channel elements and varies with the particular UWB meter design.

Consider the identification equation for a meter's impulse signal. With the transitive property of convolution, we can easily show that the determinative component of the calibration signal, that is, a signal corresponding

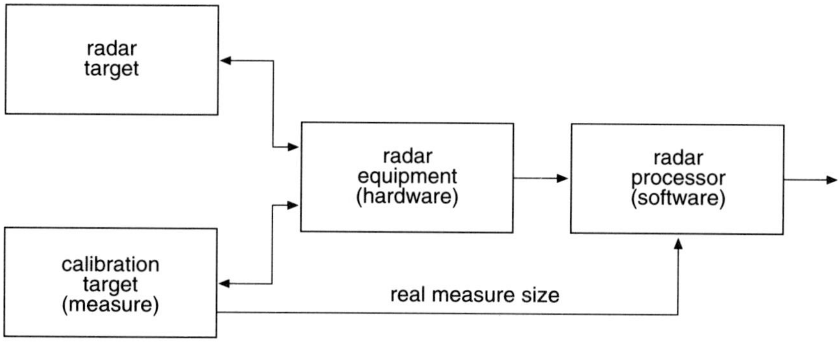

Fig. 5.1 Block diagram of UWB meter calibration

to the echo from the calibration scatterer, has the form

$$y_{c_\varepsilon}(t) = h_c(t) * w_c(t) + \varepsilon_c(t) \tag{5.3}$$

where $h_c(t)$ is the calibration target IR.

In the discrete form, eqn. 5.3, taking account of the measurement errors, is similar to eqn. 4.93

$$\mathbf{y}_{c_\varepsilon} = \mathbf{H}_c w_c + \varepsilon_c \tag{5.4}$$

Here, $\mathbf{y}_{c_\varepsilon}$ and ε_c are the sample vectors of the calibration signal and the calibration noise, respectively, and $\mathbf{H}_c$ is a matrix composed of the calibration scatterer IR counts, with a structure similar to that of the $\mathbf{W}$ matrix of eqn. 4.93.

Since eqns. 5.3 and 5.4 satisfy the regularisation conditions, they can be solved by the algorithms discussed above. The regularised solution of eqn. 5.4 yields $\hat{\mathbf{w}}_{c_\alpha}$ estimates, which permit the formation of a $\mathbf{W}$ matrix as an approximation to the real matrix. A specific feature of the identification equation for the meter IR (eqn. 5.4), as compared to that for the target IR (eqn. 4.93), is that the errors in writing the system of linear algebraic equations matrix can usually be neglected. With the meter IR estimates, we can find its FR with the discrete Fourier transform. The general ridge expressions for the meter IR and FR have the form:

$$\hat{\mathbf{w}}_{c_\alpha} = (\mathbf{H}_c^T \mathbf{H}_c + \alpha_c \mathbf{I})^{-1} \mathbf{H}_c^T \mathbf{y}_{c_\varepsilon} \tag{5.5}$$

$$\hat{\mathbf{W}}_{c_\alpha} = F\{\hat{\mathbf{w}}_{c_\alpha}\} \tag{5.6}$$

and the optimal value of the regularisation parameter can also be found from the residual principle.

One can see that the problem of simultaneous evaluation of target responses and the meter's dynamic characteristics can be solved in two steps. At first, the backscatter data from a calibration target are used to write identification eqn. 5.4 to obtain the meter's regularised IR estimate, for instance, in the form of eqn. 5.5. Then, the data on the calibration and backscatter by the test target are used to write the identification equation for the radar target (eqn. 4.93) to obtain its regularised IR estimate, for instance, the ridge (eqn. 4.158) or reduced (eqn. 4.162) estimates. Target FR evaluation is made by discrete Fourier transform with account of the time discretisation step.

5.3.2 Indirect identification of a meter

It is sometimes impossible or unreasonable to make the meter identification from a calibration scatterer. Such a situation arises with large collimator systems designed for response measurements of full-scale models or large-sized objects. The measurements take much time, sometimes a few days, and there is practically no possibility of replacing a target by a calibration object. In this case, one should employ indirect methods for the identification of the meter. These include subdivision of the radar channel into several segments and individual identification of each segment. Then the dynamic characteristics obtained are integrated into the total characteristics of the whole meter. Additional sources of UWB test signals, measuring converters and devices may be used.

5.4 Signal processing for total IR and FR estimation

5.4.1 Singularity expansion of the measurement equation matrix by the discrete Fourier transform

In order to reduce the number of operations required for solving inverse problems of the type in eqns. 4.93 and 5.4, it is useful to pass to signal processing in the frequency domain. This approach is often used in digital signal processing to solve direct problems. It is based on a fundamental property of the discrete Fourier transform (DFT), namely, its trigonometric basis invariance to the cyclic shift. (We shall restrict ourselves to discrete signals, dynamic systems and spectra. Similar results are valid for continuous signals and spectra [79].) Therefore, the DFT is an eigen transform of so called circulant matrices, to which the identification equation matrices can be converted for IR estimation of radar targets and meters.

We shall start with a few definitions [81]. Let $\mathbf{a}_1$ be a vector, $\mathbf{a}_1 = [a_1 \quad a_2 \quad \ldots \quad a_k \quad \ldots \quad a_n]^T$, and $\mathbf{a}_k$ the vector obtained from $\mathbf{a}_1$ by a $(k-1)$ step cyclic shift; $\mathbf{a}_k = [a_k \quad a_{k+1} \quad \ldots \quad a_n \quad a_1 \quad \ldots \quad a_{k-1}]^T$. The $(k-1)$ shift can be conveniently interpreted as the product of the vector $\mathbf{a}_1$ and the $\mathbf{M}_{k-1}$ matrix of the $(k-1)$ cyclic shift

$$\mathbf{a}_k = \mathbf{M}_{k-1}\mathbf{a}_1 \tag{5.7}$$

The $\mathbf{M}_{k-1}$ matrix can, in turn, be expressed by the unit M_1 matrix of a one-step shift

$$\mathbf{M}_{k-1} = \mathbf{M}_1^{k-1} \tag{5.8}$$

where

$$\mathbf{M}_1 = \begin{bmatrix} 0 & 1 & 0 & \dots & 0 \\ 0 & 0 & 1 & \dots & 0 \\ \vdots & \vdots & \vdots & & \vdots \\ 0 & 0 & 0 & \dots & 1 \\ 1 & 0 & 0 & \dots & 0 \end{bmatrix}$$

The $\mathbf{A}_c$ matrix is called circulant (periodic), if its columns are composed of the first $\mathbf{a}_1$ column by the cyclic shift procedure

$$\begin{aligned} \mathbf{A}_c &= [\mathbf{a}_1|\mathbf{a}_2|\dots|\mathbf{a}_{N-1}] \\ &= [\mathbf{a}_1|\mathbf{M}_1\mathbf{a}_1|\mathbf{M}_2\mathbf{a}_1|\dots|\mathbf{M}_{N-1}\mathbf{a}_1]^T \\ &= [\mathbf{a}_1|\mathbf{M}_1\mathbf{a}_1|\mathbf{M}_1^2\mathbf{a}_1|\dots|\mathbf{M}_1^{N-1}\mathbf{a}_1]^T \end{aligned} \tag{5.9}$$

Let us examine the DFT matrix

$$\mathbf{F} = [W_N^{kn}], \quad k, n = \overline{0, N-1} \tag{5.10}$$

where $W_N = \exp(-\mathrm{j}2\pi/N)$, permitting the conversion of a discrete time domain signal represented by the vector $\mathbf{a}$ to its discrete (complex) spectrum also in the vector form: $\mathbf{A} = [A_1 \dots A_N]^T$

$$\mathbf{A} = \mathbf{Fa} \tag{5.11}$$

The inverse DFT is carried out with the matrix

$$\mathbf{F}^{-1} = \frac{1}{N}[W_N^{-kn}], \qquad k, n = \overline{0, N-1} \tag{5.12}$$

It allows the conversion of the discrete spectrum A to the initial discrete signal $\mathbf{a}$

$$\mathbf{a} = \mathbf{F}^{-1}\mathbf{A} \tag{5.13}$$

Recall that the A_k components of the vector $\mathbf{A}$ (the discrete spectrum of the signal $\mathbf{a} = [a_1 \dots a_N]^T$) are directly related to the spectral (continuous) function, $A(\mathrm{i}\omega)$, of an analogue signal, $a(t)$, generating a discrete signal, $\mathbf{a}$ ($\mathbf{a} = [a(0) \dots a((N-1)T)]^T$)

$$A_k \equiv A[k] = \frac{1}{T}A(\mathrm{i}\omega)\bigg|_{\omega=2\pi/(NT)}, \qquad k = \overline{0, N-1} \tag{5.14}$$

There is a theorem in which the DFT is a complete transform of any circulant matrix (for the demonstration, see, e.g. [81]), namely

$$\mathbf{F}\mathbf{A}_c\mathbf{F}^{-1} = \mathbf{D}_A \tag{5.15}$$

where $\mathbf{D}_A$ is a diagonal matrix made up of the vector elements $\mathbf{A} = \mathbf{Fa}_1$

$$\begin{aligned} \mathbf{D}_A &= \mathrm{diag}\{A_1, \dots, A_N\} \\ &= \mathrm{diag}\{\mathbf{A}\} \end{aligned} \tag{5.16}$$

Eqn. 5.11 can be rewritten as the orthogonal expansion of the $\mathbf{A}_c$ matrix

$$\mathbf{A}_c = \mathbf{F}\mathbf{D}_A\mathbf{F}^{-1} \equiv \mathbf{F}\mathbf{D}_A\mathbf{F}^T \tag{5.17}$$

It follows from a comparison of eqns. 5.17 and 4.146 that, in the circulant case, the eigenvector matrices and the diagonal matrix of singular numbers (complex-valued) are replaced by those of the DFT basis vectors and of discrete spectrum counts of the vector $\mathbf{a}_1$ underlying the circulant matrix $\mathbf{A}_c$.

If the SLAE matrix

$$\mathbf{A}_c\mathbf{x} = \mathbf{y} \tag{5.18}$$

is circulant, its expansion (eqn. 5.17) can considerably reduce the computational costs for the solution of the corresponding set of equations. We shall demonstrate this with the conventional matrix inversion to obtain the solution

$$\hat{\mathbf{x}} = \mathbf{A}_c^{-1}\mathbf{y} \tag{5.19}$$

leaving aside, for the time being, the issue of its accuracy.

Conventional estimates of the vector and its DFT in the frequency domain can be obtained by expanding the matrix in eqn. 5.17

$$\begin{aligned} \mathbf{A}_c\mathbf{x} = \mathbf{y} \quad &\Leftrightarrow \\ \mathbf{F}^{-1}\mathbf{D}_A\mathbf{F}\mathbf{x} = \mathbf{y} \quad &\Leftrightarrow \\ \mathbf{D}_A(\mathbf{F}\mathbf{x}) = \mathbf{F}\mathbf{y} \quad &\Leftrightarrow \\ \mathbf{D}_A\mathbf{X} = \mathbf{Y} \quad &\Leftrightarrow \\ \hat{\mathbf{X}} = \mathbf{D}_A^{-1}\mathbf{Y} \quad & \\ \hat{\mathbf{x}} = \mathbf{F}^{-1}\hat{\mathbf{X}} \quad & \end{aligned} \tag{5.20}$$

where $\mathbf{Y} = \mathbf{F}\mathbf{y}$ is the DFT of the data vector $\mathbf{y}'$, $\mathbf{D}_A = \mathrm{diag}\{\mathbf{F}\mathbf{a}_1\} = \mathrm{diag}\{\mathbf{d}_A\}$ is the diagonal matrix composed of the vector elements $\mathbf{d}_A \equiv \mathbf{A} = \mathbf{F}\mathbf{a}_1$, which is the DFT of the vector $\mathbf{a}_1$ generating the circulant $\mathbf{A}_c$ matrix.

The reduction in the number of frequency domain operations, relative to the time domain, is attained by using the so called fast Fourier transformation (FFT) algorithm [84]. Now we shall evaluate the computational effectiveness of the frequency algorithms. It is known that the number of operations (multiplications) in a direct inversion of the $N \times N$ $\mathbf{A}_c$ matrix is of the order of N^3 [84]. In addition, N^2 operations must be made in multiplying the inverse matrix by the data vector. Therefore, the total number of operations is

$$I_{t.d.} \sim N^3 + N^2 \tag{5.21}$$

The conversion of the $\mathbf{A}_c$ matrix to the $\mathbf{D}_A$ matrix involves $N\log_2 N$ multiplications, its inversion N operations, the formation of the data vector DFT – $N\log_2 N$ operations; the multiplication of the resultant (complex-valued) vector by the complex inverse matrix involves $4N$ operations (one complex multiplication is equivalent in expenditure to four scalar ones) and the translation of the estimate obtained to the time domain needs $N\log_2 N$ operations

$$I_{f.d.} = 3N\log_2 N + 4N \tag{5.22}$$

The gain from the passage to the frequency domain processing grows rapidly with N. For example, at $N \gg 1$

$$V = \frac{I_{t.d.}}{I_{f.d.}} \sim \frac{N^3}{3N\log_2 N} \tag{5.23}$$

At $N = 1024$, $V = 3.5 \times 10^4$ and at $N = 4096$, $V = 4.7 \times 10^5$.

5.4.2 *Frequency versions of estimation algorithms*

The expressions presented allow us to transform initial eqns. 4.93 and 5.4 for the processing of time domain algorithms to the computationally more profitable frequency algorithms. For this, we must make only one additional conversion step from the low triangular matrix in eqn. 4.93 to the equivalent circulant matrix [62, 64]. The equations

$$\mathbf{Wh} = \mathbf{y} \tag{5.24}$$

and

$$\mathbf{W}_c\mathbf{h}_c = \mathbf{y}_c \tag{5.25}$$

with the low triangular $N' \times N'$ **W** matrix and the circulant $N \times N$ $\mathbf{W}_c$ matrix will be considered equivalent, if the elements of the vectors $\mathbf{x}_c$ and $\mathbf{y}_c$ coincide with those of the vectors **x** and **y** at $i \leqslant N$ $(N' \geqslant N)$.

Assume N_1 to be the duration of an impulse response of the meter or of the calibration target (see below), that is, the bandwidth in the low triangular **W** matrix; we also take N_2 to be the duration of a discontinuous signal, **x**. It is easy to show [62] that eqns. 5.24 and 5.25 are equivalent, if $N' \geqslant N_1 + N_2 - 1$. For this condition to be met, the vectors **y** and **w** (the latter generates the **W** matrix) are supplemented by zero counts and the dimension of the vector **h** is increased up to N.

Having passed from the initial eqn. 5.24 to eqn. 5.25 with the circulant matrix, we can solve the latter equation with the Fourier transform in the frequency domain, as indicated above. We obtain the unbiased estimates of the vector **x** and its DFT **X** (here and below, we drop, for simplicity, the notations for the circulant matrices and the respective vectors)

$$\begin{gathered} \mathbf{Wh} = \mathbf{y} \quad \Leftrightarrow \\ \mathbf{D}_W\mathbf{H} = \mathbf{Y} \quad \Leftrightarrow \\ \hat{\mathbf{H}} = \mathbf{D}_W^{-1}\mathbf{Y} \end{gathered} \tag{5.26}$$

$$\hat{\mathbf{h}} = \mathbf{F}^{-1}\hat{\mathbf{H}} \tag{5.27}$$

where $\mathbf{Y} = \mathbf{F}\mathbf{y}$ is the DFT of the vector $\mathbf{y}$, $\mathbf{D}_W = \text{diag}\{\mathbf{F}\mathbf{w}_1\} = \text{diag}\{\mathbf{d}_W\}$ is the diagonal matrix composed of the elements of the vector $\mathbf{d}_W = [d_{W1} \dots d_{Wn}]^T$, which is the DFT of the vector $w_1 = [w(0) \dots w((N-1)T)]^T$ of a discretised IR of the meter.

The physical sense of d_{Wi} elements is evident: they are proportional to the meter's FR samples at frequencies $\omega_i = 2\pi i/NT$

$$d_{Wi} = \frac{1}{T} W(\mathrm{i}\omega)\bigg|_{\omega = 2\pi(i-1)/(NT)} \tag{5.28}$$

We pointed out in Chapter 4 that classical methods are often unable to provide the necessary accuracy of the estimation. This is also true of the estimates obtained but we can give a clear physical interpretation of this effect. Indeed, suppose that the data vector $\mathbf{y}_\varepsilon$ contains an error vector $\mathbf{y}_\varepsilon = \mathbf{y} + \varepsilon$. Then its DFT also contains an error

$$\begin{aligned}\mathbf{Y}_\varepsilon &= \mathbf{F}\mathbf{y}_\varepsilon \\ &= \mathbf{F}\mathbf{y} + \mathbf{F}\varepsilon \\ &= \mathbf{Y} + \mathbf{E}\end{aligned} \tag{5.29}$$

Hence, taking account of eqn. 5.26, the estimate $\hat{\mathbf{H}}$ of the DFT $\mathbf{H}$ of the vector $\mathbf{h}$ contains the error

$$\begin{aligned}\Delta\hat{\mathbf{H}} &= \hat{\mathbf{H}} - \mathbf{H} \\ &= \mathbf{D}_W^{-1}\mathbf{E} \\ &= \mathbf{D}_W^{-1}\mathbf{F}\varepsilon\end{aligned} \tag{5.30}$$

The same refers to the estimate of the vector $\mathbf{h}$

$$\begin{aligned}\Delta\hat{\mathbf{h}} &= \hat{\mathbf{h}} - \mathbf{h} \\ &= \mathbf{F}^{-1}\mathbf{D}_W^{-1}\mathbf{E} \\ &= \mathbf{F}^{-1}\mathbf{D}_W^{-1}\mathbf{F}\varepsilon\end{aligned}$$

According to DFT theory, the squared norms of vector $\Delta\hat{\mathbf{h}}$ and vector $\Delta\hat{\mathbf{H}}$ are connected in the following manner [81]

$$\|\Delta\hat{\mathbf{h}}\|^2 = \|\Delta\hat{\mathbf{H}}\|^2/N \tag{5.31}$$

therefore, the accuracy of the inverse problem solution can be verified in both the time and frequency domains.

The squared error norm of the estimate (error energy)

$$\begin{aligned}\|\Delta\mathbf{H}\|^2 &= (\Delta\mathbf{H}^{T*}\Delta\mathbf{H}) \\ &= [(\mathbf{D}_W\mathbf{F}\varepsilon)^{T*}\mathbf{D}_W^{-1}\mathbf{F}\varepsilon] \\ &= [(\mathbf{F}\varepsilon)^{T*}\mathbf{D}_W^{-1^*}\mathbf{D}_W^{-1}(\mathbf{F}\varepsilon)] \\ &= [\mathbf{E}^{T*}\mathbf{D}_W^{-2}\mathbf{E}^T]\end{aligned} \tag{5.32}$$

where

$$\mathbf{D}_W^{-2} = \text{diag}\left\{\ldots, \frac{1}{d_{Wi}d_{Wi}^*}, \ldots\right\}$$

Since, $\mathbf{E}^{T*}\mathbf{E} = \|\varepsilon\|^2 N$

$$\|\hat{\mathbf{h}}\|^2 = \|\Delta\mathbf{H}\|^2/N = \left[\sum_{i=1}^{N} \frac{1}{d_{Wi}d_{Wi}^*}\right]\|\varepsilon\|^2 \tag{5.33}$$

The frequency bandwidth used for the IR and FR estimations is determined by the time discretisation step T: $\Delta F = 1/2T$, which, in turn, is chosen with Kotelnikov's theorem from the condition $T \leqslant 1/(2F_{up})$, where F_{up} is the upper FR frequency of the measuring system, that is, a reference frequency, starting with which the meter FR variation from zero can be practically neglected. Besides, for most meters, except for video pulse (pseudovideo pulse) meters, FR tend to zero at low frequencies and even have appreciable gaps in the operation frequency band due to re-emission in the radar channel. As a result, at certain frequencies, the $(d_{Wi}d_{Wi}^*)$ values, proportional to the squared AFR of the meter: $W(\omega)$ at $\omega_i = 2\pi(i-1)/(NT)$ (as follows from the physical meaning of the d_{Wi} elements), tend to zero, while the factor

$$\sum_{i=1}^{N} \frac{1}{d_{Wi}d_{Wi}^*} \equiv \text{cond}^2(\mathbf{W})$$

tends to infinity. Hence, the error energy for the unbiased estimates (eqns. 5.26 and 5.27) also tends to infinity. It is clear that the inverse filtering procedure commonly recommended for the estimation of target IR and FR and for the digital correction of the radar channel performance [1, 58, 60, 85] is not applicable either practically or theoretically. The reason for the wide application (or, rather, citation) of inverse filtering for data processing is that many authors use the term 'inverse filter' as a symbol for the response estimation algorithm, without a critical theoretical and practical analysis. On the other hand, some authors do use heuristic algorithms which can provide a certain solution stability, but they do not discuss them in depth, substituting for them with the above symbolism.

Consider some frequency versions of the effective pseudo-inverse and regularised estimates of the target FR and IR examined above. It is easy to show that the pseudo-inverse estimate of a discrete FR takes the form

$$\hat{\mathbf{H}}_\tau = \mathbf{D}_W^+ \mathbf{Y}_\varepsilon \tag{5.34}$$

where $\mathbf{D}_W^+ = \text{diag}\{\ldots, d_{W\tau i}^+, \ldots\}$.

In this case, the elements of the effective pseudo-inverse $d_{W\tau i}^+$ matrix are chosen from the condition

$$d_{W\tau i}^+ = \begin{cases} 0, & d_{Wi} < \tau \\ 1/d_{Wi}, & d_{Wi} \geqslant \tau \end{cases} \tag{5.35}$$

The parameter τ can be found by iteration from the residual principle. A much simpler way of finding it will be presented below.

Regularised estimates of discrete FR of a target are written as

$$\hat{\mathbf{H}}_\alpha = \mathbf{D}_{W_\alpha}^{-1}\mathbf{Y}_\varepsilon \tag{5.36}$$

$$\hat{\mathbf{H}}_\beta = \mathbf{D}_{W_\beta}^{-1}\mathbf{Y}_\varepsilon \tag{5.37}$$

$$\mathbf{D}_{W_\alpha}^{-1} = \mathrm{diag}\left\{\ldots, \frac{d_{Wi}^*}{d_{Wi}d_{Wi}^* + \alpha}, \ldots\right\}$$

$$\mathbf{D}_{W_\beta}^{-1} = \mathrm{diag}\left\{\ldots, \frac{1}{\beta d_{Wi}}, \ldots\right\}$$

The relationship between these estimates is quite obvious

$$\hat{\mathbf{H}}_\alpha = \hat{\mathbf{H}}_\beta|_{\beta = 1 + (\alpha/d_{Wi}d_{Wi}^*)}$$

The vector elements of the discrete FR estimates define the target FR estimates at discrete frequencies in accordance with eqn. 5.28.

In practice, the regularised estimation (eqn. 5.36) is more convenient to write and calculate in such form [1]

$$\hat{\mathbf{H}}_\alpha = \frac{W(k)Y_\varepsilon(k)}{W(k)W^*(k) + \alpha(WW^*)_{max}} \tag{5.38}$$

where $W(k)$ is the kth element of the vector calculated by the FFT from the vector of meter IR and $(WW^*)_{max}$ is the maximum value of the values $W(k)W^*(k)$, $k = \overline{0, N-1}$.

Estimates of discretised IR of a radar target are obtained from eqns. 5.34, 5.36 and 5.37 by the DFT

$$\begin{aligned} \hat{\mathbf{h}}_\alpha &= \mathbf{F}^{-1}\hat{\mathbf{H}}_\alpha \\ \hat{\mathbf{h}}_\beta &= \mathbf{F}^{-1}\hat{\mathbf{H}}_\beta \\ \hat{\mathbf{h}}_\tau &= \mathbf{F}^{-1}\hat{\mathbf{H}}_\tau \end{aligned} \tag{5.39}$$

Note that the described procedure for UWB RTR can be treated as a correction of the dynamic characteristics of the meter's hardware by a digital filter [1, 85]. Depending on the estimation procedure chosen, the discrete FR of a filter is defined by the diagonal matrix elements in eqns. 5.34, 5.36 and 5.37, while IR is defined by inverse DFT. Correction makes all general dynamic characteristics of a measuring and data processing system, characterised by the matrices

$$\begin{aligned} \mathbf{W}_{MDP_\alpha} &= \mathbf{W}\mathbf{D}_{W_\alpha}^{+} \\ \mathbf{W}_{MDP_\beta} &= \mathbf{W}\mathbf{D}_{W_\beta}^{-1} \\ \mathbf{W}_{MDP_\tau} &= \mathbf{W}\mathbf{D}_{W_\tau}^{-1} \end{aligned} \tag{5.40}$$

much closer to those of an ideal (undistorting) system. With a decreased additive noise level τ, $\alpha \rightarrow 0$, $\beta \rightarrow 1$ and $\mathbf{W}_{MDP} \rightarrow \mathbf{I}$. In the opposite case, correction provides a compromise between the levels of the dynamic (regular) and noise components of the measurement error.

5.4.3 Experimental results

Some experimental results will be presented to illustrate the effectiveness of the signal processing methods that we have described. Fig. 5.2a shows an experimental impulse response for a metallic sphere of 55.2 mm diameter, corresponding to the frequency bandwidth 0 to 8 GHz, and Fig. 5.2b depicts a sampling converter output signal which is seen to be rather different from the sought-for IR. In particular, we are unable to identify in it a creeping wave, with a time position that is uniquely related to the sphere size. After the signal processing together with a calibration signal (Section 5.3) corresponding to a sphere of a different diameter, the regularised estimate, obtained with the ridge estimates (eqns. 4.93 and 5.4), is quite close to the real value (Fig. 5.2c). Fig. 5.3 shows, respectively, the AFR of the sphere prior to (Fig. 5.3a) and after (Fig. 5.3b) the processing, together with the theoretical AFR of the sphere. The higher accuracy is due to the fact that the impulse response of a UWB meter (Fig. 5.4a) has been considerably improved by the digital signal processing (Fig. 5.4b).

Similar results have been obtained by other biased estimation algorithms, in particular, by effective pseudo-inverse estimates of the type in eqn. 5.34. We should like to note that since the correcting pseudo-inverse filter FR defined by eqn. 5.34 has sharp breaks at frequencies with a large signal to noise ratio, this FR coincides with that of the inverse filter; at frequencies with a small ratio the corrected IR of the system may have essential time sidelobes associated with the Gibbs phenomenon. These sidelobes decrease the meter's range resolution for low intensity glittering points or may even result in the detection of false (nonexistent) local scatterers. Such situations are illustrated in Figs. 5.4 and 5.5. A meter IR is shown prior to the pseudo-inverse correction (Fig. 5.5a) and after it (Fig. 5.4c). Also presented are a signal from a sphere (Fig. 5.5a) and the result of effective pseudo-inverse estimation (Fig. 5.5b). It is seen that the processing has revealed the creeping wave more clearly, but it is comparable with the sidelobe of the mirror image (Fig. 5.5b) generated by the sidelobe of the corrected IR (Fig. 5.4c). To remove this effect, we have suggested an additional, linear weighted processing in the frequency domain, which reduces the sidelobe level of the meter IR and permits reliable identification of a creeping wave (Fig. 5.5c).

5.5 Compensation of polarisation distortions in multi-dimensional processing

The above expressions for the evaluation of target FR and IR do not take into account the polarisation distortions arising in the transmitting and

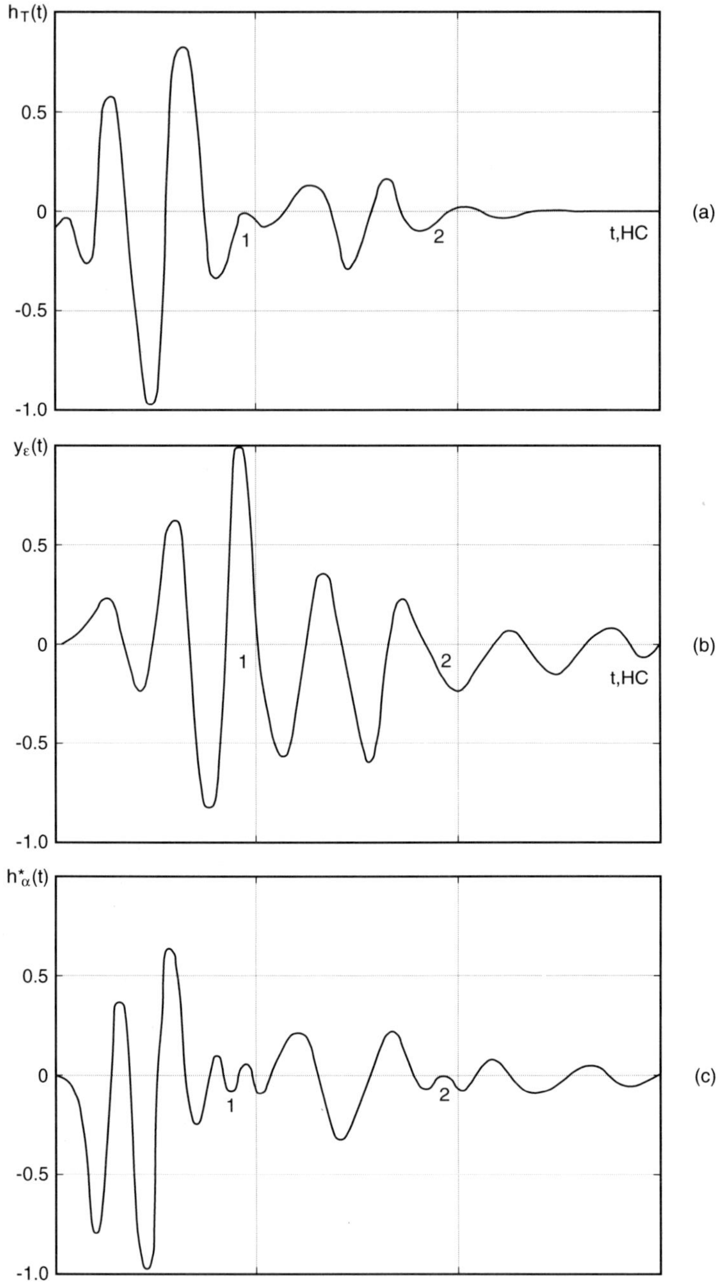

Fig. 5.2 Impulse response of the sphere

a Smoothed impulse response
b Output signal of UWB meter
c Impulse response after processing

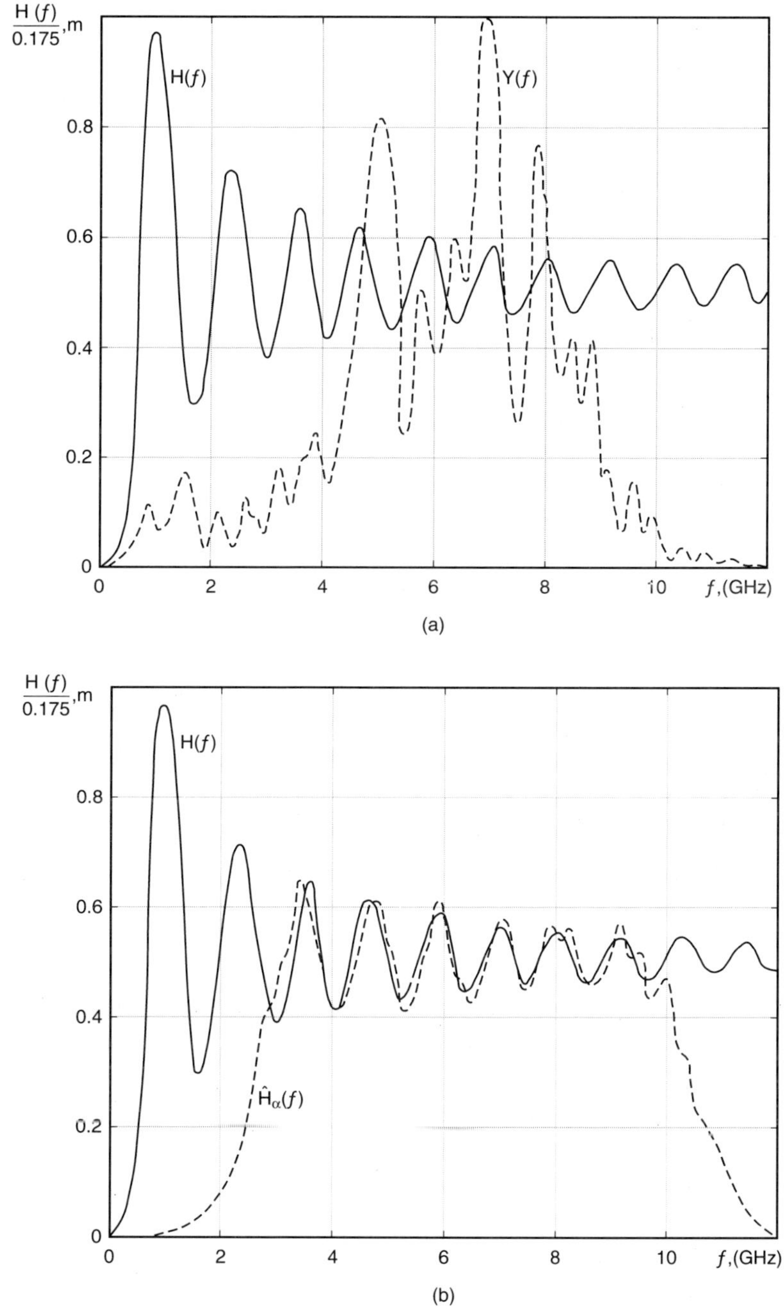

Fig. 5.3 Experimental estimations of amplitude-frequency response of the sphere

a Before processing
b After processing

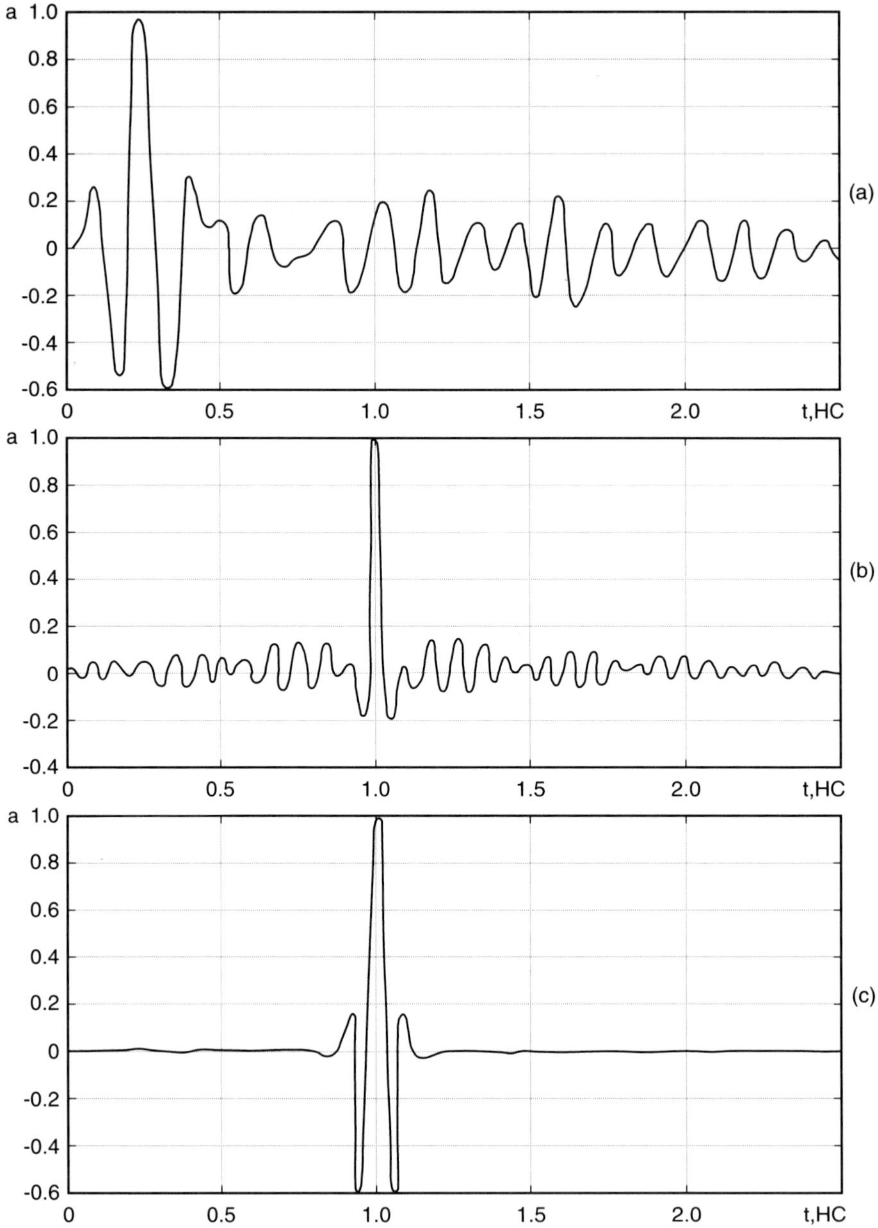

Fig. 5.4 *Impulse response of UWB meter*

a Before pseudo-inverse processing
b After pseudo-inverse processing
c After regularised processing

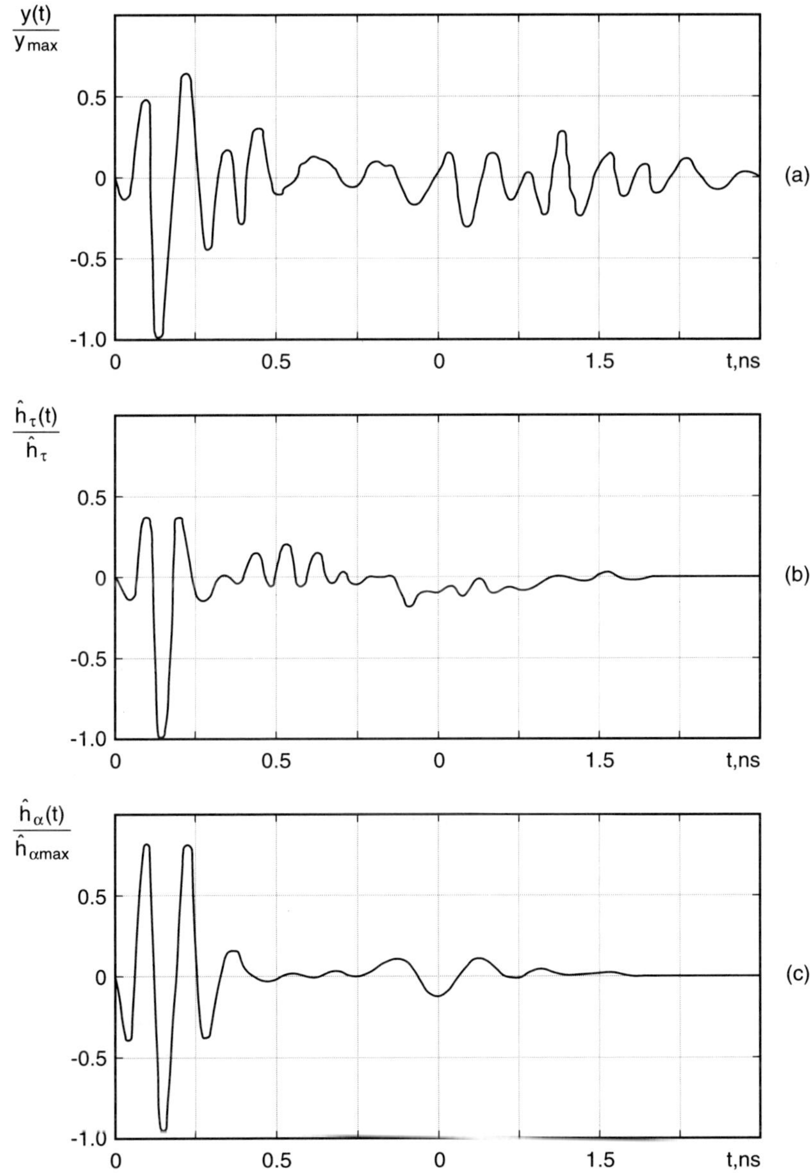

Fig. 5.5 Experimental estimations of impulse response of the sphere

a Before pseudo-inverse processing
b After pseudo-inverse processing
c After regularised processing

receiving antennas. This restricts their applicability to the case of medium-precision measurements, when an admissible reduction in precision is compensated for by the simplicity of measurement and processing algorithms. This simplicity consists in using only one probe signal for the estimation of each polarisation matrix component recorded with the desired combination of polarisations of the transmitter and receiver. Precision response measurements require a compensation of the polarisation distortions by means of identification of all polarisation components of the meter's dynamic characteristics for several calibration objects, followed by simultaneous processing of incident signals obtained for all polarisation combinations of the antennas.

Let us derive the estimation equations for the IR and FR scattering matrix elements under polarisation distortions, which are generalisations of eqns. 4.9. To simplify the derivation procedure, it is convenient to use a vector rather than a matrix representation for the target and meter IR and FR at various polarisations of the antennas. For this purpose, we introduce the vector $\mathbf{h}_v(t)$ of a target IR, consisting of all elements of the polarisation matrix $\mathbf{h}(t)$, omitting the slopes for simplicity

$$\mathbf{h}_v(t) = [h_{11}(t) \quad h_{12}(t) \quad h_{21}(t) \quad h_{22}(t)]^T \tag{5.41}$$

Let

$$\mathbf{h}_g^{(j)}(t) = [h_{g1}^{(j)}(t) \quad h_{g2}^{(j)}(t)]^T \tag{5.42}$$

be the column vector of the incident waveform generator (with variable polarisation) at its jth polarisation $(j = 1, 2)$. Also, consider the IR row vector for the receiver (with variable polarisation) at its ith polarisation $(i = 1, 2)$

$$\mathbf{h}_r^{T(i)}(t) = [h_{r1}^{(i)}(t) \quad h_{r2}^{(i)}(t)] \tag{5.43}$$

Note that the generator and receiver polarisations can be varied, for instance, by changing the orientation of linearly polarised transmitting and receiving antennas through 90° (see below).

Let us introduce the vector of the received signals (their determinate components)

$$\mathbf{y}_v(t) = [y_{11}(t) \quad y_{12}(t) \quad y_{21}(t) \quad y_{22}(t)]^T \tag{5.44}$$

where $y_{ij}(t)$ is the receiver output signal at the ith receiver polarisation and the jth transmitter polarisation.

Provided that the distortions in the propagation space can be neglected, one obtains, for a single position of the meter, the following expression for an arbitrary element of the vector $\mathbf{y}_v(t)$

$$y_{ij}(t) = \mathbf{w}^{(ij)T}(t) * \mathbf{h}_v(t) \tag{5.45}$$

where $\mathbf{w}^{(ij)T}(t)$ is the row vector of the UWB meter IR at the ith receiver and the jth transmitter polarisations.

The vector is

$$\mathbf{w}^{(ij)T}(t) = [w_{11}^{(ij)}(t) \quad w_{12}^{(ij)}(t) \quad w_{21}^{(ij)}(t) \quad w_{22}^{(ij)}(t)] \tag{5.46}$$

where

$$w_{pq}^{(ij)}(t) = \frac{1}{4\pi R_0^2} h_{pp}^{(i)}(t) * h_{gq}^{(j)}(t) \tag{5.47}$$

is the IR at the ith receiver and the jth transmitter polarisations of the pth polarisation of the scattered and the qth polarisation of the incident electromagnetic fields $(i, j, p, q = \overline{1,2})$. The resultant vector of the received signals is

$$\mathbf{y}_v(t) = \mathbf{w}(t) * \mathbf{h}_v(t) \tag{5.48}$$

where $\mathbf{w}(t)$ is the resultant meter IR matrix for all possible polarisations of the receiving and transmitting antennas and for the incident and scattered electromagnetic fields.

Polarisationally ideal antennas transmit and receive an electromagnetic wave with a polarisation coinciding with that of the respective antenna

$$h_{rp}^{(i)}(t) = \begin{cases} 0, & p \neq i \\ h_{ri}^{(i)}(t), & p = i \end{cases} \tag{5.49}$$

$$h_{gq}^{(j)}(t) = \begin{cases} 0, & q \neq j \\ h_{gj}^{(j)}(t), & q = j \end{cases} \tag{5.50}$$

Under these conditions, the $\mathbf{w}(t)$ matrix elements are

$$w_{pq}^{(ij)}(t) = \begin{cases} 0, & p \neq i, \quad q \neq j \\ w_{ij}^{(ij)}(t), & p = i, \quad q = j \end{cases} \tag{5.51}$$

where

$$w_{ij}^{(ij)}(t) = \frac{1}{4\pi R_0^2} h_{pi}^{(i)}(t) * h_{gj}^{(j)}(t)$$

Hence, the $\mathbf{w}(t)$ matrix is diagonal

$$\mathbf{w}(t) = \text{diag}\{w_{11}^{(11)}(t) \quad w_{12}^{(12)}(t) \quad w_{21}^{(21)}(t) \quad w_{22}^{(22)}(t)\}$$

and the set of eqns. 5.48 falls into four independent equations

$$y_{ij}(t) = w_{ij}^{(ij)}(t) * h_{ij}(t) \tag{5.52}$$

which coincide with eqns. 4.9. In the general case, one should use eqns. 5.48.

Consider a solution specific to this set of equations for the case of linearly polarised antennas with variable orientation. For definiteness, we shall assume that polarisation index 1 corresponds to the vertical and index 2 to the horizontal polarisations of the antennas and waves. Evidently

$$\begin{aligned} \mathbf{h}_r^{T(2)}(t) &= [\mathbf{P}\mathbf{h}_r^{(1)}(t)]^T \\ \mathbf{h}_g^{(2)}(t) &= \mathbf{P}\mathbf{h}_g^{(1)}(t) \end{aligned} \tag{5.53}$$

where

$$\mathbf{P} = \begin{bmatrix} 0 & 1 \\ 1 & 0 \end{bmatrix}$$

is the permutation matrix and $\mathbf{h}_p^{T(1)}(t) = [h_{p1}(t) \quad h_{p2}(t)]$ and $\mathbf{h}_g^{(1)}(t) = [h_{g1}(t) \quad h_{g2}(t)]^T$ are vectors, with elements which are determined by the characteristics of the meter hardware. In accordance with eqns. 5.49 and 5.50, the set of eqns. 5.48 is then written as

$$\begin{bmatrix} y_{11}(t) \\ y_{12}(t) \\ y_{21}(t) \\ y_{22}(t) \end{bmatrix} = \begin{bmatrix} w_{11}(t) & w_{12}(t) & w_{21}(t) & w_{22}(t) \\ w_{12}(t) & w_{11}(t) & w_{22}(t) & w_{21}(t) \\ w_{21}(t) & w_{22}(t) & w_{11}(t) & w_{12}(t) \\ w_{22}(t) & w_{21}(t) & w_{12}(t) & w_{11}(t) \end{bmatrix} * \begin{bmatrix} h_{11}(t) \\ h_{12}(t) \\ h_{21}(t) \\ h_{22}(t) \end{bmatrix} \tag{5.54}$$

where

$$w_{pq}(t) = \frac{1}{4\pi R_0^2} h_{rp}(t) * h_{gq}(t)$$

is the meter IR for an incident wave with the qth and a scattered wave with the pth polarisations.

It follows from the derived set of equations that every component of the vector signal is described by four terms corresponding to all polarisation components of the test object IR. For example

$$\begin{aligned} y_{11}(t) &= w_{11}(t) * h_{11}(t) + [w_{12}(t) * h_{12}(t) + w_{21}(t) * h_{21}(t) + w_{22}(t) * h_{22}(t)] \\ &= y_{11r}(t) + \Delta y_{11}(t) \end{aligned} \tag{5.55}$$

where $y_{11r}(t)$ is the useful component of the output signal defined by the corresponding component of the IR ($h_{11}(t)$) and that of the major polarisation channel of the meter ($w_{11}(t)$) and $\Delta y_{11}(t)$ is a regular error in the observed signal due to the meter polarisation distortions, which is determined by the characteristics of both the meter ($w_{12}(t)$, $w_{21}(t)$, $w_{22}(t)$) and the target ($h_{12}(t)$, $h_{21}(t)$, $h_{22}(t)$).

The expression for the regular error can be rewritten with allowance for the fact that in single-position measurements $h_{12}(t) = h_{21}(t)$

$$\Delta y_{11}(t) = [w_{12}(t) + w_{21}(t)] * h_{12}(t) + w_{22}(t) * h_{22}(t) \tag{5.56}$$

Hence, other conditions being equal, the regular error in the signal measurement will be smaller for underpolarising targets, for example, for bodies of revolution irradiated axially.

An arbitrary component of the signal has a similar structure

$$y_{ij}(t) = y_{ijr}(t) + \Delta y_{ij}(t)$$

and it also contains a regular error

$$\Delta y_{ij}(t) = \varphi\{w_{12}(t), w_{21}(t), w_{22}(t); h_{pq}(t) \quad (p \neq i, q \neq j)\}$$

The ill-posed nature of the problem of UWB response estimation from the measurements leads to a high sensitivity of classical evaluation not only to a random but also to a regular error in the initial data. Therefore, in precision measurements, one must compensate the polarisation distortions by means of simultaneous processing of all vector signal components. Consider the methods for solving this problem.

The set of integral equations (eqn. 5.54) can be written in a compact form

$$\mathbf{y}_v(t) = \mathbf{w}(t) * \mathbf{h}_v(t) \tag{5.57}$$

where $\mathbf{y}_v(t)$ and $\mathbf{h}_v(t)$ are the vectors of the observed signal and of the target IR. Since each signal, $y_{ij}(t)$, is observed against the background of additive noise, $\varepsilon_{ij}(t)$, the real vector signal obtained in the experiment represents the realisation of a random multi-dimensional process

$$\begin{aligned} \mathbf{y}_v(t) &= \mathbf{y}_v(t) + \varepsilon_v(t) \\ &= \mathbf{w}(t) * \mathbf{h}_v(t) + \varepsilon_v(t) \end{aligned} \tag{5.58}$$

where $\varepsilon_v(t) = [\varepsilon_{11}(t) \quad \varepsilon_{12}(t) \quad \varepsilon_{21}(t) \quad \varepsilon_{22}(t)]^T$ is multi-dimensional noise, and $\mathbf{y}_v(t) = [y_{11}(t) \quad y_{12}(t) \quad y_{21}(t) \quad y_{22}(t)]^T$.

By making time discretisation of the time-dependent signals and impulse responses, we can pass to a multi-dimensional regression equation, which is the generalisation of eqn. 4.9. It can be solved by the same methods as those used above for scalar signal processing. However, it is better to perform the frequency domain processing by FFT algorithms.

The Fourier transform of eqn. 5.58 is written as

$$\mathbf{Y}_v(\mathrm{i}\omega) = \mathbf{W}(\mathrm{i}\omega)\mathbf{H}_v(\mathrm{i}\omega) + \mathbf{E}_v(\mathrm{i}\omega) \tag{5.59}$$

where

$$\mathbf{Y}_v(\mathrm{i}\omega) = [Y_{11}(\mathrm{i}\omega) \quad Y_{12}(\mathrm{i}\omega) \quad Y_{21}(\mathrm{i}\omega) \quad Y_{22}(\mathrm{i}\omega)]^T$$

is the vector of the spectral functions of the signal $y_{ij}(t)$,

$$\mathbf{H}_v(\mathrm{i}\omega) = [H_{11}(\mathrm{i}\omega) \quad H_{12}(\mathrm{i}\omega) \quad H_{21}(\mathrm{i}\omega) \quad H_{22}(\mathrm{i}\omega)]^T$$

is the vector of polarisation components of the target FR and

$$\mathbf{W}(\mathrm{i}\omega) = \begin{bmatrix} W_{11}(\mathrm{i}\omega) & W_{12}(\mathrm{i}\omega) & W_{21}(\mathrm{i}\omega) & W_{22}(\mathrm{i}\omega) \\ W_{12}(\mathrm{i}\omega) & W_{11}(\mathrm{i}\omega) & W_{22}(\mathrm{i}\omega) & W_{21}(\mathrm{i}\omega) \\ W_{21}(\mathrm{i}\omega) & W_{22}(\mathrm{i}\omega) & W_{11}(\mathrm{i}\omega) & W_{12}(\mathrm{i}\omega) \\ W_{22}(\mathrm{i}\omega) & W_{21}(\mathrm{i}\omega) & W_{12}(\mathrm{i}\omega) & W_{11}(\mathrm{i}\omega) \end{bmatrix}$$

is the matrix of frequency response of the meter's polarisation channels $(W_{ij}(\mathrm{i}\omega) = F\{w_{ij}(t)\})$. $\mathbf{E}_v(\mathrm{i}\omega) = [E_{11}(\mathrm{i}\omega) \quad E_{12}(\mathrm{i}\omega) \quad E_{21}(\mathrm{i}\omega) \quad E_{22}(\mathrm{i}\omega)]^T$ is the vector of random noise spectral functions.

The discrete analogue of eqn. 5.59 has the form

$$\mathbf{Y}_v(k) = \mathbf{W}(k)\mathbf{H}_v(k) + \mathbf{E}_v(k), \qquad k = \overline{1, N} \tag{5.60}$$

in which the vectors and the matrix have the same structure as in eqn. 5.59, and their elements are interrelated, for example, by the known expression

$$\mathbf{Y}_v(k) = \frac{1}{T}\mathbf{Y}_v(\mathrm{i}\omega)\Big|_{\omega=2\pi k/(TN)}$$

For the solution of this equation, it is expedient to use a regularised, ridge estimate, which, in the present case, looks as follows

$$\hat{\mathbf{H}}_v(k) = [\mathbf{W}(k)\mathbf{W}^{*T}(k) + \alpha(k)\mathbf{I}]^{-1}\mathbf{W}^{*T}(k)\mathbf{Y}_v(k), \qquad k = \overline{1,N} \tag{5.61}$$

The regularisation parameter can be found either from the general residual ($\alpha(k) = \alpha$ over all values of $k \in [1, N]$ is global regularisation) or from the residual at certain frequencies ($\alpha(k) = \text{var}$ is local regularisation).

To obtain the estimate, one should first specify the $\mathbf{W}(k)$ matrix ($k = \overline{1,N}$) of the meter FR. As was pointed out above, one can calibrate the meter with target calibration standards. In contrast to the scalar case, it is necessary to use here several calibration targets.

Indeed, because the convolution equation is transitive, for the signal vector

$$\mathbf{y}_{vc}(t) = [y_{11c}(t) \quad y_{12c}(t) \quad y_{21c}(t) \quad y_{22c}(t)]^T$$

obtained by calibration we can write the following set of integral equations, similar to the set (eqn. 5.54)

$$\begin{bmatrix} y_{11c}(t) \\ y_{12c}(t) \\ y_{21c}(t) \\ y_{22c}(t) \end{bmatrix} = \begin{bmatrix} h^{(1)}_{11c}(t) & h^{(1)}_{12c}(t) & h^{(1)}_{21c}(t) & h^{(1)}_{22c}(t) \\ h^{(2)}_{12c}(t) & h^{(2)}_{11c}(t) & h^{(2)}_{22c}(t) & h^{(2)}_{21c}(t) \\ h^{(3)}_{21c}(t) & h^{(3)}_{22c}(t) & h^{(3)}_{11c}(t) & h^{(3)}_{12c}(t) \\ h^{(4)}_{22c}(t) & h^{(4)}_{21c}(t) & h^{(4)}_{12c}(t) & h^{(4)}_{11c}(t) \end{bmatrix} \begin{bmatrix} w_{11}(t) \\ w_{12}(t) \\ w_{21}(t) \\ w_{22}(t) \end{bmatrix} \tag{5.62}$$

where $h^{(l)}_{ijc}$ is the (i,j)th IR polarisation component of the lth calibration target. The frequency analogue, taking account of the noise effect, is

$$\mathbf{Y}_{vk}(\mathrm{i}\omega) = \mathbf{H}_k(\mathrm{i}\omega)\mathbf{W}_v(\mathrm{i}\omega) + \mathbf{E}_{vk}(\mathrm{i}\omega) \tag{5.63}$$

where $\mathbf{Y}_{vc}(\mathrm{i}\omega) = [Y_{11c}(\mathrm{i}\omega) \quad Y_{12c}(\mathrm{i}\omega) \quad Y_{21c}(\mathrm{i}\omega) \quad Y_{22c}(\mathrm{i}\omega)]^T$ is the vector of the signal spectral functions $y_{ijc}(t) = y_{ijc}(t) + \varepsilon_{ijc}(t)$, where $\varepsilon_{ijc}(t)$ is random calibration noise; $\mathbf{W}_v(\mathrm{i}\omega) = [W_{11}(\mathrm{i}\omega) \quad W_{12}(\mathrm{i}\omega) \quad W_{21}(\mathrm{i}\omega) \quad W_{22}(\mathrm{i}\omega)]^T$ is the vector of FR polarisation components of the meter;

$$\mathbf{H}_k(\mathrm{i}\omega) = \begin{bmatrix} H^{(1)}_{11c}(\mathrm{i}\omega) & H^{(1)}_{12c}(\mathrm{i}\omega) & H^{(1)}_{21c}(\mathrm{i}\omega) & H^{(1)}_{22c}(\mathrm{i}\omega) \\ H^{(2)}_{12c}(\mathrm{i}\omega) & H^{(2)}_{11c}(\mathrm{i}\omega) & H^{(2)}_{22c}(\mathrm{i}\omega) & H^{(2)}_{21c}(\mathrm{i}\omega) \\ H^{(3)}_{21c}(\mathrm{i}\omega) & H^{(3)}_{22c}(\mathrm{i}\omega) & H^{(3)}_{11c}(\mathrm{i}\omega) & H^{(3)}_{12c}(\mathrm{i}\omega) \\ H^{(4)}_{22c}(\mathrm{i}\omega) & H^{(4)}_{21c}(\mathrm{i}\omega) & H^{(4)}_{12c}(\mathrm{i}\omega) & H^{(4)}_{11c}(\mathrm{i}\omega) \end{bmatrix}$$

is the FR matrix of a set of calibration targets ($H^{(l)}_{ijc}(\mathrm{i}\omega) = F\{h^{(l)}_{ijc}(t)\}$); $\mathbf{E}_{vc}(\mathrm{i}\omega) = [E_{11c}(\mathrm{i}\omega) \quad E_{12c}(\mathrm{i}\omega) \quad E_{21c}(\mathrm{i}\omega) \quad E_{22c}(\mathrm{i}\omega)]^T$ is the vector of spectral functions of random calibration noise.

The finite-dimensional analogue of this expression has the same structure

$$\mathbf{Y}_{vc}(k) = \mathbf{H}_c(k)\mathbf{W}_v(k) + \mathbf{E}_{vc}(k) \tag{5.64}$$

For the solution of this set of equations, it is necessary that the $\mathbf{H}_c(i\omega)$ matrix (or its equivalent, the $\mathbf{H}_c(k)$ matrix) should be nonsingular at all frequencies ω or at all k, where the SNR is nonzero. This can be achieved by an adequate choice of the set of calibration targets, otherwise the matrix will be degenerate. For example, if one nondepolarising, axially symmetric object (a sphere) is used for the calibration, the matrix of the set (eqn. 5.63) takes the form

$$\mathbf{H}_c(i\omega) = \begin{bmatrix} H_{sp}(i\omega) & 0 & 0 & H_{sp}(i\omega) \\ 0 & H_{sp}(i\omega) & H_{sp}(i\omega) & 0 \\ 0 & H_{sp}(i\omega) & H_{sp}(i\omega) & 0 \\ H_{sp}(j\omega) & 0 & 0 & H_{sp}(j\omega) \end{bmatrix} \tag{5.65}$$

and is degenerate. Even an attempt to carry out multi channel calibration by choosing various nondepolarising targets (varying in shape or size) cannot change this situation. For this reason, it is also necessary to employ depolarising calibration targets. In particular, the first ($l = 1$) and the second ($l = 2$) calibration targets may represent a metallic sphere and a cylinder arranged at an angle of 30° to the vertical axis. After the measurements have been made and the vector calibration signal obtained, the vector-matrix equation (eqn. 5.64) can be solved by the regularising algorithm

$$\hat{\mathbf{W}}_v(k) = [\mathbf{H}_c(k)\mathbf{H}_c^{*T}(k) + \alpha\mathbf{I}]^{-1}\mathbf{H}_c^{*T}(k)\mathbf{Y}_{vc}(k) \tag{5.66}$$

The estimate obtained can be further used to evaluate the target FR and IR from eqn. 5.61.

The efficiency of the algorithms described is illustrated by the measurements presented in Fig. 5.6. It can be seen that the polarisation distortions can be reduced by 15–20 dB in simultaneous processing of all vector signal components.

5.6 Local estimation of total UWB radar responses

5.6.1 Local regularisation in the time domain

The effectiveness of a regularisation procedure for UWB response estimation largely depends on what additional information is invoked on the problem being solved. The above estimations used quantitative information on the error level of the initial data (to find the regularisation parameter or the reduction factor) as well as qualitative information concerning the class of validity of the problem solution. But it is also useful to employ additional information, in particular on the local backscatter of electromagnetic waves by bodies of complicated geometry.

In most cases, a signal scattered by a large (relative to the resolution unit) body represents a series of echo pulses from local scatterers – simple or

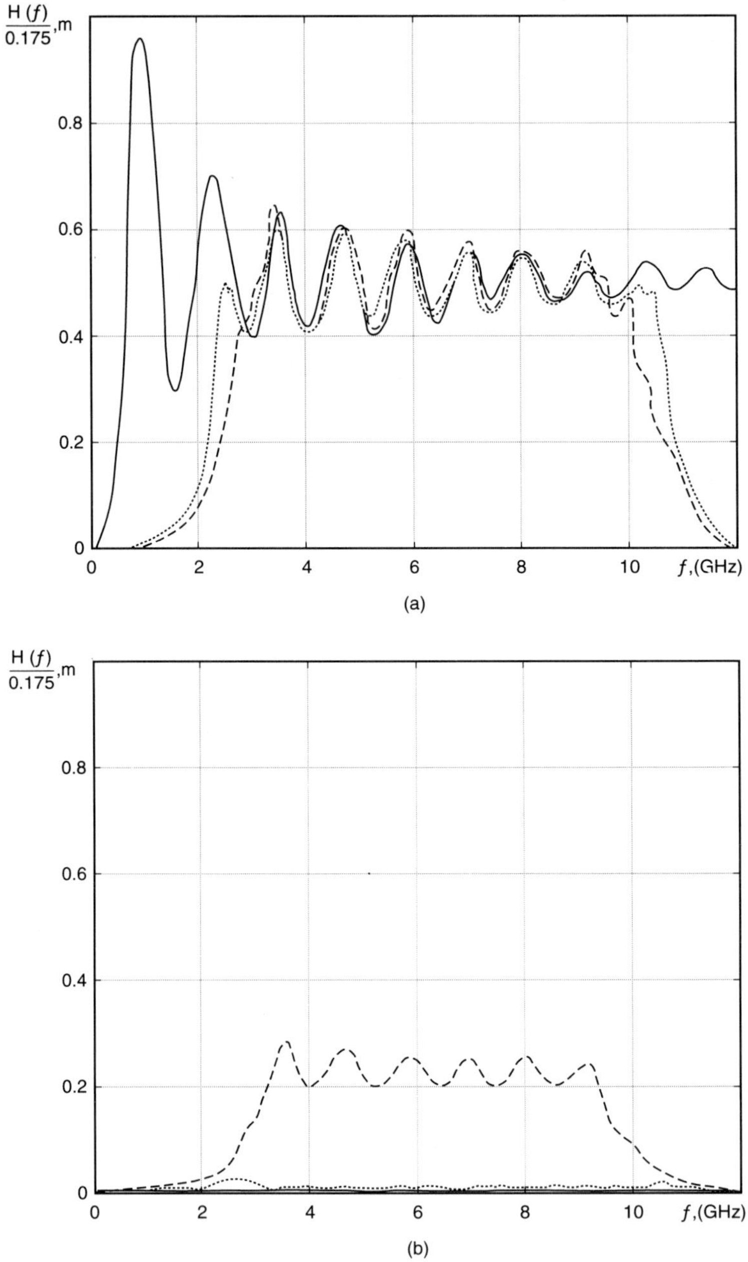

Fig. 5.6 Estimations of amplitude-frequency responses of the sphere

a For parallel polarisation
b For cross polarisation
- - - - before processing
...... after processing

complex glittering points (Chapter 3). The waveform and intensity of each local return signal may vary within wide limits. Evidently, every signal impulse (time interval) must have its own regularisation parameter. If we use a common regularisation parameter chosen, say, from the residual principle, then the characteristics of the correcting filter will be the same for all pulses, reducing the estimation accuracy.

Another reason for the presence in a backscattered signal of time-local components with different spectral structure and time variation is the existence of two qualitatively different components – forced and natural – both in total and local signals (see Chapter 2). Therefore, the use of individual regularisation parameters seems to be quite reasonable.

It is hardly possible to give a quantitative value of the gain in the estimation accuracy owing to local regularisation in the general case, but its high effectiveness can be illustrated using a few examples.

Suppose a target represents two glittering points differing in intensity. The recorded signal consists of two impulse signals with different amplitudes against a noise background. In the case of global regularisation, the correcting filter parameters (the regularisation parameter, etc.) are mainly chosen from the condition for optimal estimation of the more intense local scatterer. The transmission bandwidth of the meter after the filtering becomes large, because the powerful response from the high intensity point gives a high SNR. As a result, the residual additive noise at the filter output may screen the response from the low intensity point. Local regularisation removes this screening effect.

The method for local regularised processing in the time domain is as follows:

The total recorded signal $y_\varepsilon(n)$, $n = \overline{1, N}$, is split into M partial pulses $y_{\varepsilon m}(n)$, $m = \overline{1, M}$, corresponding to the scattering by individual local centres or their compact groups

$$y_\varepsilon(n) = \sum_{m=1}^{M} y_{\varepsilon m}(n), \qquad n = \overline{1, N} \tag{5.67}$$

where $y_{\varepsilon m}(n) = y_\varepsilon(n)$, $n = \overline{N_{Hm}, N_{km}}$.

Here, N_{Hm} and N_{km} are the beginning and the end of counting the mth partial pulse (bounds of the mth time interval).

Since the signal processing can best be made in the frequency domain, all partial pulses must have coinciding discrete frequencies, so every pulse, $y_{\varepsilon m}(n)$, is supplemented by zero counts to the right and to the left as far as the dimension N

$$Y_{\varepsilon mg}(n) = \begin{cases} 0, & n = \overline{1, N_{Hm} - 1} \\ Y_{\varepsilon m}(n), & n = \overline{N_{Hm}, N_{km}} \\ 0, & n = \overline{N_{km} + 1, N} \end{cases}$$

This procedure is followed by signal processing for each of the glittering points to estimate their FR and IR, $\hat{\mathbf{H}}_m(k)$, $\hat{\mathbf{h}}_m(n)$, $k, n = \overline{1, N}$. Then one finds the total IR and FR estimates for the radar target in question

$$\hat{h}(n) = \sum_{m=1}^{M} \hat{h}_m(n)$$
$$\hat{H}(k) = \sum_{m=1}^{M} \hat{H}_m(k) \tag{5.68}$$

5.6.2 Local regularisation in the frequency domain

The characteristics of a correcting filter (the regularisation parameter and its analogues) in the above algorithms were evaluated by simultaneous processing of the whole array of frequency samples of the initial data. Let us analyse the possibilities of separate data processing at each frequency individually. This will help us to get a better understanding of the physical sense of regularising estimation algorithms and their potentiality. We shall try to offer some simplified but computationally more efficient algorithms.

It was shown in Section 5.4 that the DFT count, $Y(k)$, of the signal, $y(n)$, at the kth discrete frequency ($k = \overline{1, N}$), is related to that of the sought-for target FR, $H(k)$, meter FR, $W(k)$, and the DFT of the noise, $E(k)$, as

$$\begin{aligned} Y_\varepsilon(k) &= W(k)H(k) + E(k) \\ &= Y(k) + E(k) \end{aligned} \tag{5.69}$$

where $Y(k)$ is the determinate component of the signal DFT.

We shall assume that the noise is centred ($\bar{E}(k) = 0$) and has known dispersion $D\{E(k)\} \equiv M\{E^*(k)E(k)\} = \sigma_E^2(k)$.

The inverse target FR estimate at the kth frequency has the form

$$\begin{aligned} \hat{H}(k) &= \frac{Y_\varepsilon(k)}{W(k)} \\ &= H(k) + \frac{E(k)}{W(k)} \end{aligned} \tag{5.70}$$

This is unbiased, because its mathematical expectation is

$$\begin{aligned} \overline{\hat{H}(k)} &= H(k) + \frac{\overline{E(k)}}{W(k)} \\ &= H(k) \end{aligned} \tag{5.71}$$

The variance of the estimate is written as

$$\begin{aligned} \sigma_H^2(k) &\equiv D\{\hat{H}(k)\} \\ &= \frac{\sigma_E^2(k)}{|W(k)|^2} \end{aligned} \tag{5.72}$$

It can easily be shown that the estimate (eqn. 5.70) is effective in the class of all unbiased estimates (on the assumption of Gaussian noise), since the dispersion (eqn. 5.72) coincides with the Cramer–Rao bound.

The relative variance of the estimate is

$$\delta_H^2(k) = \frac{\sigma_H^2(k)}{|H(k)|^2} = \frac{\sigma_E^2(k)}{|Y(k)|^2} = \frac{1}{q_0^2(k)} \tag{5.73}$$

where $q_0^2(k) = |Y(k)|^2/\sigma_E^2(k)$ is the power SNR.

Examine the properties of biased estimates with reference to a reduced estimate

$$\hat{H}_r(k) = r(k)\hat{H}(k) = r(k)\frac{Y(k)}{W(k)} = r(k)H(k) + r(k)\frac{E(k)}{W(k)} \tag{5.74}$$

It appears that it is related, in a fairly simple way, to the ridge estimation

$$\hat{H}_\alpha(k) = \hat{H}_r(k)|_{r=1/(1+\alpha)} \tag{5.75}$$

Its mathematical expectation

$$\overline{\hat{H}_r(k)} = r(k)H(k) \tag{5.76}$$

differs from the true value of $H(k)$, that is, the estimate is really a biased one. The mean estimation error

$$\Delta\hat{H}_r(k) = H(k) - \hat{H}_r(k) = H(k)(1 - r(k)) + r(k)\frac{E(k)}{W(k)} \tag{5.77}$$

known as the estimate bias, is

$$\overline{\Delta\hat{H}_r(k)} = H(k)(1 - r(k)) \tag{5.78}$$

The squared bias module has the form

$$l_0^2 = \overline{\Delta\hat{H}_r^*(k)}\,\overline{\Delta\hat{H}_r(k)} = |H(k)|^2(1 - r(k))^2 \tag{5.79}$$

The reduced estimate dispersion is described by the formula

$$\sigma_{Hr}^2(k) \equiv D\{\hat{H}_r(k)\} = r^2(k)\frac{\sigma^2(k)}{|W(k)|^2}$$
$$= r^2(k)\sigma_H^2(k) \tag{5.80}$$

and the relative dispersion is written as

$$\sigma_{Hr}^2(k) = r^2(k)\delta_H^2(k)$$
$$= r^2(k)\frac{1}{q_0^2(k)} \tag{5.81}$$

The total mean square error (module) of the estimation (TMSES) is defined as

$$l_r^2(k) = M\{\Delta\hat{H}_r^*(k)\Delta\hat{H}_r(k)\}$$
$$= |H(k)|^2(1-r(k))^2 + r^2\sigma_H^2(k)$$
$$= l_0^2(k) + \sigma_{Hr}^2(k) \tag{5.82}$$

It follows from eqn. 5.82 that the TMSES consists of two components: the squared bias and the estimate dispersion. Each component can be reduced to zero by choosing an appropriate reduction factor, although this choice will not be optimal for the TMSES in the general case (Fig. 5.7). Clearly, there is such a value, at which the TMSES value will be minimal

$$r_{opt}(k) = \min_r l_r^2(k) \tag{5.83}$$

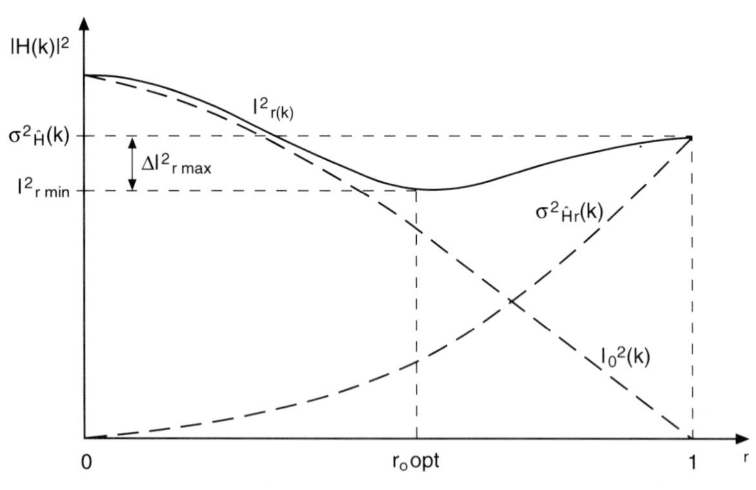

Fig. 5.7 Total mean square error of estimation versus coefficient of reduction

This value also corresponds to the maximum gain in accuracy (in the estimation error power) for the biased estimate relative to the unbiased one

$$\Delta l_r^2(k) = \sigma_H^2(k) - l_r^2(k)$$
$$= \sigma_H^2(k)(1 - r^2(k)) - |H(k)|^2(1 - r(k))^2 \tag{5.84}$$
$$\Delta l_{r\,max}^2(k) = \sigma_H^2(k)(1 - r_{opt}^2(k)) - |H(k)|^2(1 - r_{opt}(k))^2 \tag{5.85}$$

The optimal value, $r_{opt}(k)$, can be found by solving the equation

$$\frac{dl_r^2(k)}{dr(k)} = 0 \tag{5.86}$$

which readily yields the following expression

$$r_{opt}(k) = \frac{|H(k)|^2}{\sigma_H^2(k) + |H(k)|^2}$$
$$= \frac{q_0^2(k)}{1 + q_0^2(k)} \tag{5.87}$$

The expressions for the minimum TMSES and absolute gain are written as

$$l_{r\,min}^2(k) = |H(k)|^2 \left[\frac{1}{1 + q_0^2(k)}\right] \tag{5.88}$$

$$\delta_{r\,min}^2(k) = \frac{1}{1 + q_0^2(k)} \tag{5.89}$$

$$\Delta l_{r\,max}^2(k) = |H(k)|^2 \left[\frac{1}{q_0^2(k)(1 + q_0^2(k))}\right] \tag{5.90}$$

The relative gain from the application of the biased estimation method is defined as

$$V_r = \sigma_H^2(k)/l_{r\,min}^2(k) \tag{5.91}$$

With eqns. 5.72 and 5.88, we have

$$V_r = 1 + \frac{1}{q_0^2(k)} \tag{5.92}$$

It follows from the latter expression that the relative effectiveness of biased estimates increases with decreasing SNR (Fig. 5.8). Since the frequency band of an impulse meter always contains frequencies with a low SNR, the regularised estimates for these components will be appreciably more effective than the conventional ones.

The local frequency regularisation procedure carried out for all N frequencies of the signal bandwidth also optimises the meter in the integral

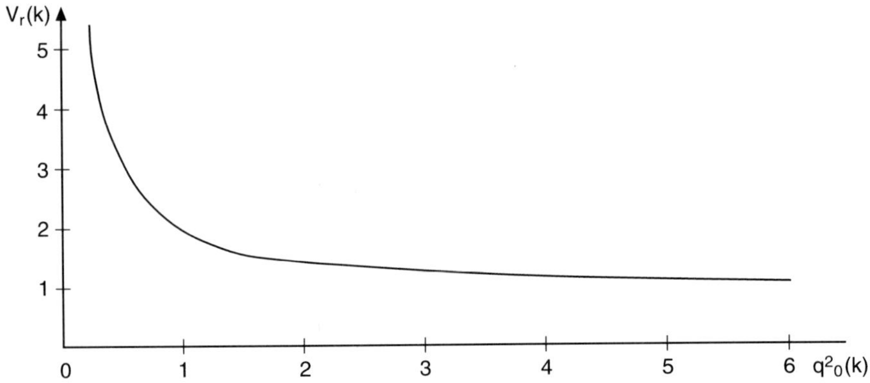

Fig. 5.8 Gain of biased estimation versus signal to noise ratio

quality indices. Indeed, since the overall (over all frequencies) TMSES is a sum of minimum possible summands, it is also minimal

$$\begin{aligned} l^2_{r\,min} &= \sum_{k=1}^{N} l^2_{r\,min}(k) \\ &= \sum_{k=1}^{N} \frac{|H(k)|^2}{1 + q_0^2(k)} \\ &= \sum_{k=1}^{N} \frac{|H(k)|^2}{1 + \dfrac{|Y(k)|^2}{\sigma^2(k)}} \\ &= \sum_{k=1}^{N} \frac{\sigma_E^2(k)|H(k)|^2}{\sigma_E^2(k) + |W(k)|^2|H(k)|^2} \end{aligned} \tag{5.93}$$

If $\sigma_E^2(k) \equiv \sigma_E^2$, then

$$l^2_{r\,min} = \sigma_E^2 \sum_{k=1}^{N} \frac{|H(k)|^2}{\sigma_E^2 + |W(k)|^2|H(k)|^2} \tag{5.94}$$

More convenient generalised (overall) characteristics are the minimum frequency-averaged TMSES

$$\begin{aligned} \overline{l^2_{r\,min}} &= l^2_{r\,min}/N \\ &= \frac{\sigma_E^2}{N} \sum_{k=1}^{N} \frac{|H(k)|^2}{\sigma_E^2 + |W(k)|^2|H(k)|^2} \end{aligned} \tag{5.95}$$

and the minimum average relative TMSES

$$\overline{\delta^2_{r\,min}} = \frac{\overline{l^2_{r\,min}}}{H^2(k)} = \frac{\sigma_E^2 \sum_{k=1}^{N} \frac{1}{\sigma_E^2 + |W(k)|^2 |H(k)|^2}}{\sum_{k=1}^{N} |H(k)|^2} \tag{5.96}$$

In practice, the estimation of the potential capabilities of a meter is often made with a given test target with constant RCS, σ_0; then $|H(k)| = \sqrt{\sigma_0}$, and eqn. 5.96 transforms to

$$\sigma^2_{r\,min\,0} = \frac{\sigma_E^2}{N} \sum_{k=1}^{N} \frac{1}{\sigma_E^2 + |W(k)|^2 \sigma_0} \tag{5.97}$$

Note that this parameter, calculated in the frequency domain, characterises, by virtue of Parseval's theorem, the relative IR error for a target with constant RCS. In this case, the target IR is a time-dependent delta function. Therefore, the meter's output response (after the digital processing), or, rather, the mean square value of the response, is closest to the delta function while its spectrum is closest to a uniform function, if the dynamic characteristics of the meter hardware were preset. Therefore, local frequency regularisation can potentially provide a reduction of the actual meter (hardware) to a synthesised analogue–digital system with optimal (in the above sense) dynamic characteristics.

By analogy with eqn. 5.97, we can derive an expression for the mean relative dispersion of the unbiased (inverse) FR estimate for a target with constant RCS

$$\delta_0^2 = \frac{\sigma_E^2}{N\sigma_0} \sum_{k=1}^{N} \frac{1}{|W(k)|^2} \tag{5.98}$$

The mean overall gain from the local frequency regularisation, as compared with inverse filtering, can be defined as

$$V_r = \frac{\delta_0^2}{\delta^2_{r\,min}} \tag{5.99}$$

Taking account of eqns. 5.97 and 5.98, we have

$$V_r = \frac{\sum_{k=1}^{N} \frac{1}{|W(k)|^2}}{\sum_{k=1}^{N} \frac{1}{\sigma_E^2 + |W(k)|^2 \sigma_0}} \tag{5.100}$$

Eqns. 5.95 and 5.100 permit the estimation of attainable mean values for the relative total square error in target IR and FR estimates for meter hardware with given dynamic and noise characteristics. They can also yield the mean value of the overall gain from the application of regularising algorithms for UWB radar response estimation. Practically, V_r may have a value of the order of 10^2–10^4, which, naturally, corresponds to the condition number of the SLAE to be solved in signal processing.

Theoretically, the efficiency of the procedure suggested is higher than that of the regularising procedures analysed earlier. But experimental tests and simulation have shown the advantages to be insignificant, and the expressions derived above may be regarded only as indications of a potential precision of a UWB meter. This is so because there is no possibility of using in practice the reduction coefficients (regularisation coefficients) which could provide the highest estimation accuracy. Indeed, the optimal values of the reduction coefficients depend on the SNR which, in turn, depends on the unknown target FR.

To get around this difficulty, one may use various approaches, including reduced stochastic estimation [61]. In this case, the quasi optimal reduction coefficient is chosen from the empirical SNR

$$r_{k\,opt}(k) = \frac{q_e^2(k)}{1 + q_e^2(k)} \tag{5.101}$$

where $q_e^2(k) = |Y(k)|^2/\sigma_E^2(k)$.

Since the function $Y(k)$ is random, also random are the values of the reduction coefficient and regularisation parameters. An analytical treatment of stochastic estimation accuracy is difficult to carry out, so it was studied by simulation. It can be shown that under low SNR conditions ($q^2 \leqslant 0.707$), the TMSE of an empirical estimate, being much smaller than the relative unbiased estimate dispersion, considerably exceeds the corresponding value for an optimal local estimate (Fig. 5.9). To raise the accuracy of empirical local frequency estimates, it is reasonable to try to take advantage of additional information on the target backscatter properties, which was implicitly done by the ridge estimation algorithm.

5.6.3 Global regularisation in the frequency domain

Since the accuracy loss in stochastic regularisation is due to errors in the empirical evaluation of the SNR at a single, isolated frequency, it would seem natural to try to use measurements made at other frequencies. The estimates obtained lose their local properties and are defined by the global information on the target backscatter characteristics.

Such types of information may come from an assumption about the type of frequency dependence of target RCS (AFR), in particular, on the possibility of describing it, for the meter operation bandwidth, by a polynomial to a certain power. We shall illustrate the effectiveness of this approach with a

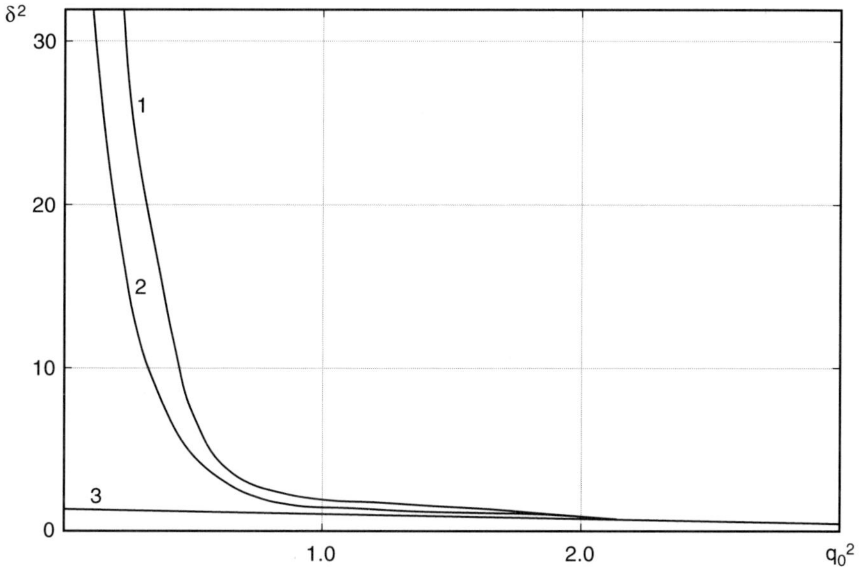

Fig. 5.9 Efficiency of the estimations

1 Stochastic local
2 Stochastic global
3 Optimal local

simple return signal measurement for a target with constant RCS, $H(k) = H$ (*a priori*, the frequency dependence of RCS is described by the zero-order polynomial).

The expressions derived above give the SNR at the kth frequency

$$q_0^2(k) \equiv Y^2(k)/\sigma_E^2(k) = \frac{|H|^2|W|^2(k)}{\sigma_E^2(k)} \tag{5.102}$$

To calculate its empirical values, we shall make use of the target AFR unbiased estimate over all frequency components of a signal (overall estimate) or from some of them (group estimate). Putting, for simplicity, $\sigma_E^2(k) = \sigma_E^2$, we can easily derive an expression for the effective unbiased FR estimate

$$\hat{H} = \frac{1}{\sum_{k=1}^{N} W(k)} \sum_{k=1}^{N} \hat{H}(k) W(k) \tag{5.103}$$

where $\hat{H}(k) = Y(k)/W(k)$ is the local FR value at a single kth frequency.

The use of this estimate for finding the SNR value at every frequency from eqn. 5.102 markedly increases the estimation accuracy and that of the

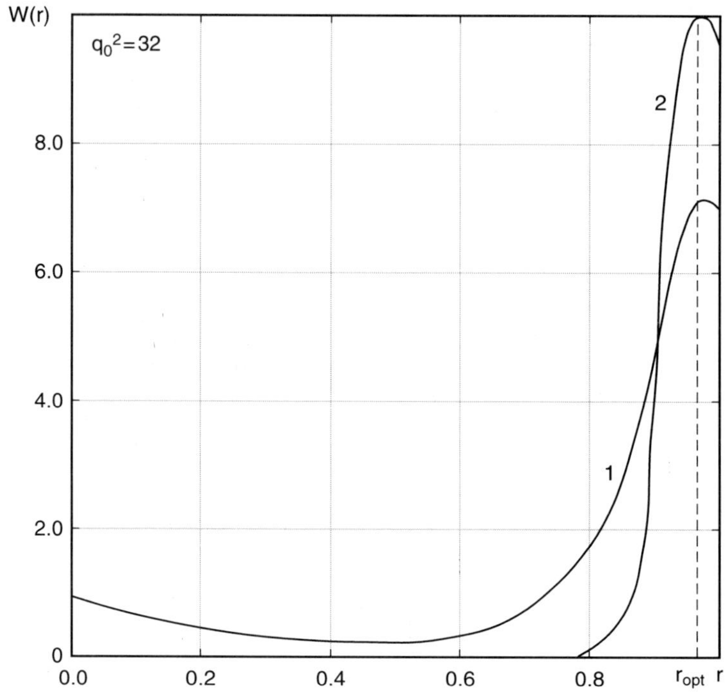

Fig. 5.10 Empirical probability density functions ($q_0^2 = 32$)

1 Stochastic local estimations
2 Global estimations

final regularised estimate of FR counts. This is clear, for example, from a comparison of the empirical probability density functions for the reduction coefficients of the local and global stochastic estimates (Fig. 5.10) obtained under the following conditions:

- the number of discrete counts, 8;
- the distribution of square AFR counts of the meter, $W^2(k) = W^2 2^{(k-1)}$, where W^2 is a constant;
- $\sigma_E^2(k) = \sigma_E^2 = \text{const}$.

The graphs show that the global reduction coefficient lies closer to the optimal value, r_{opt}, than to the local value and that the efficiency of the method rises with decreasing SNR. For this reason, the global (group) estimates have an accuracy approaching that of the optimal local estimates.

The empirical reduced estimate of the type (eqn. 5.74) can easily be transformed to a conventional, regularised form

$$\hat{H}_\alpha(k) = \frac{W^*(k)Y(k)}{W(k)W^*(k) + \alpha_q(WW^*)_{max}} \tag{5.104}$$

where α_q is the normalised regularisation parameter with the sense of the SNR estimate at the frequency of the meter maximum AFR

$$\alpha_q = \left| \frac{|H|^2 |W|^2_{max}}{\sigma_E^2} \right|^{-1} \tag{5.105}$$

Clearly, its structure is equivalent to that of the above estimate (eqn. 5.36). In contrast to the latter, however, the estimate (eqn. 5.104) contains an explicit definition of the optimal regularisation parameter; therefore, there is no need for the iteration of $\alpha = \alpha_{opt}$ to take much computational time. A simulation has shown the evaluation efficiency of both approaches to be close to one another.

An additional possibility for raising the estimation accuracy is provided by the iteration procedure suggested, which uses an unbiased target FR estimate at the first stage, as in the previous case, but at the subsequent stages it uses the biased estimates obtained at the earlier stages. Statistical modelling has demonstrated that a two-step procedure is sufficient to provide a 10–15% reduction in the TMSE values. Fig. 5.11 presents the distribution

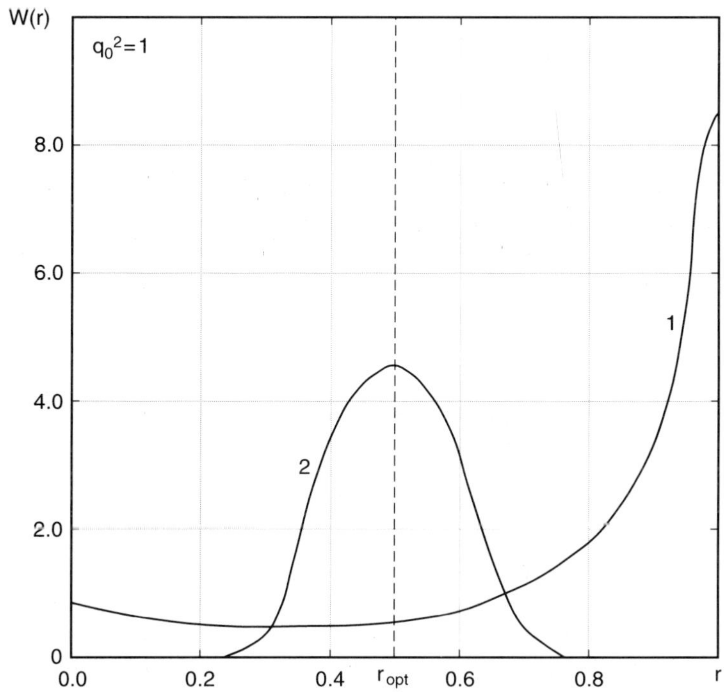

Fig. 5.11 Empirical probability density functions ($q_0^2 = 1$)

1 Stochastic local estimations
2 Global estimations

graphs for the reduction coefficient for FR estimation (1) by the least squares method and (2) after refinement by a single iterative step.

5.6.4 Time-local and frequency-global regularised estimations

Global estimates in the frequency domain have a disadvantage independent of the way of choosing the regularisation parameter: the appearance of specific errors if the *a priori* assumption of target RCS constancy within the measuring bandwidth is violated. At frequencies with a lower actual target RCS than its frequency-averaged estimate obtained from eqn. 5.38 or 5.104, the local regularisation parameter, $\alpha(k) = \alpha_q|W|^2_{max}/|W(k)|^2$, is underestimated while the reduction coefficient, $r(k)$, is overestimated. As a result, the estimate has a nonoptimal ratio between the bias and the noise error: the latter has a larger dispersion, leading to a higher TMSE than the optimal value. At frequencies with a higher actual target RCS than its average value, the situation is the reverse: the reduction coefficient turns out to be underestimated while the regularisation parameter is overestimated, leading to a greater estimate bias and lower accuracy.

There are several ways of avoiding such additional errors. One has already been mentioned; it is to make the *a priori* model of the frequency dependence of RCS more complex. But this approach is ineffective when measuring return signals from targets of large electrical dimensions and complex scattering FR, because it involves a large number of polynomial frequency dependent parameters of the RCS to be estimated, which cannot be done with a high accuracy.

Another approach, also mentioned above, is to use group estimates and to put the target RCS as constant within certain frequency subranges. This reduces the level of methodological errors, but such estimates have a lower accuracy than the optimal regularised local values because of the small number of measurements to be made in every frequency subrange.

It appears that the most reasonable solution is to combine the concepts of time-local and frequency-global regularisations. By localising and cutting out responses from individual glittering points at the first stage, one can, in most cases, identify the echoes from local scatterers with small longitudinal electrical dimensions and a weak frequency dependence of the RCS. Frequency-global regularisation can be employed at the second stage, because then it will not entail excessive methodological errors. Radar response estimates for individual local centres can be reduced to the total responses for the whole target in both the time and frequency domains with the procedures discussed.

An additional possibility for an increase in estimate accuracy is provided by the following suggested iterative procedure, in which at the first stage of regularisation parameter determination (as in the previous case) an unbiased target FR estimate is used, and at subsequent stages biased estimates received during the previous steps are used. Statistical modelling has

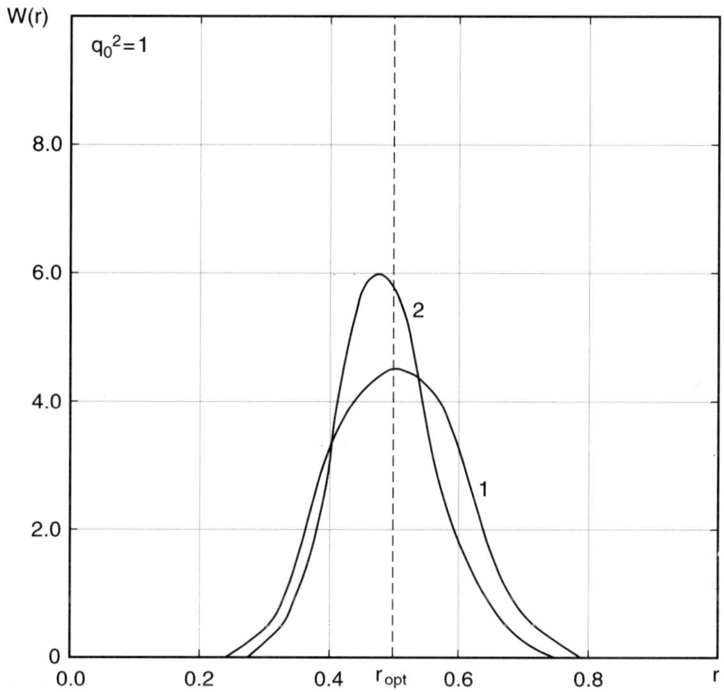

Fig. 5.12 Efficiency of the iterative global estimation

1 Stochastic global estimation
2 Stochastic global estimation with one iteration

shown that it is enough to use a two-step procedure. It provides TNSES decrease of about 10–15% (see Fig. 5.12 where diagrams of reduction factor distribution on FR estimation by the least square method (1) and by use of close definition after a single iteration (2) are presented).

5.7 Secondary signal processing by probe meters

5.7.1 Focused aperture synthesis

One of the techniques used for local response measurement and radar pulse imaging by UWB probe systems is focused aperture synthesis. For simplicity, we shall consider the problem of two-dimensional imaging, using a linear focused synthetic aperture (FSA). An extension to the case of three-dimensional imaging by a plane FSA is obvious.

A linear FSA (the space region where a transmit–receive probe can be placed) is located in the interval $x[-L/2, L/2]$ at $Y = D$. The meter's coordinate reference $X0Y$ is superimposed with the construction coordinate reference $X^{\Theta}0Y^{\Theta}$, around which a target can be rotated; here, Θ is the object's aspect relative to the FSA. The antenna is moved at a step Δl, point by point through N_L points; $N_L = (L/\Delta l) + 1$, $L = (N_L - 1)\Delta l$. The coordinates of the ith antenna point are (l_i, D), where $l_i = i\Delta l$, $i = \overline{-I, I}$, $I = (N_L - 1)/2$. The antenna has a normalised directivity pattern (DP), $d(\varphi)$, where φ is the angle from the aperture surface normal. It is assumed that the impulse and frequency responses of the antenna depend on the regular coordinate as

$$\begin{aligned} h_a(t, \varphi) &= d(\varphi) h_{a0}(t) \\ K_a(\mathrm{i}\omega, \varphi) &= d(\varphi) K_{a0}(\mathrm{i}\omega) \end{aligned} \tag{5.106}$$

where $h_{a0}(t)$ and $K_{a0}(\mathrm{i}\omega)$ are the antenna IR and FR in the axial direction. For low directivity antennas, these relations hold fairly well for a wide range of the angle φ.

Aperture focusing is performed at space points within the region

$$\begin{aligned} x &\in [x_{min}, x_{max}] = [-\Delta X, \Delta X] \\ y &\in [y_{min}, y_{max}] = [-\Delta Y, \Delta Y] \end{aligned}$$

which is discretised along both coordinates with a step Δr taken to be constant. The processes of pulse imaging and aperture focusing occur at discrete points $(x_m, y_k) \triangleq (m, k)$, where $x_m = m\Delta r$, $y_k = kr$; $m = \overline{-M, M}$, $k = \overline{-K, K}$. Here, $M = \{\Delta X/\Delta r\}$, $K = \{\Delta Y/\Delta r\}$, $\Delta r \ll \lambda_0$, $\{\cdot\}$ is the integer separation operator. The image matrix is a $(2M+1) \times (2K+1)$ matrix with the total number of elements $P = (2M+1)(2K+1)$.

The imaging involves the measurement of so-called trajectory signals $s_e(t, l) = s_e(t, i)$ at each position of the synthetic aperture (Fig. 4.2). We shall initially suppose that all N_L signals are measured and recorded by a storage device prior to the imaging procedure. However, this will be shown later to be unnecessary: to save on the storage capacity and the imaging time, they can be obtained recursively, as a trajectory signal appears, without storing the previous ones.

The signals are stored in a time-discretised form, namely, a continuous signal $s_\varepsilon[t, i]$; $t[0, T_{ob}]$ is converted to a discrete signal $s_\varepsilon[n, i]$; $n = \overline{0, N_T - 1}$, where T_{ob} is the observation time interval, $s_\varepsilon[n, i] = s_\varepsilon(t, i)|_{t=nT}$, $N_T = (T_{ob}/T) + 1$ is the number of signal counts. The discretisation step must meet Kotelnikov's theorem: $T \leqslant 1/(2 \div 5) f_{up}$, where f_{up} is the upper frequency of the incident signal. The observation time interval, T_{ob}, and the dimension of the count array, N_T, are determined by the frame size

$$\begin{aligned} T_{ob} &\approx \frac{2}{c}\sqrt{\left(\frac{L}{2} + \Delta X\right)^2 + (D + \Delta Y)^2} \\ N_T &= (T_{ob}/T) + 1 \end{aligned} \tag{5.107}$$

Signal imaging involves three types of operation:

1 Time processing (filtering) of trajectory signals provides correction of the dynamic characteristics of the transmit–receive channel and, partly, those of the transmit–receive antenna. As a result, the received signals are compressed (squeezed in time) to decrease the level of sidelobes, thus increasing the radar resolution on both coordinates. The correction algorithm was discussed in Section 5.4. It is based on the IR, $w(t)$, of the measuring channel, with the probe occupying the central position relative to the reference point of the measuring coordinates. The IR, $w_0(t)$, is defined by placing a calibration scatterer at this point. The regularisation parameter of the algorithm can be computed in advance – it was found to be equal to the SNR for the threshold (minimum) RCS of a local scatterer placed at the reference point.

When the IR of the meter has an acceptable form, that is, when it has a low level of sidelobes and its extent, $\Delta R_w = 2\Delta\tau/c$, is essentially less than the required resolution spacing in depth, ΔR_y, the time processing procedure can be omitted to reduce the computational costs. The signal used in the synthesis has the form

$$s[n,i] = s_\varepsilon[n,i] \tag{5.108}$$

otherwise $s[n,i] = s_\alpha[n,i]$.

2 The phasing (time delay) of every trajectory signal obtained at the previous stage is performed with matched positions of the focal point and the respective array element. Signal phasing is a process of eliminating the time shifts of trajectory signals by sweeping them from the ith element to the focal point (x_m, y_k) and back. The distance $R(l_i, x_m, y_k)$ from the point (l_i, D) to the point (x_m, y_k) is defined as

$$\begin{aligned} R(l_i, x_m, y_k) &\triangleq R(i,m,k) \\ &= \sqrt{(l_i - x_m)^2 + (D - y_k)^2} \\ &= \sqrt{(i\Delta l - m\Delta r)^2 + (D - k\Delta r)^2} \end{aligned} \tag{5.109}$$

where $i = \overline{-I, I}$, $m = \overline{-M, M}$ and $k = \overline{-K, K}$.

Then the time delay (time shift) between the radiated signal and the signal received at the ith point, associated with the travel to the point (x_m, y_k) and back, is described by the expression

$$\begin{aligned} \tau(l_i, x_m, y_k) &\triangleq \tau(i,m,k) \\ &= \frac{2R(i,m,k)}{c} \end{aligned} \tag{5.110}$$

3 Additional correction of distortions is made. These are primarily amplitude distortions arising in the propagation space due to the directive

properties of transmit–receive antennas, different path lengths of signals, measurement clutter, etc. The quality of the various imaging procedures depends directly on how well the distortions can be compensated. Whether the measurement noise has been taken into account in the spatial processing or not is a criterion for grouping the algorithms into two classes – those for inverse filtering and pseudo-inverse regulating algorithms.

5.7.2 Spatial inverse processing

This procedure allows for the fact that the signal radiated by an antenna and that backscattered by a local glittering point both have spherical wavefronts, leading to the inverse dependence of the signal amplitude on the radar range $R(i, m, k)$. It is necessary to consider the amplitude dependence on the angle $\varphi(i, m, k)$, at which the point (m, k) is viewed from the ith array node, and on the antenna directivity, $d(\varphi)$

$$\varphi(i, m, k) = \arctan\left(\frac{m\Delta r - i\Delta l}{D - k\Delta r}\right) \tag{5.111}$$

Inverse filtering does not remove the measurement noise, so the trajectory signal, $s[n, i]$, at the ith antenna output with the field strength gain, G_a (when a point target with the ESL $l_i(m, k)$ is located at the point (m, k)), will be described by the expression:

$$s[n, i] \cong \frac{G_a^2 \lambda_0 U_0}{8\pi\sqrt{\pi}} \frac{d^2(\varphi(i, m, k))}{R^2(i, m, k)} l(m, k) w_0[n - N_t(i, m, k)] \tag{5.112}$$

where λ_0 is the average wavelength of the signal, $w_0[n]$ is the discretised, normalised IR of the meter reduced to the antenna position (with or without a preliminary correction), $N_t(i, m, k)$ is the discrete time shift

$$N_t(i, m, k) = \left\{\frac{\tau(i, m, k)}{T}\right\}$$

and U_0 is the maximum IR of the meter. Hence, the maximum value, $S_0(i, m, k)$, of the ith trajectory signal, $s[n, i]$, corresponding to the scattering at the point (m, k) and having a time shift $N_t(i, m, k)$, will be defined as

$$S_0[i, m, k] = \beta[i, m, k] l(m, k) \tag{5.113}$$

where

$$\beta[i, m, k] = \frac{G_a^2 \lambda_0 U_0}{8\pi\sqrt{\pi}} \frac{d^2(\varphi(i, m, k))}{R^2(i, m, k)}$$

In spatial inverse filtering, imaging without a regularising time domain processing is carried out with the expression

$$\hat{l}(x_m, y_m) \triangleq \hat{l}(m,k) = \sum_{i=-I}^{I} \frac{1}{\beta(i,m,k)} s_\varepsilon[N(i,m,k), i]$$

$$= k_v \sum_{i=-I}^{I} \frac{R^2(i,m,k)}{d^2(\varphi(i,m,k))} s_\varepsilon[N(i,m,k), i] \tag{5.114}$$

while, after the processing, in accordance with

$$\hat{l}(m,k) = k_v \sum_{i=-I}^{I} \frac{R^2(i,m,k)}{d^2(\varphi(i,m,k))} s_\alpha[N(i,m,k), i] \tag{5.115}$$

Here, $N(i,m,k) = N_t(i,m,k) + N_0$, where $N_0 = \{\tau/T\}$ is the maximum count number for the normalised IR of the meter or the impulse count of the transmitter, $k_v = 1/(N_L\zeta)$ is the normalisation coefficient and $\zeta = G_a^2\lambda_0 U_0/(8\pi\sqrt{\pi})$ is a constant factor.

In the ideal case of zero measurement noise and infinite duration of the meter IR, $w_0[n]$, or of the IR after time regularising processing, $w_0[n]$, the images obtained with the algorithm (eqn. 5.114 or eqn. 5.115) represent the dependence of the effective scattering length on the coordinates

$$\hat{l}(m,k) = l(m,k) \tag{5.116}$$

In reality, however, the IR waveform differs from the delta function (a single impulse), and the observable image, $\hat{l}(m,k)$, will represent a convolution of the actual target image, $l(m,k)$, with the two-dimensional IR of the meter, that is, with the point scattering function, $\rho(x_v, y_w, x_m, y_m) = \rho(v,w,m,k)$. The point scattering function is an image mapped by a given radar system when the local scatterer with a unit ESL is located at a point with the coordinates $(x_v, y_w) \triangleq (v,w)$. In accordance with the physical sense of this function, the image determinate component should look like

$$\hat{l}(m,k) = \sum_v \sum_w l(v,w)\rho(v,w,m,k)$$

$$= l(m,k) * \rho(v,w,m,k) \tag{5.117}$$

After the regularising processing, the time domain IR of the meter becomes squeezed, so that the point scattering function becomes more peaked.

We shall now take into account the noise component of the trajectory signals used in the aperture synthesis

$$s[n,i] = s_p[n,i] + n[n] \tag{5.118}$$

where $s_p[n,i]$ is the determinate (exact) component and $n[n]$ is random observation noise.

If a trajectory signal was not corrected by a regularised pseudo-inverse filter prior to the aperture synthesis, then

$$s[n,i] \equiv s_{\varepsilon}[n,i] \\ = s_p[n,i] + \varepsilon[n] \tag{5.119}$$

that is, $n[n] \equiv \varepsilon[n]$, where $\varepsilon[n]$ is the noise at the sampling converter output.

After preliminary regularising processing, the noise $n[n]$ is the result of the $\varepsilon[n]$ noise passage through a pseudo-inverse filter.

In both cases, the characteristics of $n[n]$ can be defined; for example, its dispersion, σ_n^2, may be considered to be known. For definiteness, it is assumed that no preliminary correction of the transmit–receive channel has been made.

The image $\hat{l}(m,k)$ obtained with the spatial inverse filtering algorithm (eqn. 5.114) contains the determinate and the noise components

$$\begin{aligned}\hat{l}(m,k) &= \frac{1}{N_L}\sum_{i=-I}^{I}\frac{1}{\beta(i,m,k)}[s_p[N(i,m,k)] + n[i]] \\ &= \frac{1}{N_L}\sum_{i=-I}^{I}\frac{1}{\beta(i,m,k)}\beta(i,m,k)l(m,k) + \frac{1}{N_L}\sum_{i=-I}^{I}\frac{1}{\beta(i,m,k)}n[i] \\ &= l(m,k) + \Delta l(m,k)\end{aligned} \tag{5.120}$$

The noise error dispersion for the image at the point (m,k) can be described by the formula

$$\sigma_{\Delta l}^2(m,k) = \frac{\sigma_n^2}{N_L^2}\sum_{i=-I}^{I}\frac{1}{\beta^2(i,m,k)} \tag{5.121}$$

and the relative image dispersion by the expression

$$\begin{aligned}\delta_{\Delta l}^2(m,k) &\triangleq \frac{\sigma_l^2(m,k)}{l^2(m,k)} \\ &= \frac{\sigma_n^2}{N_L^2 l^2(m,k)}\sum_{i=-I}^{I}\frac{1}{\beta^2(i,m,k)}\end{aligned}$$

By introducing the relative transmittance, $\delta\beta(i,m,k) \triangleq \beta(i,m,k)/\beta_{max}(m,k)$, we can rewrite this expression in the form

$$\begin{aligned}\delta_{\Delta l}^2(m,k) &= \frac{\sigma_n^2}{N_L^2}\sum_{i=-I}^{I}\frac{1}{S_{0\,max}^2(m,k)\delta\beta^2(i,m,k)} \\ &= \frac{1}{N_L^2 q_{S\,max}^2(m,k)}\sum_{i=-I}^{I}\frac{1}{\delta\beta^2(i,m,k)}\end{aligned} \tag{5.122}$$

where $S_{0\,max}$ is the maximum amplitude of the signal reflected by the point (m,k), and $q_{S\,max}^2$ is the maximum SNR in the observable signal.

An approximate value of the quantity (eqn. 5.122) can be found from the expression

$$\delta^2_{\Delta l}(m,k) \approx \frac{1}{N_L q^2_{S\,max}(m,k)} \cdot \frac{1}{[\delta\beta(m,k)]^2_{min}} \tag{5.123}$$

where $[\delta\beta(m,k)]_{min} = min_i\,[\delta\beta(i,m,k)]$. Hence, the image SNR $q^2_l(m,k) \triangleq 1/\delta\Delta l_n(m,k)$ will be defined by

$$q^2_l(m,k) \approx N_L q^2_{S\,max}(m,k)[\delta\beta(m,k)]^2_{min} \tag{5.124}$$

Thus, it is clear that the image quality, in terms of the minimum noise error, grows with the number of signals being measured because of the noise averaging. However, as the relative aperture size is increased, leading generally to a higher system resolution in the transverse direction, the image quality may become low because of the observation noise effects.

The analysis shows that the inverse filtering algorithm cannot provide a high quality image, since an attempt to obtain a high resolution in the longitudinal coordinate by increasing the aperture size results in greater noise interference.

5.7.3 Regularising spatial processing

To decrease the image susceptibility to measurement noise effects, one can use the reduced estimation algorithm suggested by James and Stein, in which

$$\hat{l}_r(m,k) = \frac{1}{N_L}\sum_{i=1}^{N_L} \frac{r(i,m,k)}{\beta(i,m,k)} s(i,m,k) \tag{5.125}$$

where $r(i,m,k)$ is the reduction coefficient. The total estimation error in this case will be

$$\begin{aligned}
\Delta\hat{l}(m,k) &= l(m,k) - \hat{l}_r(m,k) \\
&= l(m,k) - \frac{1}{N_L}\sum_{i=1}^{N_L} \frac{r(i,m,k)}{\beta(i,m,k)} s(i,m,k) \\
&= l(m,k) - \frac{1}{N_L}\sum_{i=1}^{N_L} \left\{ \frac{r(i,m,k)}{\beta(i,m,k)} [\beta(i,m,k)l(m,k) + n(i)] \right\} \\
&= l(m,k) - \frac{1}{N_L}\sum_{i=1}^{N_L} \left[r(i,m,k)l(m,k) + \frac{r(i,m,k)}{\beta(i,m,k)} n(i) \right] \\
&= l(m,k) - \frac{1}{N_L} l(m,k) \sum_{i=1}^{N_L} r(i,m,k) - \frac{1}{N_L}\sum_{i=1}^{N_L} \frac{r(i,m,k)}{\beta(i,m,k)} n(i) \\
&= \left[1 - \frac{1}{N_L}\sum_{i=1}^{N_L} r(i,m,k)\right] l(m,k) - \frac{1}{N_L}\sum_{i=1}^{N_L} \frac{r(i,m,k)}{\beta(i,m,k)} n(i)
\end{aligned} \tag{5.126}$$

and the mean square error at the point (m, k) will have the form

$$\overline{\Delta\hat{l}_{av}^{2}}(m,k) = \left[\left(1 - \frac{1}{N_L}\sum_{i=1}^{N_L} r(i,m,k)\right) l(m,k) - \frac{1}{N_L}\sum_{i=1}^{N_L} \frac{r(i,m,k)}{\beta(i,m,k)} n(i)\right]^2$$

$$= \left[1 - \frac{1}{N_L}\sum_{i=1}^{N_L} r(i,m,k)\right]^2 l^2(m,k) + \frac{1}{N_L}\sum_{i=1}^{N_L} \frac{r^2(i,m,k)}{\beta^2(i,m,k)}\sigma_n^2 \quad (5.127)$$

The mean square error in eqn. 5.127 represents the sum of the squared bias and the noise component of the ESL estimation error for a point target having the coordinates (m, k). If one examines the whole of the $M \times M$ region, one has to consider noise at all points (m, k). Then the total mean square error over the region of radar impulse imaging (the indices m and k are omitted for simplicity) will be described by the expression

$$\overline{\Delta\hat{l}_r^2} = \left(1 - \frac{1}{N_L}\sum_{i=1}^{N_L} r_i\right)^2 l^2 + \frac{\sigma_n^2}{N_L^2}\sum_{j=1}^{M\times M}\sum_{i=1}^{N_L}\frac{r_i^2}{\beta_i^2}$$

$$= \left(1 - \frac{1}{N_L}\sum_{i=1}^{N_L} r_i\right)^2 l^2 + \frac{M^2\sigma_n^2}{N_L^2}\sum_{i=1}^{N_L}\frac{r_i^2}{\beta_i^2} \quad (5.128)$$

Define now the optimal coefficients minimising the value of the quantity (eqn. 5.128). In the expanded form, this expression can be written as

$$\Delta\overline{\hat{l}^2} = l^2 - \frac{2l_t^2}{N_L}(r_1 + r_2 + \ldots + r_n)$$

$$+ \frac{2l^2}{N_L^2}(r_1r_2 + r_1r_3 + \ldots + r_1r_{N_L} + r_2r_3 + \ldots + r_2r_{N_L} + \ldots + r_{N_{L-1}}r_{N_L})$$

$$+ \frac{l^2}{N_L^2}\left(1 + \frac{M^2\sigma_N^2}{\beta_1^2 l_t^2}\right)r_1^2 + \ldots + \frac{l^2}{N_L^2}\left(1 + \frac{M^2\sigma_N^2}{\beta_N^2 l_t^2}\right)r_N^2 \quad (5.129)$$

By differentiating the function (eqn. 5.129) with respect to the variables r_i and equating them to zero, we obtain the following set of equations

$$-\frac{2l^2}{N_L} + \frac{2l^2}{N_L^2}r_2 + \frac{2l^2}{N_L^2}r_3 + \ldots + \frac{2l^2}{N_L^2}r_{N_L} + \frac{2l^2}{N_L^2}\left(1 + \frac{M^2\sigma_N^2}{\beta_1^2 l^2}\right)r_1 = 0$$

$$-\frac{2l^2}{N_L} + \frac{2l^2}{N_L^2}r_1 + \frac{2l^2}{N_L^2}r_3 + \ldots + \frac{2l^2}{N_L^2}r_{N_L} + \frac{2l^2}{N_L^2}\left(1 + \frac{M^2\sigma_N^2}{\beta_2^2 l^2}\right)r_2 = 0$$

$$\cdots\cdots\cdots\cdots\cdots\cdots\cdots\cdots\cdots\cdots$$

$$-\frac{2l^2}{N_L} + \frac{2l^2}{N_L^2}r_1 + \frac{2l^2}{N_L^2}r_2 + \ldots + \frac{2l^2}{N_L^2}r_{N_{L-1}} + \frac{2l^2}{N_L^2}\left(1 + \frac{M^2\sigma_N^2}{\beta_N^2 l^2}\right)r_{N_L} = 0 \quad (5.130)$$

In the matrix form, the set (eqn. 5.130) is

$$\mathbf{A}\cdot\mathbf{r}=\mathbf{N}_L \tag{5.131}$$

where

$$\mathbf{A}=\begin{bmatrix} 1+\dfrac{a}{\beta_1^2} & 1 & \dots & 1 \\ 1 & 1+\dfrac{a}{\beta_2^2} & \dots & 1 \\ \vdots & \vdots & & \vdots \\ 1 & 1 & \dots & 1+\dfrac{a}{\beta_{N_L}^2} \end{bmatrix}$$

and $\mathbf{r}=[r_1 \quad r_2 \quad \dots \quad r_N]^T$, $\mathbf{N}_L=[N_L \quad N_L \quad \dots \quad N_L]^T$. Here, $a=M^2\sigma_n^2/l_t^2$. Eqn. 5.131 yields

$$\mathbf{r}=\mathbf{A}^{-1}\mathbf{N}_L \tag{5.132}$$

In order to calculate the inverse $\mathbf{A}^{-1}$ matrix, the $\mathbf{A}$ matrix should be represented as

$$\mathbf{A}=\mathbf{B}+\mathbf{C}\cdot\mathbf{C}^T \tag{5.133}$$

where

$$\mathbf{B}=\begin{bmatrix} \dfrac{a}{\beta_1^2} & 0 & \dots & 0 \\ 0 & \dfrac{a}{\beta_2^2} & \dots & 0 \\ \vdots & \vdots & & \vdots \\ 0 & 0 & \dots & \dfrac{a}{\beta_{N_L}^2} \end{bmatrix}$$

$\varepsilon=[1 \quad 1 \quad \dots \quad 1]^T$. Then, using the relation

$$\mathbf{A}^{-1}=\mathbf{B}^{-1}-\frac{\mathbf{B}^{-1}\mathbf{C}\mathbf{C}^T\mathbf{B}^{-1}}{\mathbf{I}+\mathbf{C}^T\mathbf{B}^{-1}\mathbf{C}} \tag{5.134}$$

whcrc

$$\mathbf{B}^{-1}=\begin{bmatrix} \dfrac{\beta_1^2}{a} & 0 & \dots & 0 \\ 0 & \dfrac{\beta_2^2}{a} & \dots & 0 \\ \vdots & \vdots & & \vdots \\ 0 & 0 & \dots & \dfrac{\beta_{N_L}^2}{a} \end{bmatrix}$$

we obtain the $\mathbf{A}^{-1}$ matrix in the following form

$$\mathbf{A}^{-1} = \frac{1}{a^{N_L} + a^{N_L - 1} + \sum_{i=1}^{N_L} \beta_i^2} \times \begin{bmatrix} \left(a^{N_L - 1} + a^{N_L - 2} \sum_{i=1}^{N_L} \beta_i^2\right)\beta_1^2 & -a^{N_L - 2}\beta_1^2\beta_2^2 & \cdots & -a^{N_L - 2}\beta_1^2\beta_{N_L}^2 \\ -a^{N_L - 2}\beta_1^2\beta_2^2 & \left(a^{N_L - 1} + a^{N_L - 2} \sum_{i=1}^{N_L} \beta_i^2\right)\beta_2^2 & \cdots & -a^{N_L - 2}\beta_2^2\beta_{N_L}^2 \\ \vdots & \vdots & & \vdots \\ -a^{N_L - 2}\beta_1^2\beta_{N_L}^2 & -a^{N_L - 2}\beta_2^2\beta_{N_L}^2 & \cdots & \left(a^{N_L - 1} + a^{N_L - 2} \sum_{i=1}^{N_L} \beta_i^2\right)\beta_{N_L}^2 \end{bmatrix} \tag{5.135}$$

Multiplying the matrix (eqn. 5.135) by the vector $\mathbf{N}_L = [N_L \quad N_L \quad \ldots \quad N_L]^T$ gives the solution to the set of equations (eqn. 5.130) as the vector of the coefficients $\mathbf{r}_{opt} = [r_{1\,opt} \quad r_{2\,opt} \quad \ldots \quad r_{N_L\,opt}]$ minimising the total mean square error (eqn. 5.128). The vector elements are defined by the expression

$$r_{i\,opt} = C\beta_i^2 \tag{5.136}$$

where

$$C = \frac{N_L}{a + \sum_{i=1}^{N_L} \beta_i^2}$$

The reduction coefficient $r(i, m, k)$ is the weight of the inverse estimate $\hat{l}(i, m, k) = s(i, m, k)/\beta(i, m, k)$ obtained from the ith measurement and used in the integral imaging of the point (m, k). Measurements, for which the values of $r(i, m, k)$ are small, make practically no contribution to the point imaging, so they may be left aside.

To reduce the amount of computation, one can apply a quasi-optimal algorithm based on separate optimisation of the reduction coefficients in each measuring channel. When deriving the algorithm, we should keep in mind that the total estimation error for the ith channel is

$$\begin{aligned} \Delta\hat{l}_r(i, m, k) &= l(m, k) - \hat{l}_r(i, m, k) \\ &= (1 - r(i, m, k))l(m, k) - \frac{r(i, m, k)}{\beta(i, m, k)} n(i) \end{aligned} \tag{5.137}$$

It contains both components – the bias and the noise – and its mean squared value is the sum of the bias squared value and the estimate dispersion

$$\begin{aligned} \overline{\Delta\hat{l}_r(i, m, k)} &= \overline{\left[(1 - r(i, m, k))l(m, k) + \frac{r(i, m, k)}{\beta(i, m, k)} n(i)\right]^2} \\ &= (1 - r(i, m, k))^2 l^2(m, k) + \frac{r^2(i, m, k)}{\beta^2(i, m, k)} \sigma_n^2 \end{aligned} \tag{5.138}$$

Naturally, the minimum total mean square error for the ith channel can be obtained with the following optimal value of the reduction coefficient

$$
\begin{aligned}
r(i,m,k) &\equiv r_{opt}(i,m,k) \\
&= \frac{l^2(m,k)}{l^2(m,k) + \sigma_n^2/\beta^2(i,m,k)} \\
&= \frac{1}{1 + \dfrac{\sigma_n^2}{\beta^2(i,m,k)l^2(m,k)}} \\
&= \frac{1}{1 + \dfrac{\sigma_n^2}{s_T(i,m,k)}} \\
&= \frac{1}{1 + \dfrac{1}{q_s^2(i,m,k)}}
\end{aligned}
\tag{5.139}
$$

Hence, the reduction coefficient will be defined by the SNR value, $q_s^2(i,m,k)$. At large SNR, it will be close to unity, while at small SNR it will tend to zero.

The total mean square error of the estimation with the optimal reduction coefficient is much less than the inverse estimate dispersion $\hat{l}(i,m,k) = s(i,m,k)/\beta(i,m,k)$, and the gain from the application of this estimate grows with decreasing $\beta(i,m,k)$. In other words, the gain is especially great for the most remote array elements having a low SNR. Unfortunately, it is practically impossible to calculate the optimal value of the reduction coefficient, because it depends on the *a priori* unknown ESL value of the local scatterer $l(m,k)$. Therefore, one can take advantage of the concept of minimum, or guaranteed, approach to the evaluation of the effective scattering length of the target $l_{min}(m,k)$, which can be taken to be constant over the whole period of data acquisition.

As a result, we have

$$
r(i,m,k) = \frac{l_{min}^2(m,k)}{l_{min}^2(m,k) + \sigma_n^2/\beta^2(i,m,k)} \tag{5.140}
$$

With eqn. 5.140, the quasi optimal regularising imaging algorithm takes the form

$$
l(m,k) = k_N(m,k)\sum_{i=-I}^{I}\frac{r(i,m,k)}{\beta(i,m,k)}s[N(i,m,k),i] \tag{5.141}
$$

where $k_N(m,k) = 1/\sum_{i=-I}^{I} r(i,m,k)$ is a normalisation factor.

The quantity $\sum_{i=-I}^{I} r(i,m,k)$ characterises the effective number of elements in a synthetic array

$$
N_{L_{ef}} = \sum_{i=-I}^{I} r(i,m,k) \tag{5.142}
$$

and the quantity $L_{ef} = \Delta l(N_{L_{ef}} - 1)$ describes the effective length of the synthetic aperture.

To perform the processing in real time and to save computer storage capacity, it is reasonable to use a recursive procedure of image shot mapping

$$\hat{l}^i(m,k) = l^{(i-1)}(m,k) + \Delta l^{(i)}(m,k) \tag{5.143}$$

where $i = \overline{-I, I}$, $m = \overline{-M, M}$ and $k = \overline{-K, K}$.

Here $\Delta l^{(i)}(m,k)$ is a supplementary summand from the ith signal counts. Its form depends on the particular algorithm used

$$\Delta \hat{l}^{(i)}(m,k) = k_N \frac{R^2(i,m,k)}{d^2(\varphi(i,m,k))} s[N(i,m,k), i] \tag{5.144}$$

$$\Delta \hat{l}_r^{(i)}(m,k) = k_N(m,k) r(i,m,k) \frac{1}{\beta(i,m,k)} s[N(i,m,k), i] \tag{5.145}$$

The original form of the image in the case of zero iteration $(i = -I - 1)$ is defined as

$$\hat{l}^{(-I-1)}(m,k) = 0, \quad \forall m, \quad \forall k$$

We have mentioned above that a UWB probe system can make precise measurements (taking account of eqn. 5.117) of local target responses independent of the aspect and frequency. In actual practice, however, target return signals do depend on both aspect and frequency, so a radar meter measures local effective scattering lengths averaged over all aspects, at which the focusing points (m,k) are visible from the array nodes i, and over all frequencies of the signal bandwidth. In the latter case, the averaging weights are corrected or uncorrected AFR values of the meter.

We shall discuss the capabilities of, and problems in, local response estimation with the algorithms described, using mathematical modelling data on the probe system. The initial data and assumptions used were:

(i) signal frequency bandwidth, 4–8 GHz (duration of an impulse signal, $\tau_i = 0.3$ ns, $f_0 = 6$ GHz);
(ii) synthetic aperture length, $L = 6$ m;
(iii) distance between the central aperture element and the radar target centre, $D = 10$ m;
(iv) number of elements N_L of the array may vary;
(v) trajectory signals are superimposed by noise of a given intensity.

The modelling involved the imaging of a point scatterer, for which the point scattering function and its principal cross section (longitudinal and transverse) were calculated.

Fig. 5.13 shows the cross section of the point scattering function at $N_L = 11$ $(\Delta l = 0.6$ m). No additional time domain regularisation was made

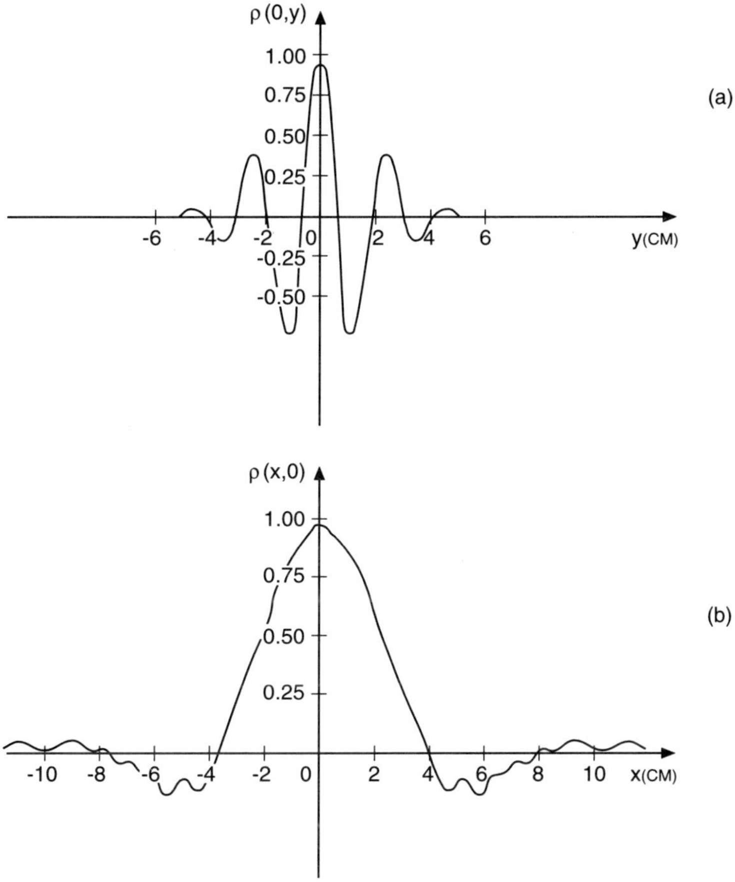

Fig. 5.13 Sections of PSF ($N_L = 11, \Delta l = 0.6\, m$)

a Longitudinal
b Cross

in this case. One can see that the UWB radar meter has a high range resolution on both coordinates. A disadvantage of inverse spatial processing is the presence of appreciable sidelobes, especially outside the principal cross sections. This reduces the image quality and RCS and decreases the meter sensitivity. The reason for this is that the use of this algorithm is, in fact, equivalent to expansion with a uniform field distribution, which is known to give a directivity pattern with a high level of sidelobes. A way of controlling this limitation is well known – additional weight processing with window functions, for example, the Hamming function, the Hann function, the Dolph–Chebyshev function, etc. But there is probably no need to perform this procedure, since additional signal weighting in the regularising algorithm suggested is made by introducing a reduction coefficient. Another

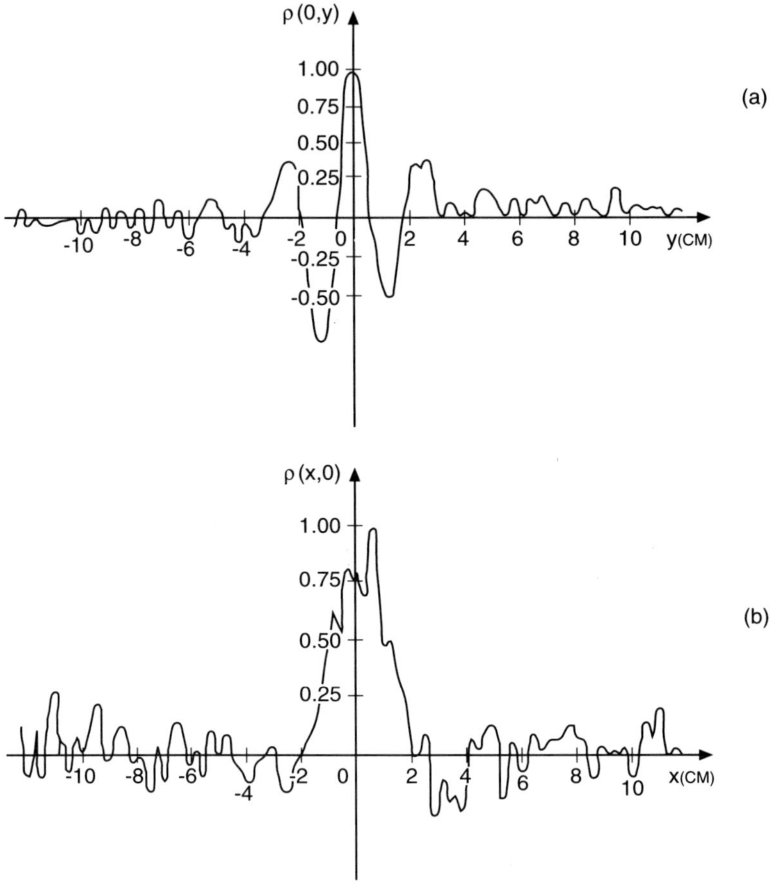

Fig. 5.14 Sections of PSF after inverse spatial processing

a Longitudinal
b Cross

disadvantage of the inverse processing algorithm is its high sensitivity to measurement noises (Fig. 5.14).

The time domain regularising processing of trajectory signals not only increases the longitudinal resolution (τ_i drops to 0.15 ns) but also improves the characteristics of the point scattering function cross sections, although some of these still remain rather high because of the sidelobes. A combination of regularising algorithms of time and spatial processing provides high quality imaging (Fig. 5.15).

The array with a synthetic aperture using UWB probing signals can be made more sparse without a noticeable decrease in the image quality (a manifestation of the time–space equivalence). Fig. 5.16 shows point

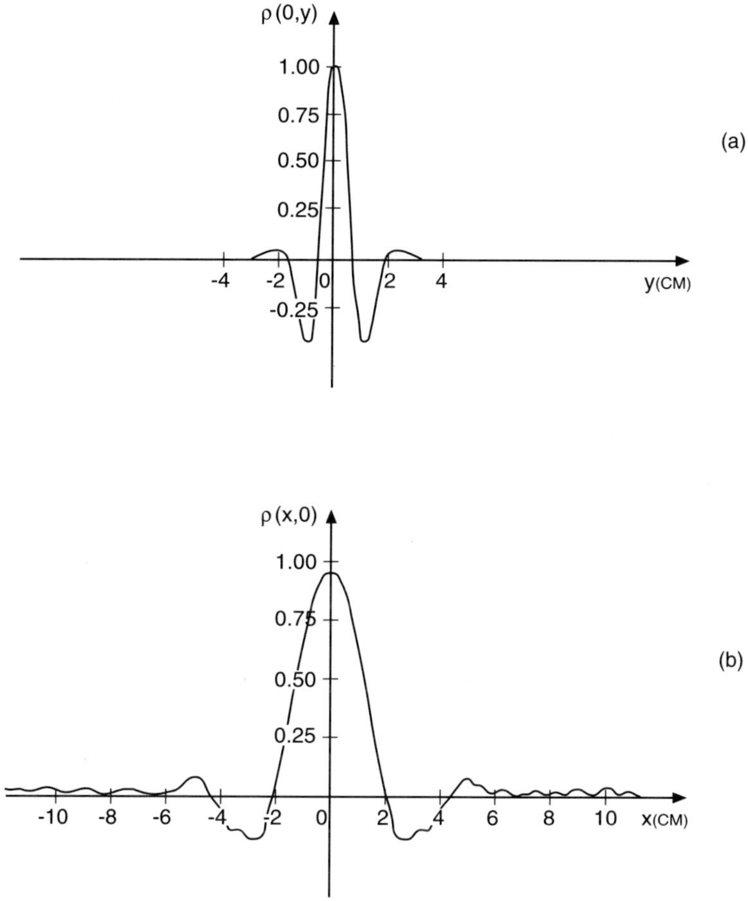

Fig. 5.15 Sections of PSF after regularisation time and spatial procession

a Longitudinal
b Cross

scattering function cross sections at $N_L = 5$ ($\Delta l = 1.5\,\mathrm{m}$) for regularising time processing. Further decrease in the number of array elements deteriorates the transverse resolution. However, the application of sparse arrays may considerably decrease the measurement time.

5.8 Secondary signal processing by collimator meters

The classical foundation for the restoration of local responses from their projections is the inverse Radon transform [39, 55, 58, 106]. For the case in

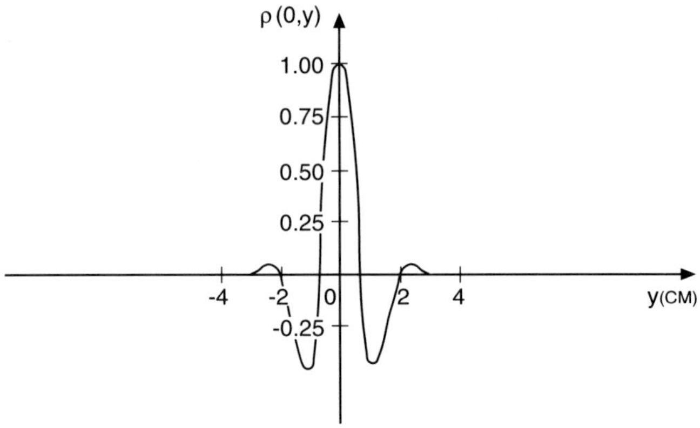

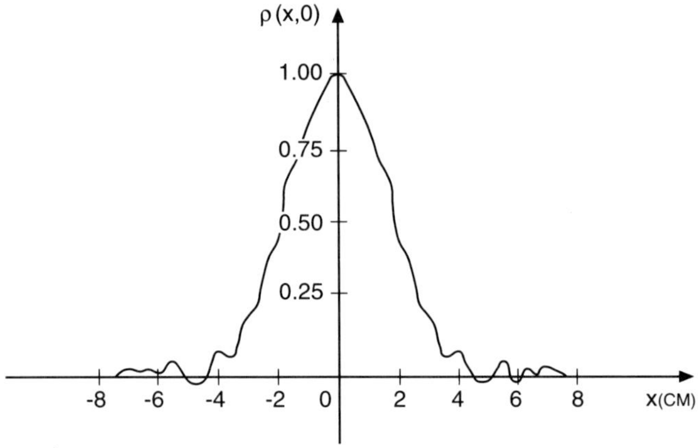

Fig. 5.16 Sections of PSF ($N_L = 5, \Delta l = 1.5\,m$)

a Longitudinal
b Cross

question, it can be written in the operator form

$$l(x_m, y_k) = \mathbf{B}_{\Theta_M}\{\mathbf{D}_{\tau_k}\{\mathbf{G}_{\tau_k}\{\tilde{y}(\tau_k, \Theta_n)\}\}\} \tag{5.146}$$

where $\mathbf{G}_{\tau_k}\{\cdot\}$ is the Hilbert transformation operator, $\mathbf{D}_{\tau_k}\{\cdot\}$ is the operator of differentiation with respect to the variable τ_k,

$$\mathbf{B}_{\Theta_M}\{\cdot\} = \frac{1}{\Theta_M} \sum_{\Theta_n = -(\Theta_M/2)}^{\Theta_M/2} \ldots (y_k \cos\Theta_n + x_m \sin\Theta_n) \tag{5.147}$$

is the inverse projection operator in the aspect sector Θ_M and $\tilde{y}$ is a signal defined by eqn. 4.33.

The effect of the operator $\mathbf{B}_{\Theta_M}$ on the signal $\tilde{y}(\tau_k, \Theta)$ is equivalent to the scale change of the signal $(\tau_k = (2/c)y_k^{\Theta})$, to the transformation of its coordinates ($y_k^\Theta = y_k \cos\Theta_n + x_m \sin\Theta_n$) and to the aspect averaging in the sector Θ_n. Simultaneous action of these operations leads to a dramatic smoothing of the result. For this reason, the operators $\mathbf{G}_{\tau_k}$ and $\mathbf{D}_{\tau_k}$ are employed for introducing the appropriate predistortions of the signal $\tilde{y}_r$, by which the smoothing due to the inverse projection can be compensated.

We can show that the frequency response of the linear system making the Hilbert transformation and differentiation is described by the expression

$$K_{GD}(if) = |f| \tag{5.148}$$

In the total aspect sector, $\Theta_m = 2\pi$, the normalised spectrum of the inverse projection operator kernel is

$$\begin{aligned} K_B(f) &\equiv K(f) \\ &= 1/|f| \end{aligned} \tag{5.149}$$

Comparison of eqns. 5.148 and 5.149 shows that the sequential action of the operators $\mathbf{G}_{\tau_k}$, $\mathbf{D}_{\tau_k}$ and $\mathbf{B}_{\Theta_M}$ is equivalent to the transformation of the signal $\tilde{y}_r$ by a nondistorting factor. Therefore, the point scattering function of the system evaluating the local responses of an object with the inverse Radon transform turns out to be radially symmetric and spatially uniform. Each of its central cross sections reproduces the incident signal waveform.

However, there are some factors in practice which distort the point scattering function of actual meters, interfering with the estimation of local ESL. The most serious challenges are:

(i) the effect of measurement noise (errors) on the signal $\tilde{y}_r(\tau_k, \Theta)$;
(ii) the value $\Theta_M < \pi$ bounding the aspect sector;
(iii) methodological errors of the local response model.

Consider the implications of each of these factors and some ways of avoiding them.

Noise affects the measurement of instantaneous values of the return signal $s_r(t, \Theta)$, introducing errors in the signal $\tilde{y}(\tau_k, \Theta)$. Taking the noise to be additive and homogeneous, we can write the observation equation for the object's local responses in the operator form

$$\tilde{y}_r(\tau_k, \Theta) = k_2 \mathbf{R}_d\{l(x, y)\} + n(t) \tag{5.150}$$

where $\mathbf{R}_d\{\cdot\}$ is the discrete Radon transformation operator, $k_2 = \text{const}$ and $n(t)$ is the noise realisation with the known dispersion σ_n^2. For the estimation of the point scattering function, we assume $l(x, y) = l_0\delta(x, y) \equiv l_h(x, y)$, and the signal reflected by an object with this local response will be designated as $h(t, \Theta)$. Then eqn. 5.150 will take the form

$$h(t, \Theta) = k_2 \mathbf{R}_d\{l_h(x, y)\} + n(t) \tag{5.151}$$

The problem of restoring $l_h(x, y)$ from the known signal $h(t, \Theta)$ in the presence of noise $n(t)$ is ill posed mathematically because of the unstable solution of $\hat{l}_h(x, y)$ obtained by the classical inverse Radon transformation at small values of $h(t, \Theta)$ due to the noise. Physically, this may be accounted for by the frequency response of the predistorting filter (eqn. 5.148), which leads to an unlimited enhancement of the high frequency noise component. In order to get a stable solution at $\Theta_M = 2\pi$, the filter with a frequency response of the type in eqn. 5.148 should be replaced by a regularising filter with the following frequency response

$$K_f(f) = \frac{1}{K(f)} \cdot K_\alpha(f) \tag{5.152}$$

where $K(f)$ is the frequency response of the stabiliser.

Quite often, the function $K(f)$ is replaced by

$$K_{1\alpha}(f) = \exp\left[-\alpha^2 2\pi \left(\frac{f}{\Delta f}\right)^2\right] \tag{5.153}$$

where $\alpha > 0$ is the regularisation parameter, and Δf is the bandwidth in which the local response is being restored. Or, it may be replaced by the function

$$K_{2\alpha}(f) = \frac{K(f)}{K(f) + \alpha K_{max}} \tag{5.154}$$

where $K_{max} = max_f \{K(f)\}$.

The optimal value of the regularisation parameter α_0 can be obtained by iteration of the residual equation

$$\frac{\Delta t}{T_v} \sum_{T_v} \frac{\Delta\Theta}{\Theta_M} \sum_{\Theta_M} [h_\alpha(t, \Theta) - h(t, \Theta)]^2 \Big|_{\alpha = \alpha_0} = \sigma_n^2 \tag{5.155}$$

where $h(t, \Theta) = \mathbf{R}_d\{\hat{l}_h(x, y)\}$ and $\Delta\Theta$ is the aspect variation step.

One should keep in mind, however, that with the use of eqn. 5.155, the amount of computation needed for obtaining only one value of α will greatly exceed that necessary for the imaging of the sought for $\hat{l}_h(x, y)$ estimate. So the problem of finding a simpler way of determining α_0 still remains.

To resolve this difficulty, one commonly uses *a priori* information about the solution and the initial data. Suppose that the AFR of the object in the bandwidth Δf determined by the given spatial resolution satisfies the relation

$$K_0(f) \geqslant K_0(\Delta f) \tag{5.156}$$

and that its measurement has been made with the desired accuracy over the whole bandwidth Δf. Then the normalised SNR at the bandwidth edge will be described by the ratio

$$q_n = q_{\Delta f} / q_{max} \tag{5.157}$$

where $q_{\Delta f}$ is the SNR at a frequency corresponding to the bandwidth edge, Δf.

Obviously, the regularisation parameter should be chosen in terms of a given accuracy at a given resolution. Considering the prominent shape of the function $K_\alpha(f)$, $\alpha = \alpha_0$ should be found starting with the condition

$$K_\alpha(f)|_{f=\Delta f/2} = q_n$$

For example, for the function $K_{1\alpha}(f)$ described by the formula in eqn. 5.153, we have

$$\alpha_0 \approx \sqrt{\frac{2}{\pi}\ln\left(\frac{1}{q_n}\right)} \tag{5.158}$$

Hence, with decreasing SNR at the given bandwidth edge, Δf, the value of α_0 grows, providing considerable noise smoothing. As α_0 increases at constant q_n, one observes range resolution losses; therefore, the dynamic errors of local response estimation grow. At $\alpha < \alpha_0$ and $q_n = \text{const}$, the processing reveals such a strong noise effect which makes it difficult to interpret the change in local responses. This can be illustrated by modelling the data obtained with the regularised inverse Radon transform (Fig. 5.17). The following initial data were employed in the modelling: the aspect sector $\Theta_M = 2\pi$; the signal was a Gaussian radio pulse with the spectral bandwidth $\Delta f = t_i^{-1} = 5\,\text{GHz}$ and centre frequency $f_0 = 7\,\text{GHz}$; the normalised mean square root deviation of the noise $\delta_n = 0.2$. Fig. 5.17a shows the values of the radial variable $r = \sqrt{x^2 + y^2}$ plotted on the abscissa.

If the aspect sector ($\Theta_M < \pi$) becomes smaller, the PSF radial symmetry is disturbed, since the reciprocal projection operator kernel loses its isotropy and homogeneity. In this case, the resolution can be estimated with the

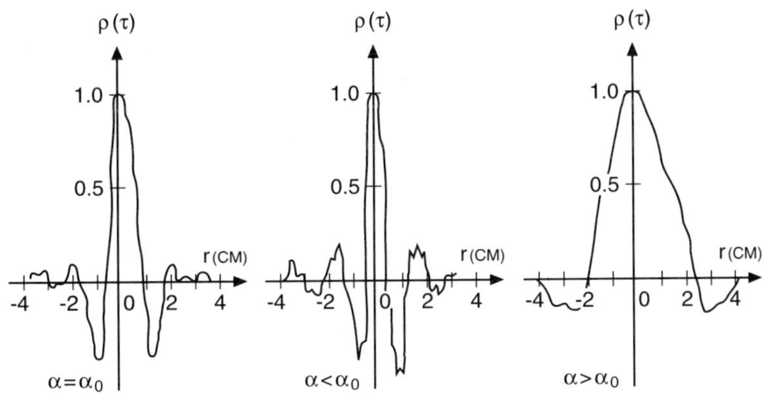

Fig. 5.17 Radial sections of PSF by $\Theta_m = 360°$ *and various values of* α

relations

$$d_x = \begin{cases} \dfrac{c}{4f_0}[(1+\mu/2)\sin\Theta_M/2]^{-1}, & \text{at } \Theta_M \leqslant \pi \\ \dfrac{c}{4f_0}(1+\mu/2)^{-1}, & \text{at } \Theta_M > \pi \end{cases} \tag{5.159}$$

$$d_y = \begin{cases} \dfrac{c}{2f_0}[(1+\mu/2)(1-\mu/2)\cos\Theta_M/2]^{-1}, & \text{at } \Theta_M \leqslant \pi \\ \dfrac{c}{2f_0}(2\sin^2\Theta_M/4+\mu\cos^2\Theta_M/4)^{-1}, & \text{at } \Theta_M > \pi \end{cases} \tag{5.160}$$

where $\mu = \Delta f/f$ is the bandwidth factor and d_x and d_y are the sizes of the resolvable element along the x and y axes, respectively.

Calculations with eqns. 5.159 and 5.160 are in good agreement with the modelling results presented in Fig. 5.18 which shows the point scattering function cross sections obtained at the previous parameters of the signal and the noise in the sector $\Theta_M = 60°$.

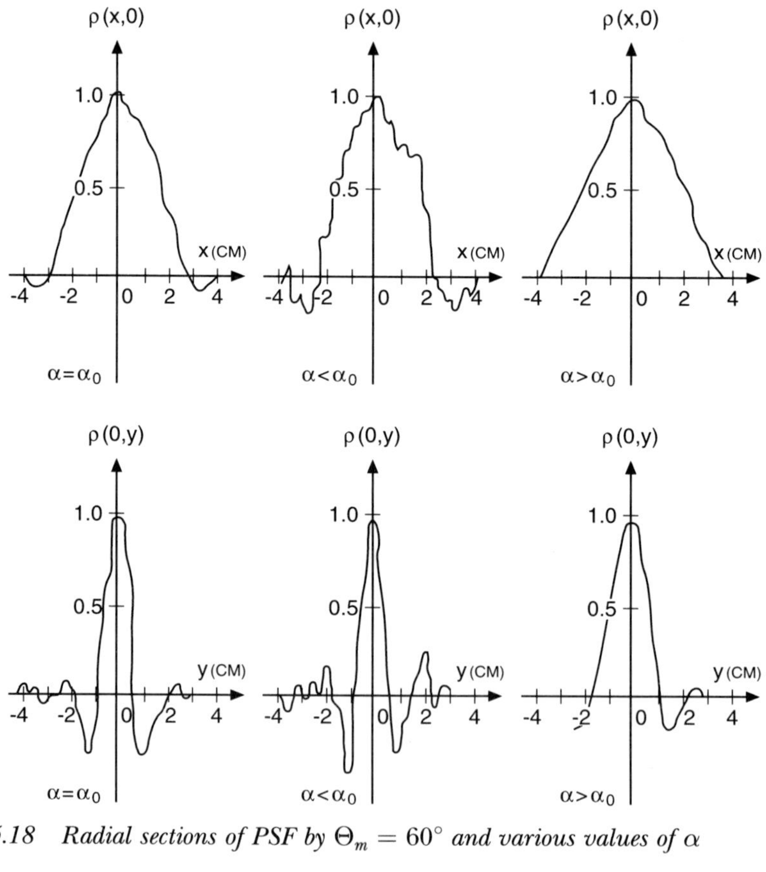

Fig. 5.18 Radial sections of PSF by $\Theta_m = 60°$ *and various values of* α

To make allowance for the reduction in the aspect sector size, the signal processing algorithm should be somewhat modified. For example, the filter characteristics will now depend on the aspect and have the form

$$K_f(f,\Theta) = \frac{1}{K(f,\Theta)} K_\alpha(f,\Theta) \tag{5.161}$$

where $K(f,\Theta)$ is the normalised spectrum of the reciprocal projection operator kernel for the aspect sector $\Theta_M < 2\pi$.

Note that the analytical calculation of the function $K(f,\Theta)$ is generally impossible. In practice, one can either compute a set of $K(f,\Theta)$ values and store them or calculate this function during the processing. For this, it is possible to use the reciprocal projection algorithm usually applied to a signal of the type $\delta(t_k,\Theta_n)$ with subsequent Fourier transformation. Preference of this or that way of computing the function $K(f,\Theta)$ depends on the acceptable compromise between the time expenditure and the computer storage required for the processing.

The modelling has shown (Fig. 5.18) that the quality of local response restoration strongly depends, as in the case with $\Theta_M = 2\pi$, on the value of α. It should be chosen from the residual equation, because recalculation of q_n with varying $K(f,\Theta)$ involves time consuming operations of spectral interpolation.

From eqns. 5.146 and 5.147 it follows that the estimates obtained in the secondary processing of local responses represent local aspect-averaged (within Θ_M) and time-averaged (in the interval t_i) ESL values for the illuminated part of the object. It is possible to show that the geometrical interpretation of the estimates, namely, the ability to attribute them to particular elements of the object's surface, generally requires knowledge of the surface boundary equation presented in the coordinates (x,y). If an object has a rather smooth, conducting surface, the local ESL estimates may give a general idea of its shape.

We would like to emphasise that actual objects do not always satisfy the assumptions used in the formulation of the problem discussed. The most important among these assumptions is that of the absence of interaction of the local scatterers with one another. One of the ways of overcoming difficulties associated with verification of the local ESL model is to describe local responses using a suitable macroparameter, for example, dielectric permittivity, and use more general versions of the Radon transform.

References

1 ASTANIN, L.Y., and KOSTYLEV, A.A.: 'Fundamentals of ultrawideband radar measurement' (Radio i Svyas, Moscow, 1989 (in Russian))

2 ASTANIN, L.Y., and KOSTYLEV, A.A.: 'The current status of research into ultrawideband radars in Russia', Digest of IEE colloquium no. 004 on *Antenna and propagation problems of ultrawideband radars*, London, 1993, pp. 3/1–3/6

3 ASTANIN, L.Y., and KOSTYLEV, A.A.: 'Ultrawideband signals – a new step in radar development', *IEEE Aerosp. Electron. Syst. Mag.*, 1992, **March**, pp. 12–15

4 ASTANIN, L.Y., KOSTYLEV, A.A., MILYAEV, P.V., and NOZDREVATYKH, Y.I.: 'Metrology and software of automatic measurement systems for impulse microwave signals and fields'. Proceedings of All-Union scientific and technical conference on *Radiotechnical measurements in high-frequency and microwave ranges*, Novosibirsk, 1984 (in Russian)

5 ASTANIN, L.Y., PROSYPKIN, S.E., and STEPANOV, A.V.: 'Equipment and means for UWB RTR measurements', *Zarubeznaja Radioelectron.*, 1991, (1) (in Russian)

6 BAUM, C.E.: 'Impulse radiating antennas', *in* BERTRONI, N. (Ed.): 'Ultra wideband and short-pulse electromagnetics' (Plenum Press, New York, 1993)

7 BAUM, C.E.: 'Toward an engineering theory of electromagnetic scattering: the singularity and eigenmode expansion methods', in USLENGHI, P.L.E. (Ed.): 'Electromagnetic scattering' (Academic Press, New York, 1978)

8 BELCIN, V.S., and SHULZENCO, G.I.: 'Formers for powerful nanosecond and picosecond pulses' (Institute of Nuclear Physics, Novosibirsk, 1991 (in Russian))

9 BENNET, C.L., and ROSS, G.F.: 'Time-domain electromagnetics and its applications', *Proc. IEEE*, 1978, **66**, (3)

10 BERNI, A.J.: 'Target identification by natural resonant estimation', *IEEE Trans. Aerosp. Electron. Syst.*, 1975, **AES-11**, (2), pp. 147–154

11 BOCHCAROV, A.M., and DOLGOV, M.N.: 'Radiolocation of low visibility aircraft', *Zarubeznaja Radioelectron.*, 1989, (2) (in Russian)

12 BOJARSKI, N.N.: 'A survey of the physical optics inverse scattering identity', *IEEE Trans. Antennas Propag.*, 1982, **AP-31**, (5), pp. 980–989

13 BOJARSKI, N.N.: 'Low frequency inverse scattering', *IEEE Trans. Antennas Propag.*, 1982, **AP-30**, (4), p. 775

14 BOROVIKOV, V.A., and KINBER, B.E.: 'Geometrical theory of diffraction', (Radio i Svyas, Moscow, 1978 (in Russian))

15 CHAMBERS, C., CLOUDE, S.R., SMITH, P.D., MILNE, A., TRAFFORD, K., and PARKES, D.M.: 'Wavelet processing of ultrawideband radar signals'. Presented at IEE Colloquium no. 004 on *Antenna and propagation problems of ultrawideband radar*, January, 1993

16 CHEN, K.M., NYQUIST, D.F., ROTHWELL, E.J., WEBB, L., and DRACHMAN, B.: 'Radar target discrimination by convolution of radar return with extinction-pulses and single-mode extraction signals', *IEEE Trans. Antennas Propag.*, 1986, **AP-34**, (7), pp. 896–904

17 CHEN, K.M., and WESTMORLAND, D.: 'Radar waveform synthesis for exciting single-mode backscatters from a sphere and application for target discrimination', *Radio Sci.*, 1982, **17**, (6), pp. 574–588

18 CHEN, T.J., and CHU, T.H.: 'Calibration of wideband polarimetric measurement system using three perfectly polarization-isolated calibrators', *IEEE Trans. Antennas Propag.*, 1992, **AP-40**, (12), pp. 1573–1577

19 COOK, C.E., and BERNFELD, M.: 'Radar signals: an introduction to the theory and application' (Academic Press, New York, 1967)

20 DAS, Y., and BOERNER, W.M.: 'On radar target shape estimation using algorithms for reconstruction from projections', *IEEE Trans. Antennas Propag.*, 1978, **AP-26**, (2), pp. 274–279

21 DASARATHY, B.V.: 'Recognition of targets characterized by point ensembles'. Proceedings of IEEE SOUTHEASTCON'81, Huntsville, Alabama, USA, 5–8 April 1981, pp. 655–659

22 DEMIDENKO, E.Z.: 'Linear and nonlinear regression' (Finansy i Statistika, Moscow, 1981 (in Russian))

23 DRACHMAN, B., and ROTHWELL, E.A.: 'Continuation method for identification of the natural frequencies of an object using a measured response', *IEEE Trans. Antennas Propag.*, 1985, **AP-33**, (4), pp. 445–450

24 DUDLEY, D.G.A.: 'State-space formulation of transient electromagnetic scattering', *IEEE Trans. Antennas Propag.*, 1985, **AP-33**, (10), pp. 1127–1130

25 DULEVICH, V.E. (Ed.): 'Theoretical foundations of radar' (Sov. Radio, Moscow, 1978 (in Russian))

26 DYBDAL, R.B.: 'Radar cross section measurements', *Proc. IEEE*, 1987, **75**, (4)

27 FEYNMAN, R.: 'The Feynman lectures of physics – vol. 2' (Addison-Wesley, Reading, MA, 1964)

28 FINKILSHTEIN, M.I. (Ed.): 'Subsurface radar' (Radio i Svyas, Moscow, 1994 (in Russian))

29 FOCK, V.A.: 'Diffraction problems and electromagnetic wave propagation' (Sov. Radio, Moscow, 1970 (in Russian))

30 FOK, F.Y.S.: 'K-pulse estimation for a right-angled bent wire using more than one impulse response', *IEEE Trans. Antennas Propag.*, 1990, **AP-38**, (7), pp. 1092–1098

31 FOK, F.Y.S., MOFFATT, D.L., and WANG, N.: 'K-pulse estimation from the impulse response of a target, *IEEE Trans. Antennas Propag.*, 1987, **AP-35**, (8), pp. 926–933

32 FOO, B.Y., CHAUDHURI, S.K., and BOERNER, W.M.: 'Polarization correction and extension of the Kennaugh–Cosgriff target-ramp response equation to the bistatic case and applications to electromagnetic inverse scattering', *IEEE Trans. Antennas Propag.*, 1990, **AP-38**, (7), pp. 964–971

33 GHARSALLAH, N., ROTHWELL, E.J., CHEN, K.M., and NYQUIST, D.P.: 'Identification of the natural frequencies of a conducting sphere from a measured transient response', *IEEE Trans. Antennas Propag.*, 1990, **AP-38**, (1), pp. 141–143

34 GLEBOVICH, G.V. (Ed.): 'Investigation of objects using picosecond pulses' (Radio i Svyas, Moscow, 1964 (in Russian))

35 GONOROVSKY, I.S.: 'Radiotechnical circuits and signals' (Radio i Svyas, Moscow, 1986 (in Russian))

36 GORELIC, A.L. (Ed.): 'Selection and recognition on the basis of location information' (Radio i Svyas, Moscow, 1990 (in Russian))

37 GRAUPE, D.: 'Identification of systems' (Robert E. Krieger, Huntington, N.Y., 1976)

38 HARMUT, H.: 'Theory of sequential analysis' (Mir, Moscow, 1986 (in Russian))

39 HERMAN, G.T.: 'Image reconstruction from projections, the fundamentals of computerized tomography' (Wiley, New York, 1980)

40 HILL, D.A.: 'Near-field detection of buried dielectric objects', *IEEE Trans. Geosci. Remote Sens.*, 1989, **GRS-27**, (4), pp. 364–368

41 HOORFAR, A., and CHANG, D.C.: 'Analytic determination of the transient response of a thin-wire antenna based upon a SEM-representation', *IEEE Trans. Antennas Propag.*, 1982, **AP-30**, (6), pp. 1145–1152

42 HUA, Y., and SARKAR, T.K.A.: 'Discussion of E-pulse method and Prony's method for radar target resonance retrieval from scattered field', *IEEE Trans. Antennas Propag.*, 1989, **AP-37**, (7), pp. 944–946

43 HUA, Y., and SARKAR, T.K.: 'Generalized pencil-of-function method for extracting poles of an EM system from its transient response', *IEEE Trans. Antennas Propag.*, 1989, **AP-37**, (2), pp. 229–234

44 HUA, Y., and SARKAR, T.K.: 'Perturbation analysis of TK method for harmonic retrieval problems', *IEEE Trans. Acoust. Speech Signal Process.*, 1988, **ASSP-36**, (2), pp. 228–240

45 IVANOV, V.V.: 'Methods of computer calculation' (Naukova Dumka, Kiev, 1986 (in Russian))

46 IVERSON, D.E.: 'Coherent processing of ultrawideband radar signals', *IEE Proc. F, Commun. Radar and Signal Process.*, 1993

47 IVERSON, D.E.: 'Measuring moving target speed and angle in ultrawideband radar'. Presented at second international conference on *Ultrawideband short pulse electromagnetics*, Politechnic Univ., 5–7 April 1994

48 KALININ, Y.N., KONONOV, A.A., KOSTYLEV, A.A., and LEVCHENKO, V.K.: 'Ultrawideband methods and means of radar object visibility control', *Zarubeznaja Radioelectron.*, 1994, (6) (in Russian)

49 KANAREIKIN, D.B., PAVLOV, N.E., and POTEKHIN, V.A.: 'Polarisation of radar signals' (Sov. Radio, Moscow, 1966 (in Russian))

50 KE, Y.A., and MESE, E.D.: 'Advances on radar target classification'. Proceedings of CIE international conference on *Radar*, Nanjing, China, 4–7 November 1986, pp. 226–232

51 KELLER, J.B.: 'Geometrical theory of diffraction', *J. Opt. Soc. Am.*, 1962, **52**, p. 116

52 KENNAUGH, E.M.: 'The K-pulse concept', *IEEE Trans. Antennas Propag.*, 1981, **AP-29**, (2), pt. 2, pp. 329–331

53 KENNAUGH, E.M., and COSSGRIFF, R.L.: 'The use of impulse response in electromagnetic scattering problems', *Nat. Conv. Rec.*, 1958, **6**, (1), pp. 72–77

54 KHARKEVICH, A.A.: 'Selected works – vol. 1, Transient wave phenomena' (Nauka, Moscow, 1986 (in Russian))

55 KONONOV, A.F.: 'Application of tomographic methods for radar imaging with ultrawideband signals', *Zarubeznaja Radioelectron.*, 1991, (1) (in Russian)

56 KORN, G.A., and KORN, Th.M.: 'Mathematical handbook for scientists and engineers' (McGraw-Hill, New York, 1968)

57 KOSTYLEV, A.A.: 'Identification and applications of resonance models of scatterers and antennas', *Zarubeznaja Radioelectron.*, 1991, (1) (in Russian)

58 KOSTYLEV, A.A.: 'Identification of radar targets using UWB signals – methods and applications', *Zarubeznaja Radioelectron.*, 1984, (4) (in Russian)

59 KOSTYLEV, A.A.: 'Optimization and adaptation of radar response meters', *in* 'Methods and devices of radioholography and acoustic optics' (Nauka, Moscow, 1995 (in Russian))

60 KOSTYLEV, A.A.: 'Signal processing in experimental investigation of scattering of short radio pulses on a conducting sphere', *Radiotekhnika*, 1984, (8) (in Russian)

61 KOSTYLEV, A.A., and KALININ, Y.N.: 'Methods for experimental recognition patterns determination with UWB signals', *Zarubeznaja Radioelectron.*, 1992, (10) (in Russian)

62 KOSTYLEV, A.A., MILYAYEV, P.V., and DORSKY, Yu.D., et al.: 'Statistical processing of experimental results using micro-computers and programmed calculators' (Energoatomizdat, Leningrad, 1991 (in Russian))

63 KOSTYLEV, A.A., and SHALDAEV, S.E.: 'The ultra wideband signals processing algorithms for target recognition'. Proceedings of second international conference on *Development directions of the radio communication systems and means*, Voronezh, May 1995, pp. 62–68

64 LOWSON, C.L., and HANSON, R.J.: 'Solving least square problems' (Prentice-Hall, Englewood Cliffs, New Jersey, 1974)

65 LUKE, Y.: 'Mathematical functions and their approximations' (Academic Press, New York, 1975)

66 MACKAY, A.J., and McCOWEN, A.: 'An improved pencil-of-functions method and comparison with transitional method of pole extraction', *IEEE Trans. Antennas Propag.*, 1987, **AP-35**, (4), pp. 435–441
67 MAJDA, G., STRAUSS, W.A., and WEI, M.: 'Computation of exponentials in transient data', *IEEE Trans. Antennas Propag.*, 1989, **AP-37**, (10), pp. 1284–1290
68 MAJDA, G., and WEI, M.A.: 'Simple procedure to eliminate known poles from a time series', *IEEE Trans. Antennas Propag.*, 1989, **AP-37**, (10), pp. 1343–1344
69 MARPLE, S., Jr.: 'Digital spectral analysis with applications' (Prentice-Hall, Englewood Cliffs, New Jersey, 1987)
70 'Mathematical encyclopaedic dictionary' (Sov. Encyclopedia, Moscow, 1988 (in Russian))
71 MAX, G.: 'Méthodes et techniques de traitement du signal et applications aux mesures physiques' (Masson, Paris, 1981)
72 MECKELBURG, H.: 'Scatterer reconstruction from multistatic far-field data', *Electron. Lett.*, 1982, **18**, (8), pp. 341–343.
73 MESE, E.D., MANCIANTI, M., and VERRAZZANI, L.: 'Target identification by means of radar', *Microw. J.*, 1984, (12), pp. 85–86, 88, 92, 94, 96, 98, 100, 102–103
74 MOFFATT, D.L.: 'Ramp response radar imagery spectral content', *IEEE Trans. Antennas Propag.*, 1981, **AP-29**, (2), pp. 399–401
75 MOFFATT, D.L., and MAINS, R.K.: 'Detection and discrimination of radar targets', *IEEE Trans. Antennas Propag.*, 1975, **AP-23**, (3), pp. 358–367
76 MOFFATT, D.L., and RHOADS, C.M.: 'Radar identification of naval vessels', *IEEE Trans. Aerosp. Electron. Syst.*, 1982, **AES-18**, (2), pp. 182–187.
77 MOFFATT, D.L., YOUNG, J.D., KSIENSKI, A.A., LIN, H.C., and RHOADS, C.M.: 'Transient response characteristics in identification and imaging', *IEEE Trans. Antennas Propag.*, 1981, **AP-29**, (2), pp. 192–205
78 NICOLSON, A.M., BENNETT, C.L., LAMENSDORF, D., and SUSMAN, L.: 'Application of time-domain metrology to the automation of broadband microwave measurements', *IEEE Trans. Microw. Theory Tech.*, 1972, **MTT-20**, (1), pp. 3–9
79 PAPOULIS, A.: 'Systems and transforms with applications in optics' (McGraw-Hill, New York, 1968)
80 POGGIO, A.J., VAN BLARICUM, M.L., MILLER, E.K., and MITTRA, R.: 'Evaluation of the processing technique for transient data', *IEEE Trans. Antennas Propag.*, 1978, **AP-26**, (1), pp. 165–173
81 POIDA, V.N.: 'Spectral analysis in discrete orthogonal bases' (Nauka i Tekhnica, Minsk, 1978 (in Russian))
82 'Polarization of signals in large complicated transport radioelectronic systems' (Transport Academy, St. Petersburg, 1994)
83 PROSYPKIN, S.E., and KOSTYLEV, A.A.: 'Algorithm of the measure information processing for investigation of local radar responses of objects', *Radioteckn.*, 1994, (3) (in Russian)
84 RABINER, L., and GOLD, B.: 'Theory and application of digital signal processing' (Prentice-Hall, Englewood Cliffs, New Jersey, 1985)
85 'Radar target characteristics: measurements and applications' (ASTANIN, L.Yu, KOSTYLEV, A.A., ZINOVIEV, Yu.S., and PASMUROV, A.Ya., Eds.) (CRC Press, Boca Raton, Ann Arbor, London, Tokyo, 1994)
86 RIEGGER, S., and WIESBEK, W.: 'Wide-band polarimetry and complex radar cross section signatures', *Proc. IEEE*, 1989, **77**, (5)
87 ROTHWELL, E.J., CHEN, K.M, and NYQUIST, D.P.: 'Extraction of natural frequencies of a radar target from a measured response using E-pulse techniques', *IEEE Trans. Antennas Propag.*, 1987, **AP-35**, (6), pp. 715–720
88 ROTHWELL, E.J., CHEN, K.M., NYQUIST, D.P., and SUN, W.: 'Frequency domain E-pulse synthesis and target discrimination', *IEEE Trans. Antennas Propag.*, 1987, **AP-35**, (4), pp. 426–434
89 ROTHWELL, E.J., NYQUIST, D.P., CHEN, K.M., and DRACHMAN, B.: 'Radar target discrimination using the extinction-pulse technique', *IEEE Trans. Antennas Propag.*, 1985, **AP-33**, (9), pp. 929–936

90 ROUTTI, J.T., and SANDBERG, J.V.: 'General purpose unfolding program with linear and nonlinear regularizations', *Comput. Phys. Commun.*, 1980, **21**, (1), pp. 119–144

91 RUCK, G.T., BARRIC, D.E., STUART, W.D., and KRICHBAUM, C.K.: 'Radar cross section handbook – vol. 1–2' (Plenum Press, New York, 1970)

92 RUSHFORTH, C.K., CROWFORD, A.E., and ZHOU, Y.: 'Least-squares reconstruction of objects with missing high-frequency components', *J. Opt. Soc. Am.*, 1982, **72**, (2), pp. 204–211

93 SARKAR, T.K., WEINER, D., NEBAT, J., and LAIN, V.K.: 'A discussion of various approaches to the identification/approximation problem', *IEEE Trans. Antennas Propag.*, 1981, **AP-30**, (1), pp. 89–98

94 SHIRMAN, J.D. et al.: 'On the first domestic investigations on UWB radar', *Radiotekn.*, 1991, (1) (in Russian)

95 SHTAGER, E.A.: 'Radiowave scattering on the bodies of complex shape' (Radio i Svyaz, Moscow, 1986)

96 SHUBERT, K.K., YOUNG, J.D., and MOFFATT, D.L.: 'Synthetic radar imagery', *IEEE Trans. Antennas Propag.*, 1977, **AP-25**, (4)

97 'Special issue on radar cross section of complex objects', *IEEE Trans. Antennas Propag.*, 1989, **AP-37**, (5)

98 SWARNER, W.G., and BRYAN, R.H.: 'Radar cross-section and imaging measurement systems', *MSN Microw. Syst. News*, 1988, **18**, (4), pp. 14–18

99 SUZUKI, T., and ARAI, I.: 'Advanced subsurface radars'. Proceedings of CIE international conference on *Radar*, Nanjing, China, 1986, pp. 215–219

100 TAYLOR, E.D. (Ed.): 'Introduction to ultra-wideband radar systems' (CRC Press, Boca Raton, FL, 1995)

101 TICE, T.: 'An overview of radar cross section measurement techniques'. Proceedings of IEEE *Instrumentation and measurement technology* conference, Washington, D.C., USA, 25–27 April 1989, pp. 344–346

102 TIKHONOV, A.N., and ARSENIN, V.Ya.: 'Methods for solving noncoherent problems' (Nauka, Moscow, 1986, 3rd edn. (in Russian))

103 TIKHONOV, A.N., ARSENIN, V.Ya., and TIMONOV, A.A.: 'Mathematical problems of computer tomography' (Nauka, Moscow, 1987 (in Russian))

104 TIKHONOV, A.N., and UFIMTSEV, M.V.: 'Statistical processing of experimental results' (Moscow State University, 1988)

105 'Transient electromagnetic fields', *in* FELSEN, L.B. (Ed.): 'Topics in applied physics – vol. 10' (Springer, New York, 1976)

106 TROITSKY, I.N.: 'Statistical theory of the tomography' (Radio i Svyaz, Moscow, 1989 (in Russian))

107 TSENG, F.I., and SARCAR, T.K.: 'Experimental determination of resonant frequencies by transient scattering from conducting spheres and cylinders', *IEEE Trans. Antennas Propag.*, 1984, **AP-32**, (9), pp. 914–918

108 UFIMTSEV, P.Ya: 'Method of waves in the theory of physical diffraction' (Sov. Radio, Moscow, 1962 (in Russian))

109 VAN BLARICUM, M.L., and MITTRA, R.A.: 'Technique for extracting the poles and residues of a system directly of its transient response', *IEEE Trans. Antennas Propag.*, 1975, **AP-23**, (6), pp. 777–781

110 VAN BLARICUM, M.L., and MITTRA, R.: 'Correction to "Problems and solutions associated with Prony's method for transient data"', *IEEE Trans. Antennas Propag.*, 1980, **AP-28**, (11), pp. 949–959

111 VAN BLARICUM, M.L., and MITTRA, R.: 'Problems and solutions associated with Prony's method for processing transient data', *IEEE Trans. Antennas Propag.*, 1978, **AP-26**, (1), pp. 174–182

112 VARGANOV, M.E., ZINOVIEV, Y.S., ASTANIN, L.Yu., KOSTYLEV, A.A., PASMUROV, A.Ya., et al.: 'Radar responses of aircraft' (TUCHKOV, L.T. (Ed.)) (Radio i Svyaz, Moscow, 1985 (in Russian))

113 VECHINSKI, S.R., and SHUMPERT, T.H.: 'Natural resonances of conducting bodies of revolution', *IEEE Trans. Antennas Propag.*, 1990, **AP-38**, (7), pp. 1133–1136

114 WAINSTEIN, L.A., and VAKMAN, D.E.: 'Frequency separation in the theory of fluctuations' (Nauka, Moscow, 1983 (in Russian))

115 WALTON, E.K., and YOUNG, J.D.: 'Radar scattering measurements of cones and computation of transient response', *IEEE Trans. Antennas Propag.*, 1981, **AP-29**, (4), pp. 595–599
116 WALTON, E.K., and YOUNG, J.D.: 'The Ohio State University compact radar cross-section measurement range', *IEEE Trans. Antennas Propag.*, 1984, **AP-32**, (11), pp. 1218–1223
117 YOUNG, J.D.: 'An automated multiple-frequency radar system'. International IEEE/AP-S symposium program and digest, Atlanta, Ga, USA, 1974, pp. 347–349
118 YOUNG, J.D.: 'Radar imaging from ramp response signatures', *IEEE Trans. Antennas Propag.*, 1976, **AP-24**, (3), pp. 276–282
119 YU, F.T.: 'Introduction to diffraction information processing, and holography' (MIT Press, 1973)

List of abbreviations

AEC	anechoic chamber
AFR	amplitude-frequency response
DEIOV	differential equations in input–output variables
DFT	discrete Fourier transform
DP	directivity pattern
ERL	effective radar length
FFT	fast Fourier transform
FR	frequency response
FSA	focused synthetic aperture
GTD	geometrical theory of diffraction
IR	impulse response
MDP	measuring and data processing
PFR	phase-frequency response
PM	polarisation matrix
PSF	point scattering function
RAM	radio absorbing material
RCS	radar cross section
RR	ramp response
RTR	radar target response
SEM	singularity expansion method
SLAE	system of linear algebraic equations
SNR	signal to noise ratio
SR	step response
STFT	short time Fourier transform
TF	transfer function
TMSES	total mean square error of the estimation
TR	total response
TWT	travelling wave tube
UWB	ultrawideband
WT	wavelet transform

Index